Copepods in Aquaculture

Copepods in Aquaculture

Edited by
Cheng-Sheng Lee
Patricia J. O'Bryen
Nancy H. Marcus

Blackwell
Publishing

Cheng-Sheng Lee, PhD, is the Director of the Aquaculture Interchange Program, Oceanic Institute in Hawaii and the Executive Director for the Center for Tropical and Subtropical Aquaculture, Cooperative State Research, Education, and Extension Service, U.S. Department of Agriculture.

Patricia J. O'Bryen is the project manager for the Aquaculture Interchange Program, Oceanic Institute in Hawaii.

Nancy H. Marcus, PhD, is the Robert O. Lawton Distinguished Professor and Mary Sears Professor of Oceanography, Florida State University.

Blackwell Publishing Professional
2121 State Avenue, Ames, Iowa 50014, USA

Orders: 1-800-862-6657
Office: 1-515-292-0140
Fax: 1-515-292-3348
Web site: www.blackwellprofessional.com

Blackwell Publishing Ltd.
9600 Garsington Road, Oxford OX4 2DQ, UK
Tel.: +44 (0)1865 776868

Blackwell Publishing Asia
550 Swanston Street, Carlton, Victoria 3053, Australia
Tel.: +61 (0)3 8359 1011

First edition, 2005

Library of Congress Cataloging-in-Publication Data
Copepods in aquaculture / edited by Cheng-Sheng Lee, Patricia J. O'Bryen, Nancy H. Marcus.—1st ed.
p. cm.
Papers presented at a workshop held in Honolulu, Hawaii, May 5–8, 2003,
Includes index.
ISBN-13: 978-0-8138-0066-0
ISBN-10: 0-8138-0066-8
1. Fishes—Feeding and feeds—Congresses.
2. Copepoda—Congresses. I. Lee, Cheng-Sheng.
II. O'Bryen, P. J. (Patricia J.) III. Marcus, Nancy H.

SH156.C67 2005
639.3′2—dc22

2004029918

The last digit is the print number: 9 8 7 6 5 4 3 2

Contents

Contributors

Chapter numbers in brackets follow each name. *Indicates the corresponding author.

Donald N. Barnes [17] University of Southern Mississippi-Gulf Coast Research Laboratory, P.O. Box 7000, Ocean Springs, Mississippi 39564-7000, USA.

Stephen C. Battaglene [15] Marine Research Laboratories, Tasmanian Aquaculture and Fisheries Institute and Aquafin Cooperative Research Centre, University of Tasmania, Hobart, Tasmania 7001, Australia.

Jessica L. Beck [10] Department of Biological Sciences, Florida Institute of Technology, 150 West University Blvd., Melbourne, Florida 32901, USA.

Leon Boyd [6] Department of Food Science, North Carolina State University, Raleigh, North Carolina 27695, USA.

Matthew P. Bransden [15] Marine Research Laboratories, Tasmanian Aquaculture and Fisheries Institute and Aquafin Cooperative Research Centre, University of Tasmania, Hobart, Tasmania 7001, Australia.

Carol A. Burkart [5] Mountain Empire Community College, Big Stone Gap, Virginia 24219, USA.

Edward J. Buskey [8*]The University of Texas at Austin, Marine Science Institute, 750 Channel View Drive, Port Aransas, Texas 78373, USA. buskey@utmsi.utexas.edu

Edward J. Buskey is a professor of marine science in the Department of Marine Science and a senior research scientist at the Marine Science Institute of the University of Texas at Austin. Dr. Buskey's research interests focus on the behavioral ecology of zooplankton and include studies of planktonic predator—prey interactions, aggregative behavior of zooplankton, adaptive value of marine bioluminescence, and the role of zooplankton grazers in harmful algal blooms.

He received his A.B. in biology from Brown University, his M.Sc. in zoology from the University of British Columbia, and his Ph.D. in biological oceanography from the University of Rhode Island.

Tzyy-Ing Chen [14] Tungkang Marine Laboratory, Fisheries Research Institute, Council of Agriculture R.O.C., Tungkang, Pingtung, 92845 Taiwan.

Shin-Hong Cheng [14] Tungkang Marine Laboratory, Fisheries Research Institute, Council of Agriculture R.O.C., Tungkang, Pingtung, 92845 Taiwan.

Edward J. Chesney [11*] Louisiana Universities Marine Consortium, Chauvin, Louisiana 70344, USA. echesney@lumcon.edu

Edward J. Chesney is an associate professor at the Louisiana Universities Marine Consortium. Dr. Chesney has been involved in research and management related to fisheries, coastal oceanography, and aquaculture for over 30 years in such diverse locations as the Gulf of Mexico, and the east and west coasts of the United States, and Spain.

He received his B.S. from the University of Michigan School of Natural Resources and his Ph.D. from the University of Rhode Island, Graduate School of Oceanography in 1984. He is currently collaborating directly with the aquaculture industry to improve hatchery and rearing techniques of marine finfish. Recent collaborative projects related to aquaculture include studies of the role of gamete quality in the spawning of fishes and the subsequent effects on the quality of fish eggs and larvae, ways to improve hatchery production techniques, the use of recirculating systems for the spawning and culture of marine finfish, studies to improve production techniques for live feeds, and a feasibility study for offshore aquaculture in the Gulf of Mexico.

Allen D. Davis [12] Department of Fisheries and Allied Aquacultures, Auburn University, Auburn, Alabama 36849, USA.

Graeme A. Dunstan [15] CSIRO Marine Research and Aquafin Cooperative Research Centre, Hobart, Tasmania, 7001 Australia.

John W. Fleeger [2*] Department of Biological Sciences, Louisiana State University, Baton Rouge, Louisiana 70803, USA. zoflee@lsu.edu

John W. Fleeger holds the George C. Kent Chair in Biological Sciences at Louisiana State University, Baton Rouge. Dr. Fleeger has been an active researcher specializing in the ecology, evolution, and ecotoxicology of harpacticoid copepods for more than two decades. His research with harpacticoids and other benthic invertebrates includes topics in systematics (traditional and molecular), culture procedures, behavior, toxicology, and diet. Field-related research includes examination of factors (including nutrients, predation, hydrodynamics, and competition) that influence distribution and abundance patterns.

He received his Ph.D. from the University of South Carolina in Columbia. He is a former chair of the International Association of Meiobenthologists and President of the Gulf Estuarine Research Association. He served as chair of the Department of Zoology and Physiology for 6 years and has been a co-organizer of the Benthic Ecology Meetings (in 1999) and an International Conference of Meiobenthology (in 2001).

Ma. Salvacion Golez [13] Tokyo University of Fisheries, 4-5-7 Konan, Minato-ku, Tokyo 108-8477, Japan.

Sarah E. Hazzard [5] Institute of Ecosystem Studies, Millbrook, New York 12545, USA.

Ju-shey Ho [3*] Department of Biological Sciences, California State University, 1250 Bellflower Blvd., Long Beach, California 90840-3702, USA. jsho@csulb.edu

Ju-shey Ho is a professor of biology at California State University, Long Beach. Dr. Ho has worked on symbiotic copepods since 1960 and published more than 200 articles on the taxonomy, systematics, and biogeography of symbiotic copepods. He has received several honors connected to his work, including an *Award in Science and Technology* from the Taiwanese-American Foundation and a *Life Time Achievement Award* from The 8th International Conference on Copepoda. He is also frequently called upon for help in the identification of copepods parasitic on various types of marine animals ranging from sponges to mammals.

He received his B.S. at National Taiwan University in Taipei, Taiwan, in 1958 and obtained advanced degrees (M.S. and Ph.D.) in biology from Boston University. Dr. Ho has traveled to many parts of the world to collect and study symbiotic copepods. That long-term experience in search for the symbiotic copepods led him to conceive the idea of using symbiotic copepods for mass production of nauplii to obtain live feed for marine finfish larvae rearing.

Adrianna Ianora [4*] Ecophysiology Laboratory, Stazione Zoologica A. Dohrn, Villa Comunale 80121, Naples, Italy. iannora@alpha.szn.it

Adrianna Ianora is senior scientist of the Ecophysiology Laboratory at the Stazione Zoologica in Naples, Italy. Dr. Ianora received her degree at the University of Naples in 1980 and has been working at the Stazione Zoologica since 1983 on copepod ecology, reproduction, development, nutrition, and physiology. She is an editorial board member of the *Journal of Plankton Research*, *Scientia Marina*, and Marine Ecology: PSZNI, and review editor of the journal, *Marine Ecology Progress Series*. She is author of over 90 scientific publications and coeditor of three books published by Springer-Verlag on zooplankton taxonomy and Antarctic biology. Currently, she is studying the effects of phytoplankton chemical defenses on copepod reproduction and development and screening for the presence of new antiproliferative compounds in diatoms, dinoflagellates, and other microalgal groups.

Gary S. Kleppel [5*] Department of Biological Sciences, University at Albany, State University of New York, 1400 Washington Ave., Albany, New York 12222, USA.

Gary S. Kleppel is a professor of biological sciences at the University at Albany, State University of New York. His interests are in the nutritional ecology of zooplankton and sustainable development. Dr. Kleppel's research in zooplankton nutrition began in the early 1980s with studies on ichthytoplankton off southern California and has continued with research on the role of nutrition in defining ecosystems and ecological interactions within and among copepod populations in coastal and ocean waters. Dr. Kleppel is director of the Biodiversity, Conservation and Policy Program at the University at Albany and the Land Use-Coastal Ecosystem Study (NOOA/COP), which seeks sustainable solutions to the impacts of development along rapidly urbanizing coastlines.

Dr. Kleppel received his Ph.D. from Fordham University in 1979 and conducted postdoctoral research at the University of Southern California from 1981 to 1986. He taught at the Oceanographic Center of Nova Southeastern University, in Fort Lauderdale, Florida, and the University of South Carolina, Columbia prior to coming to the University at Albany.

Jon Kobashigawa [16] The Oceanic Institute, 41-202 Kalanianaole Hwy., Waimanalo, Hawaii 96795, USA.

Tomonari Kotani [16] Industrial Promotion Foundation, Ikeda 2-1303-8, Oomura 856-0026, Japan.

Justin M. Krebs [10] Department of Biological Sciences, Florida Institute of Technology, 150 West University Blvd., Melbourne, Florida 32901, USA.

Hsiang-Pin Lan [12] Department of Fisheries and Allied Aquacultures, Auburn University, Auburn, Alabama 36849, USA.

Cheng-Sheng Lee [Editor, 18] The Oceanic Institute, 41-202 Kalanianaole Hwy., Waimanalo, Hawaii 96795, USA. cslee@oceanicinstitute.org

Cheng-Sheng Lee is the Director of the Aquaculture Interchange Program at the Oceanic Institute in Hawaii and the Executive Director for the Center for Tropical and Subtropical Aquaculture, Cooperative State Research, Education, and Extension Service, U.S. Department of Agriculture. His research interests have focused on control of reproduction and the early life history of marine finfish and shrimp, rotifer biology and culture, and general aquaculture management.

Jason T. Lemus [17] University of Southern Mississippi-Gulf Coast Research Laboratory, P.O. Box 7000, Ocean Springs, Mississippi 395-7000, USA.

Jason D. Licamele [10] Department of Biological Sciences, Florida Institute of Technology, 150 West University Blvd., Melbourne, Florida 32901, USA.

Emily E. Lipman [12] Department of Fisheries and Allied Aquacultures, Auburn University, Auburn, Alabama 36849, USA.

Jeffrey M. Lotz [17] University of Southern Mississippi-Gulf Coast Research Laboratory, P.O. Box 7000, Ocean Springs, Mississippi 39564-7000, USA.

Nancy H. Marcus [Editor, 1*] Department of Oceanography, Florida State University, Tallahassee, Florida 32306, USA. marcus@ocean.fsu.edu

Nancy H. Marcus is the Robert O. Lawton Distinguished Professor and Mary Sears Professor of Oceanography at Florida State University. Her research has focused on the induction, maintenance, and termination of diapause in marine copepods, specifically coastal species that produce diapause eggs.

Kimo Marion [16] The Oceanic Institute, 41-202 Kalanianaole Hwy., Waimanalo, Hawaii 96795, USA.

Ephrime B. Metillo [15] Department of Biological Sciences, Mindanao State University-Iligan Institute of Technology, A. Bonifacio Ave., Iligan City, Philippines 9200.

Augustin Molnar [16] The Oceanic Institute, 41-202 Kalanianaole Hwy., Waimanalo, Hawaii 96795, USA.

David T. Morehead [15*] Marine Research Laboratories, Tasmanian Aquaculture and Fisheries Institute and Aquafin Cooperative Research Centre, University of Tasmania, Hobart, Tasmania 7001, Australia.

David T. Morehead is involved in striped trumpeter research at the Tasmanian Aquaculture and Fisheries Institute's Marine Research Laboratories (University of Tasmania). He received his bachelor of applied science degree at Curtin University (Western Australia) in 1991 and a graduate diploma in Antarctic and southern ocean studies with honors at the University of Tasmania in 1992. In 1997, Dr. Morehead completed his Ph.D. in the management of reproduction in striped trumpeter, focusing on reproductive endocrinology. Since that time, Dr. Morehead has concentrated on larviculture, with emphasis on nutrition, health, and system design.

Komarey Kao Moss [12] Myron B. Thompson Academy, PCS, 629 Pohukaina St., Suite 3, Honolulu, Hawaii 96813, USA.

L. Casey Nicholson [17] University of Southern Mississippi-Gulf Coast Research Laboratory, P.O. Box 7000, Ocean Springs, Mississippi 39564-7000, USA.

Patricia J. O'Bryen [Editor, 18*] The Oceanic Institute, 41-202 Kalanianaole Hwy., Waimanalo, Hawaii 96795, USA. pobryen@oceanicinstitute.org

Patricia J. O'Bryen is the Project Manager for the Aquaculture Interchange Program at the Oceanic Institute in Hawaii. She has co-edited the proceedings of five previous workshops held by the Oceanic Institute. Ms. O'Bryen has a B.A. in Zoology and an M.A. in English as a Second Language from the University of Hawaii.

John T. Ogle [17*] University of Southern Mississippi-Gulf Coast Research Laboratory, P.O. Box 7000, Ocean Springs, Mississippi 39564-7000, USA.

John T. Ogle is a senior research scientist with The University of Southern Mississippi, Gulf Coast Research Laboratory at Ocean Springs, Mississippi, USA. He received his B.S. and M.S. from Texas A&M University and has worked in aquaculture for the past 30 years.

Atsushi Ohno [13] Tokyo University of Fisheries, 4-5-7 Konan, Minato-ku, Tokyo 108-8477, Japan.

Michael F. Payne [9] Seahorse Sanctuary, Red Bluff Road, Kalbarri, Western Australia 6536, Australia.

Ronald P. Phelps [12*] Department of Fisheries and Allied Aquacultures, Auburn University, Auburn, Alabama 36849, USA.

Ronald P. Phelps is an associate professor in the Department of Fisheries and Allied Aquacultures at Auburn University. Dr. Phelps has worked in the area of fish reproduction and hatchery management since 1977. Much of his recent research efforts have concentrated on the reproduction of the marine fish red snapper *Lutjanus campechanus*, its larval rearing, and production of live feeds. He received his B.S. at Auburn University in fisheries science in 1969 and his Ph.D. in 1975.

Adelaide Rhodes [6*] Essential Live Feeds, 343 Soquel Ave. #197, Santa Cruz, California 95062, USA.

Adelaide C.E. Rhodes recently received her Ph.D. in zoology from North Carolina State University. Her dissertation research focused on developing a simple technique for rearing copepods to use in aquaculture. Dr. Rhodes started her career in aquaculture in 1993, when she joined the U.S. Peace Corps program in Honduras as a freshwater fish culture extension worker. In 1996, she was selected as a Knauss Sea Grant Marine Policy Fellow in the National Sea Grant Office in the National Oceanic and Atmospheric Administration (NOAA), where she assisted in the development of the NOAA and Department of Commerce aquaculture policies. She received her B.A. in physics from the University of Virginia in 1990 and her M.S. in oceanography from the Florida Institute of Technology in 1992.

Robert J. Rippingale [9*] 15 Stuart Crescent, Lesmurdie, Western Australia 6076, Australia.

Robert J. Rippingale retired from Curtin University of Technology in Perth, Western Australia, where he was a senior teaching fellow. Dr. Rippingale taught aquatic ecology and focused research activity on estuarine ecosystems. Under a grant from the Australian Fisheries Research and Development Corporation, his particular interest in estuarine copepods led to work with copepods in cultivation and eventually linking copepod production to providing food for larval fish.

He received his Ph.D. from the University of Western Australia and has authored numerous articles in *Aquaculture* and the *Australian Journal of Marine and Freshwater Research*. In 2001, he and coauthor Michael F. Payne published a guide to procedures for cultivating *Gladioferens imparipes*, a calanoid copepod from the Swan River Estuary in Australia.

Robin J. Shields [16*] School of Biological Sciences, University of Wales Swansea, Singleton Park, Swansea SA2 8PP, UK.

Robin J. Shields is director of research for the Aquaculture Wales sustainable development group, located at the University of Wales, Swansea, UK. He was previously finfish program manager at the Oceanic Institute in Hawaii. Dr. Shields has been actively researching the biology and dietary requirements of larval marine fish for more than 10 years, since obtaining his Ph.D. from the University of Wales Bangor. This research is directed at developing reliable production techniques for juvenile marine fish, ranging from temperate to tropical species. He is a technical reviewer for several scientific journals and has acted as an advisor to organizations, including the U.S. Center for Tropical and Subtropical Aquaculture and the U.S. National Marine Fisheries Service.

Huei-Meei Su [14*] Fisheries Research Institute, Council of Agriculture R.O.C., Tungkang, Pingtung, 92845 Taiwan.

Huei-Meei Su is a senior researcher at the Tungkang Marine Laboratory, Fisheries Research Institute, Council of Agriculture, Republic of China. Dr. Su has worked on live food culture and utilization for marine larviculture since 1978 and since 1985 on harmful microalgae occurring in aquaculture ponds. She received her B.S. in botany at National Taiwan University in 1972 and has advanced degrees from the Institute of Oceanography (M.S. and Ph.D.). Dr. Su maintains a culture collection of microalgae and rotifers and has provided organisms as starting cultures to farmers and researchers in Taiwan, and has also served to extend education for farmers.

Mao-Sen Su [14] Tungkang Marine Laboratory, Fisheries Research Institute, Council of Agriculture R.O.C., Tungkang, Pingtung, 92845 Taiwan.

Gede S. Sumiarsa [12] Department of Fisheries and Allied Aquacultures, Auburn University, Auburn, Alabama 36849, USA.

Larren Tang [16] The Oceanic Institute, 41-202 Kalanianaole Hwy., Waimanalo, Hawaii 96795, USA.

Joebert D. Toledo [13*] Aquaculture Department, Southeast Asian Fisheries Development Center, Tigbauan, Iloilo 5021, Philippines.

Joebert D. Toledo is a scientist at the Southeast Asian Fisheries Development Center (SEAFDEC) Aquaculture Department (AQD) in Iloilo, Philippines. He now heads the grouper research program at ADQ, where he documented the natural spawning of grouper in concrete tanks and floating net cages. His technical skills significantly contributed in the generation of technologies for the breeding and seed production of milkfish *Chanos chanos*, sea bass *Lates calcarifer*, and mangrove red snapper *Lutjanus argentimaculatus*. He often serves as a resource person in local and international training courses and in-situ seminars and is a consultant for several fish hatcheries.

Dr. Toledo received his B.Sc. in marine fisheries at the University of the Philippines in 1980 and his M.Sc. from Hiroshima University in 1990. As a Ronpaku Fellow, he earned his Ph.D. from Hiroshima University in 2002.

Ralph G. Turingan [10*] Department of Biological Sciences, Florida Institute of Technology, 150 West University Blvd., Melbourne, Florida 32901, USA.

Ralph G. Turingan is an associate professor of biological sciences at Florida Institute of Technology in Melbourne, Florida. He has taught fish biology and finfish aquaculture. His research centers on the ecology and evolution of functional design in fishes. He recently applied his field of research into aquaculture. He is now conducting research on the development of feeding functional morphology and prey-capture performance in hatchery-reared fish larvae. The central hypothesis of his research is that the development of the larva's prey-capture mechanism influences its ability to capture prey. Understanding how the feeding mechanism constrains the type of prey a fish larva is able to capture and ingest underlies the ability of hatchery managers to select the optimal prey for a given ontogenetic stage of fish larva.

Shin-ichi Uye [7*] Graduate School of Biosphere Sciences, Hiroshima University, 4-4 Kagamiyama 1 Chome, Higashi-Hiroshima 739-8528, Japan. suye@hiroshima-u.ac.jp

Shin-ichi Uye is a Professor of Biological Oceanography at Hiroshima University. Dr. Uye has worked on zooplankton production ecology in coastal waters of Japan, mainly in the Inland Sea of Japan (Seto Inland Sea). His research field has expanded from the copepod autoecology to trophodynamics of the plankton community in relation to anthropogenic environmental impacts. One of his current interests is jellyfish blooms and their impact on fisheries.

He received his D.Agr. from Tohoku University in 1981. He is currently the president of the Plankton Society of Japan, and vice-president of the World Association of Copepodologists.

Preface

Cheng-Sheng Lee

Rotifers and brine shrimp nauplii are the two common live-food organisms for early life stages of marine finfish in hatcheries. Copepods provide additional desirable characteristics such as size and nutritional value to finfish larvae and, until now, have played a supplemental role in larval rearing. Nauplii of some copepod species have been used to successfully raise fish species that cannot use rotifers as a first feed. Copepodids of some species provide greater amounts of the highly unsaturated fatty acids eicosapentaenoic acid and docosahexaenoic acid than rotifers or brine shrimp. Unless, however, mass-production technology for copepods is established, copepods will not be a major live-feed organism for finfish larvae in hatcheries. Limited information about copepod culture is available, but it is generally found from among diverse sources.

A workshop, Culture of Copepods and Applications to Marine Finfish Larval Rearing, was held in Honolulu, Hawaii, from May 5 to 8, 2003, to gather information about the culture of copepods in hatcheries and basic biological studies on various types of copepods with a potential application to larval rearing in hatcheries. Funding for the workshop was provided under a grant from the National Oceanic and Atmospheric Administration (NOAA; grant no. NA07RG0579, Amendment 2) to The Oceanic Institute (OI), Hawaii. With the assistance of Dr. Nancy H. Marcus, copepod biologists and scientists working on copepod culture were invited to present their work and discuss key issues at the workshop. Copepod biologists and ecologists covered the production of resting eggs, reproductive performance, behavior, and natural productivity of various types of copepods, and identified potential species for aquaculture purposes. Scientists specializing in live feed and early life stages of marine finfish presented their studies on feeding behavior, culturing copepods, and applying copepods to larviculture. Ample time was allowed for information exchange and discussion among the participants from a diversity of disciplines represented at the meeting.

This book is a peer-reviewed collection of papers that were presented at the workshop, with a final chapter summarizing the discussions that took place. It is hoped that these studies and the knowledge gained thereby will help advance the techniques for culturing copepods and applying them to larviculture. The opinions expressed in these papers reflect those of the authors and not necessarily those of NOAA or OI. Warmest thanks are extended to the contributors and reviewers for the quality of the papers and the valuable information presented.

Copepods in Aquaculture

1
Calanoid Copepods, Resting Eggs, and Aquaculture

Nancy H. Marcus

ABSTRACT

A major bottleneck in the cultivation of many marine fish species for commercial purposes is the lack of a suitable food for the first-first-feeding larval stages. The standard feeds, *Artemia* and rotifers (*Brachionus* spp.), are not always effective foods. This could be due to several factors, including size, biochemical composition, and swimming behavior. Studies of wild-caught fish larvae report that calanoid copepod nauplii are an important item in the diet of various fish species. Hence, establishing a protocol for the large-scale cultivation of calanoid copepods would appear to be a fruitful direction for further research. Indeed, a few studies have already been conducted, demonstrating the suitability of species of *Acartia* and *Gladioferens* for raising some fish species. In view of the fact that people are trying to cultivate a fairly wide variety of fish species, e.g., small ornamentals to large consumables, from estuarine to open ocean habitats, a single calanoid species, however, may not solve the needs of the aquaculture industry. This paper discusses several of the calanoid copepods, especially in terms of the phenomenon of dormancy, and makes suggestions as to species that hold promise for meeting the needs of aquaculture. Promising species are found in genera including *Centropages, Labidocera, Acartia,* and *Eurytemora.*

INTRODUCTION

During the last decade a number of excellent reviews and manuals addressing the use of copepods as a live feed by aquaculturists have been produced. These include *Manual on the Production and Use of Live Food for Aquaculture* (Lavens and Sorgeloos 1996), *Intensive Cultivation of a Calanoid Copepod for Live Food in Fish Culture* (Rippingale and Payne 2001), and *Live Feeds in Marine Aquaculture* (Støttrup and McEvoy 2003). Nevertheless copepods are still not being used routinely by the aquaculture industry.

Of the ten orders of copepods, the Calanoida, Harpacticoida, Cyclopoida, and Mormonilloida have pelagic representatives in marine systems. Of these four taxa, the calanoid copepods are particularly abundant members of the pelagic realm in estuaries and other coastal habitats and generally represent an important link between the phytoplankton and fish in these inshore nursery systems. Largely for these reasons, the calanoid copepods have received considerable attention by researchers (see review by Mauchline 1998). Most species are approximately 1.0 mm in total length with some being as small as 0.4 mm and some as large as 10.0 mm (see Fig. 3 in Mauchline 1998). Based on citations in the scientific literature, *Acartia clausi* and *Calanus finmarchicus* are the most widely studied species (Mauchline 1998), followed by *Temora longicornis, Paracalanus parvus, Calanus helgolandicus, Pseudocalanus elongatus, Acartia tonsa, Centropages hamatus, Centropages typicus*, and *Temora stylifera*.

Since most of this basic research on the calanoid copepods has been limited to only a few species, it is difficult to make informed choices about the best candidates for aquaculture because, as noted by Støttrup (2000), "A basic knowledge of physiological processes and population dynamics of a species is a prerequisite for the development of rearing techniques." Calanoid copepods are consumed by many larval fish in the wild (e.g., Pepin and Penney 1997) and some species (e.g., *A. tonsa*, *C. hamatus*, *Eurytemora affinis*, and *Gladioferens imparipes*), have already been used for the cultivation of fish (see review by Støttrup and McEvoy 2003). A feature that this author believes makes most of these species particularly good candidates for finfish aquaculture is that they undergo a dormant phase during their life cycle.

DORMANCY

Dormancy or suppressed development is a characteristic feature of the life cycle of many copepod species (Grice and Marcus 1981; Dahms 1995; Williams-Howze 1997). Depending on the species, the suppression of development may occur at the embryonic, naupliar, copepodid, or adult phase. Dormancy has been described as representing a spectrum of suppressed development, ranging from quiescence (retarded) to diapause (arrested) (see Grice and Marcus 1981). Quiescence occurs as an immediate response to adverse conditions in the environment; diapause typically occurs in response to a cue (e.g., photoperiod) that occurs prior to the onset of deteriorating environmental conditions (Grice and Marcus 1981). Generally, quiescent individuals resume development as soon as the immediate environmental conditions improve. For example, the development rate of many copepods, especially warm-water species, is slowed when they are exposed to cold temperatures, but speeds up when temperatures increase again. In the case of diapause, an organism generally undergoes biochemical, physiological, and/or endocrinal changes and enters diapause itself or produces diapause eggs (see Grice and Marcus 1981). In diapausing individuals development resumes only after completion of a refractory phase that might last days to months, depending on the species (Marcus 1979; Grice and Marcus 1981). During the refractory phase individuals do not resume development even if conditions are favorable (Watson and Smallman 1971). The distinction between quiescence and diapause is important because for at least one copepod species, *C. hamatus,* diapause eggs are capable of long-term survival (several months) even when exposed to toxic chemicals such as hydrogen sulfide (Marcus and Lutz 1998).

The existence of arrested development in copepods has been recognized for decades. The dormant CV copepodid stage of *Calanus* spp. attracted the attention of researchers early on in the 20th century, but the factors controlling this developmental delay still remain elusive (see Marcus and Boero 1998). While dormancy during the CV stage has been noted for several open-ocean copepod genera, dormancy during the egg (embryonic) phase is characteristic of many coastal and estuarine copepod species (Grice and Marcus 1981). In fact, most of the copepod species mentioned previously in the context of rearing larval fish are also known to exist as dormant/resting eggs in the environment. Indeed, the list of copepod species that exist as resting eggs in the environment has gradually grown over the last three decades as more geographic regions throughout the world's oceans have been studied. In the first review of the topic, Grice and Marcus (1981) noted that since the initial observation of copepod resting eggs in *A. tonsa* (Zillioux and Gonzalez 1972), 15 more marine calanoid copepod species had been shown to have a dormant egg phase as part of the life cycle. More recently, Mauchline (1998) noted the existence of a resting egg phase in 44 calanoid copepod species. In many cases this categorization was based on observations of nauplii hatching from samples of sediments that were collected from the seabed and incubated in the laboratory. While the nauplii presumably came from resting eggs existing in the sediments, the time at which the eggs were originally spawned and settled to the seabed in the field was not known. Thus it cannot be concluded with certainty that the nauplii developed from diapause eggs, quiescent nondiapause eggs, or some other intermediate egg type (Chen and Marcus 1997). A few studies (see review Marcus 1996), however, have either raised copepods in the laboratory and revealed conditions that led to the production of diapause and nondiapause eggs, collected eggs from wild females and revealed different hatching responses under existing field conditions and after incubation in the laboratory, or documented morphological features that can be used to distinguish diapause and nondiapause eggs that are found in the field. *Labidocera aestiva*, *C. hamatus*, *A. clausi*, *A. tonsa,* and *E. affinis* are among the calanoid copepod species that have been studied in this regard (e.g., Marcus 1980; Uye 1985; Ban 1992; Chen and Marcus

1997; Castro-Longoria 2001), that is, species for which diapause eggs have been clearly shown to exist.

RESTING EGGS AND AQUACULTURE

The remainder of this chapter will address the relevance of copepod resting eggs to aquaculture. Naess (1991a) believed that resting eggs of copepods were important for the maintenance of zooplankton populations in outdoor "extensive" pond systems that were used to rear fish. As a result, he examined the tolerance of copepod eggs to freezing, desiccation, and the insecticide rotenone, since the procedures for rearing fish involved annual draining of the ponds and exposure to rotenone. Rotenone was applied to kill parasites and predators. Based on the results, Naess concluded that subjecting ponds to these treatments led to reduced hatching of copepod resting eggs from the bottom sediments. At the time of this study the eggs were not identified as to species, but previous work (Naess 1991b) had shown that *A. clausi*, *E. affinis*, and *C. hamatus* were present in the ponds. Despite the fact that these treatments resulted in reduced egg hatching, Naess suggested that differences in egg type (quiescent nondiapause versus diapause) and differences in the tolerance of the various species to environmental stressors would eventually lead to dominance in the ponds of resistant copepod species by directional selection. If the surviving species were still suitable for the fish larvae, this evolutionary response would be beneficial, because it would allow treatment of the pond systems with compounds that reduce harmful parasites and predators.

In a subsequent study Naess (1996) quantified the abundance of copepod resting eggs in the sediments of outdoor marine ponds used for raising fish. The ponds were located along the coast of Norway. Five of the ponds were used exclusively for raising zooplankton. The bottoms of these systems were described as muddy, anoxic, and rich in sulfide. The other two ponds had more sandy, oxygenated seabeds and were used for raising cod (*Gadus morhua*). Copepod eggs were found in the seabed of all ponds, but the densities varied both within and between ponds. Densities ranging from 10^4 to 10^6 eggs/m^2 were found. Eggs of *E. affinis*, *Acartia teclae*, *A. clausi*, and *T. longicornis* were all found at densities exceeding 10^4 eggs/m^2. *Centropages hamatus* was represented in the form of resting eggs, but the densities were generally less than 10^3 eggs/m^2. The hatching success of the eggs varied. Naess suggested that one reason for reduced egg abundance was the presence of predators in the sediments. Hence, in ponds that were subjected to rotenone, densities were higher because presumably the benthic predators had been reduced. This suggestion is supported by more recent studies that indicate benthic amphipods can have a negative impact on the survival of benthic resting eggs of planktonic copepods (Albertsson and Leonardsson 2001). More recently, Engell-Sørensen et al. (2004) reported on the rearing of flounder (*Platichthys flesus*) juveniles in semiextensive systems associated with a Danish fjord. Some of the copepods used for food were derived from resting eggs obtained from the bottom of rearing ponds.

In the book, *Manual on the Production and Use of Live Feeds for Aquaculture,* Lavens and Sorgeloos (1996) suggested that copepod resting eggs could be used as an inoculum to initiate cultures of copepods. The suggestion is reasonable, but this author believes that the value of copepod resting eggs to aquaculture extends beyond their use as an inoculum. Since it is possible to store copepod resting eggs for days to years and obtain newly hatched nauplii, the risk of having no nauplii available to feed to fish larvae can be reduced. A second advantage is that since the resting stages of some taxa are resistant to surface disinfection agents commonly used in aquaculture, it may be possible to develop protocols that reduce the risk of parasitic contamination of larval fish cultures when nauplii derived from copepod resting eggs are used as a source of food.

VIABILITY OF RESTING EGGS

Research to date has shown that survival of copepod eggs with storage varies with egg type (nondiapause or diapause), species, and conditions of storage. When held under conditions that prevent hatching, nondiapause eggs of copepods generally do not remain viable for as long as diapause eggs. For example, Marcus et al. (1997) collected eggs from wild-caught females of *C. hamatus*, *A. tonsa*, and *L. aestiva*. Control eggs were incubated under conditions suitable for development, and within a few days, hatching was generally 84% or greater, indicating that the eggs were nondiapause. Experimental eggs were stored under sulfidic and/or anoxic (283–352 μM) conditions to prevent hatching. When the experimental eggs were returned to normoxic and nonsulfidic conditions within 7 days, viability (as evidenced by hatching) ranged between 50% and 80% (depending on the species). Longer exposures to sulfide and/or anoxia (e.g., 4 weeks) resulted in further reductions in hatching for all species. In a subsequent study, Marcus and Lutz (1998) collected eggs from wild-caught females of *C. hamatus* at a time of the year when the species was producing mostly diapause eggs (Chen and Marcus 1997). The eggs were incubated in the laboratory for 5 days to allow nondiapause eggs to hatch. The remaining unhatched presumptive diapause eggs were exposed to sulfide and/or anoxia as great as 1,363 μM. These eggs survived 20 weeks of such exposure with little or no reduction in viability, as shown by the hatch of more than 80% of the eggs when they were returned to normoxic and nonsulfidic conditions. In fact, some eggs survived exposure to anoxia for 437 days (Marcus and Lutz 1998). Studies of *E. affinis* also indicate that diapause eggs survive storage longer than nondiapause eggs (Ban and Minoda 1992). Generally, studies such as these are designed to gain insight into how long copepod resting eggs survive in the field. Hence, they do not typically consider storage conditions that may not have any relevance to survival of eggs in the field but that might extend shelf life in the laboratory and thus be of great importance to aquaculturists. There is a need for studies of the combined effects of anoxia and cold on shelf life. For example, nondiapause eggs of *Acartia steuri* and *A. clausi* that were covered by muddy sediments survived 100 and 165 days, respectively, when stored at 5°C, but only 75 and 70 days at 20°C (Uye 1980). Similar results were reported by Ban and Minoda (1992) for eggs of *E. affinis* stored in sediments at 15°C and 4°C. It is likely that dissolved oxygen concentrations were reduced in the sediments.

While the focus of this chapter is on resting eggs, it is important to note a study by Payne and Rippingale (2001). They stored nauplii of *G. imparipes* in static containers of seawater at 8°C. The nauplii were not fed. When these nauplii were held under these conditions for up to 12 days and then transferred to 20°C and fed, survival was excellent (> 99%), but for nauplii held longer, survival declined. Payne and Rippingale (2001) emphasized the minimal maintenance effort required for storing the nauplii in this way and hence the advantage this afforded in holding nauplii to feed fish larvae when needed. The authors did not indicate whether or not the nauplii developed while held at 8°C, but it is unlikely, since they were not fed.

Nauplii from Resting Eggs

If copepod eggs (and other stages as well) are to be stored for later use as a food source, not only must they survive storage, but also the nutritional quality of the nauplii that hatch from the eggs must be maintained. Highly unsaturated fatty acids (HUFAs), especially docosahexaenoic acid (DHA) and eicosapentaenoic acid (EPA), are essential components of the diet of fish (Sargent et al. 1997). Støttrup et al. (1999) and Støttrup and McEvoy (2003) reported that copepod nauplii are excellent sources of these essential fatty acids, unlike *Artemia* and rotifers (*Brachionus* spp.), which must be treated with fatty acid supplements to achieve a good nutritional status for feeding fish larvae. Moreover, Støttrup et al. (1999) showed that there was no change in the EPA content of *A. tonsa* eggs that were stored for 12 weeks at 4°C, but the amount of DHA decreased.

Resistance of Resting Stages to Surface Disinfection Agents

A problem associated with using live feeds in aquaculture is the introduction of pathogens and other contaminants into larval fish cultures in association with the food. This risk could be minimized with copepod resting eggs if surface disinfection protocols could be developed for their use. There is evidence that resting stages of several taxa, including the eggs of copepods, survive exposure to surface disinfection agents. Naess and Bergh (1994) subjected resting eggs of *A. clausi* and *E. affinis* to three disinfection agents (glutaraldehyde, FAM-30, or Buffodine). Eggs were washed with 0.2 μm filtered sterile seawater and monitored for hatching and survival. Bacterial growth was reduced on the eggs following treatment, but hatching and survival differed among the treatments. Lavens and Sorgeloos

(1996) suggested that nondiapause eggs may experience higher mortality than diapause eggs following such treatment because the outer covering of nondiapause eggs is thinner than the covering of diapause eggs (Ianora and Santella 1991; Fanelli et al. 1992). Other disinfection agents have been tested with resting stages of other taxa. Douillet (1998) reported that sodium hypochlorite was an effective disinfection agent for rotifer cysts. When cysts were exposed for 3 minutes to 0.5% sodium hypochlorite, hatching remained high, and in 97% of the trials the resulting populations were bacteria free. Higher concentrations reduced viability. Similarly Pati and Belmonte (2003) exposed cysts of *Artemia*, a rotifer, and a ciliate to five disinfectant agents commonly used in aquaculture. They reported that some cysts were able to survive commonly used dosage levels of these chemicals. While the goal of the study had been to determine levels of the disinfectants that would successfully kill cysts because they could be the source of contamination in aquaculture tanks, the fact that some survived suggests that certain concentrations might be effective for surface disinfection of live feeds that are derived from resting cysts or eggs.

One of the limitations of using the diapause eggs of some species (e.g., *L. aestiva* and *C. hamatus*) for aquaculture is that based on research to date, they must be stored for a lengthy period (weeks to months) before they can be induced to hatch, and eggs stored for a very long term (years) require more time to hatch when placed under conditions suitable for hatching (personal observation). For example diapause eggs of *L. aestiva* must be stored at 5°C for approximately 10–30 days before they can be induced to hatch at 19°C (Marcus 1987). Conversely, diapause eggs of *C. hamatus* must be stored for approximately 4 months at 25°C before they can be induced to hatch at temperatures less than 20°C (personal observation). In addition diapause eggs of *C. hamatus* that have been stored for approximately 4 months at 25°C normally hatch within 5–7 days when placed at 15°C. Diapause eggs that were stored for 22 months at 25°C, however, required 10 days of subsequent incubation at 15°C to hatch (personal observation). Research that elaborates the hatching response of dormant eggs under various conditions is needed so that better control of the process is achieved. For example Jo and Marcus (2004) subjected diapause eggs of *C. hamatus* to different temperature regimens following storage under anoxic conditions at 25°C for approximately 8 months. They found that for eggs stored for this period of time, the time to hatching could be shortened by exposing the eggs to 15°C for only 36 hours followed by re-exposure to 25°C. It may also be possible to reduce the time to hatching by exposing eggs to certain chemical conditions. For example, normally dehydration breaks the diapause of *Artemia* cysts, and cysts must be rehydrated for development to resume, followed by hatching. Clegg and Jackson (1998), however, showed that diapause could be broken in *Artemia* cysts, which were prevented from dehydrating, by exposing them to ammonium chloride.

CONCLUSION

Clearly, the production of large numbers of copepod resting eggs, especially diapause eggs, could facilitate the availability of copepod nauplii for aquaculture. Sufficient information on the cultivation requirements of several copepod species (e.g., *L. aestiva*, *C. hamatus*, *Acartia spp*., and *E. affinis*) that produce diapause eggs or other egg types capable of surviving lengthy periods of suppressed development exists to support continued exploration of this approach. *Labidocera aestiva* is a temperate warm-water free-spawning species. Diapause and nondiapause egg production are primarily controlled by photoperiod and temperature (Marcus 1980). Both egg types are smooth. Generally the diapause eggs must be stored for at least 1 month at 5°C before they can be induced to hatch by placing them at 25°C (Marcus 1987). Diapause eggs survive storage at 5°C for several months. *Centropages hamatus* is a temperate cold-water free-spawning species. Diapause and nondiapause egg production are controlled by photoperiod and temperature (Marcus and Murray 2001). Both egg types are spiny, which can be somewhat of a problem, because the spines provide a surface to which debris readily clings (personal observation). Diapause eggs must be stored at 25°C for approximately 4 months before placing the eggs at temperatures less than 20°C induces hatching (Chen and Marcus 1997). Diapause eggs of *C. hamatus* can survive as long as 22 months when held at 25°C (Marcus and

Murray 2001). *Eurytemora affinis* is a widespread species that occurs in brackish environments. Photoperiod plays a role in the type of eggs produced, as does population density (Ban 1992). Females carry eggs in sacs, but sacs with diapause eggs are released in the water column and sink to the seabed. Diapause eggs of *E. affinis* survive storage for as long as 20 months (Ban unpublished, as cited in Ban and Minoda 1992). Since *E. affinis* tolerates reduced salinity, and some populations occur in freshwater lakes (Ban and Minoda 1991), it may be possible to maintain cultures free of marine parasites by rearing animals in freshwater. *Acartia spp.* spawn eggs freely into the water column, and several species produce diapause eggs (see Marcus 1996; Castro-Longoria and Williams 1999). *A. clausi* and *A. bifilosa* are temperate cold-water species; *A. tonsa*, *Acartia californiensis*, and *Acartia latisetosa* are temperate warm-water species. Diapause egg production by *A. clausi* is controlled by photoperiod (Uye 1985); the factors controlling diapause egg production by *A. tonsa* are not well understood (see William-Howze 1997). Nevertheless *A. clausi* and *A. tonsa* have been cultivated extensively and in some cases for feeding larval fish. Nondiapause eggs of species of *Acartia* are amenable to some manipulation aimed at suppressing development. For example, eggs stored in the cold remain viable for several weeks (Støttrup et al. 1999).

In conclusion, dormancy during the egg (embryonic) phase is a feature that potentially makes certain copepod species particularly suitable for aquaculture. If eggs can be stored for later use, then it is no longer necessary to maintain simultaneous cultures of copepods and fish larvae. Moreover, having a stock of eggs facilitates recovery from population crashes that may occur during cultivation phases. Hence, research that determines the most effective storage conditions to prolong survival and maintain nutritional quality is needed. In addition, dormant eggs may lend themselves more to disinfection than later stages, thus reducing the chances for contamination of fish cultures.

REFERENCES

Albertsson, J. and Leonardsson, K. 2001. Deposit-feeding amphipods (*Monoporeia affinis*) reduce the recruitment of copepod nauplii from benthic resting eggs in the northern Baltic Sea. *Marine Biology* 138:793–801.

Ban, S. 1992. Effects of photoperiod, temperature, and population density on induction of diapause egg production in *Eurytemora affinis* (Copepoda: Calanoida) in Lake Ohnuma, Hokkaido, Japan. *Journal of Crustacean Biology* 12:361–367.

Ban, S. and Minoda, T. 1991. The effect of temperature on the development and hatching of diapause and subitaneous eggs in *Eurytemora affinis* (Copepoda:Calanoida) in Lake Ohnuma, Hokkaido, Japan. *Bulletin of the Plankton Society of Japan. Proceedings of the 4th International Conference on Copepoda*:299–308.

Ban, S. and Minoda, T. 1992. Hatching of diapause eggs of *Eurytemora affinis* (Copepoda:Calanoida) collected from lake bottom sediments. *Journal of Crustacean Biology* 12:51–56.

Castro-Longoria, E. 2001. Comparative observations on the external morphology of subitaneous and diapause eggs of *Acartia* species from Southampton water. *Crustaceana* 74:225–236.

Castro-Longoria, E. and Williams, J.A. 1999. The production of subitaneous and diapause eggs: a reproductive strategy for *Acartia bifilosa* (Copepoda:Calanoida) in Southampton Water, UK. *Journal of Plankton Research* 21:65–84.

Chen, F. and Marcus, N.H. 1997. Subitaneous, diapause, and delayed-hatching eggs of planktonic copepods from the northern Gulf of Mexico: morphology and hatching success. *Marine Biology* 127:587–597.

Clegg, J.S. and Jackson, S.A. 1998. The metabolic status of quiescent and diapause embryos of *Artemia franciscana* (Kellogg). *Advances in Limnology. Stuttgart* 52:425–439.

Dahms, H. 1995. Dormancy in Copepoda—an overview. *Hydrobiologia* 306:199–211.

Douillet, P. 1998. Disinfection of rotifer cysts leading to bacteria free populations. *Journal of Experimental Marine Biology and Ecology* 224:183–192.

Engell-Sørensen, K., Støttrup, J.G., and Holmstrup, M. 2004. Rearing of flounder (*Platichthys flesus*) juveniles in semiextensive systems. *Aquaculture* 230:475–491.

Fanelli, G., Ianora, A., and Santella, L. 1992. Produzione de uova de diapausa nel ciclo biologico del copepode *Acartia latisetosa* (Kriczaguin). *Oebelia Supplement* XVII:295–302.

Grice, G. and Marcus, N.H. 1981. Dormant eggs of marine copepods. *Oceanography and Marine Biology Annual Review* 19:125–140.

Ianora, A. and Santella, L. 1991. Diapause embryos in the neustonic copepod *Anomlocera patersoni*. *Marine Biology* 108:387–395.

Jo, S.-G. and Marcus, N.H. 2004. The effect of temperature and duration of incubation on the hatching of diapause eggs of *Centropages hamatus* (Copepoda, Calanoida). *Aquaculture Research* 35:432–435.

Lavens, P. and Sorgeloos, P. 1996. Manual on the Production and Use of Live Food for Aquaculture. FAO Fisheries Technical Paper 361. Food and Agriculture Organization of the United Nations, Rome, Italy.

Marcus, N.H. 1979. On the population biology and nature of diapause of *Labidocera aestiva* (Copepoda:Calanoida). *Biological Bulletin* 157:297–305.

Marcus, N.H. 1980. Photoperiodic control of diapause in the marine calanoid copepod *Labidocera aestiva*. *Biological Bulletin* 159:311–318.

Marcus, N.H. 1987. Differences in the duration of egg diapause of *Labidocera aestiva* (Copepoda: Calanoida) from the Woods Hole region, Massachusetts. *Biological Bulletin* 173:169–177.

Marcus, N.H. 1996. Ecological and evolutionary significance of resting eggs in marine copepods: past, present, and future studies. *Hydrobiologia* 320:140–152.

Marcus, N.H and Boero, F. 1998. Minireview: The importance of benthic-pelagic coupling and the forgotten role of life cycles in coastal aquatic systems. *Limnology and Oceanography* 43:763–768.

Marcus, N.H. and Lutz, R.V. 1998. Longevity of subitaneous and diapause eggs of *Centropages hamatus* (Copepoda:Calanoida) from the northern Gulf of Mexico. *Marine Biology* 131:249–257.

Marcus, N.H., Lutz, R.V., and Chanton, J.P. 1997. Impact of anoxia and sulfide on the viability of eggs of three planktonic copepods. *Marine Ecology Progress Series* 146:291–295.

Marcus, N.H. and Murray, M. 2001. Copepod diapause eggs: a potential source of nauplii for aquaculture. *Aquaculture* 201:49–60.

Mauchline, J. 1998. *The Biology of Calanoid Copepods*. San Diego, California, USA: Academic Press. 710 pp.

Naess, T. 1991a. Tolerance of marine calanoid resting eggs: effects of freezing, desiccation and Rotenone exposure—a field and laboratory study. *Marine Biology* 111:455–459.

Naess, T. 1991b. Marine calanoid copepod resting eggs in Norway: abundance and distribution of two copepod species in the sediment of an enclosed marine basin. *Marine Biology* 110:261–266.

Naess, T. 1996. Benthic resting eggs of calanoid copepods in Norwegian enclosures used in mariculture: abundance, species, composition, and hatching. *Hydrobiologia* 320:161–168.

Naess, T. and Bergh, O. 1994. Calanoid copepod resting eggs can be surface-disinfected. *Aquacultural Engineering* 13:1–9.

Pati, A.C. and Belmonte, G. 2003. Disinfection efficacy on cyst viability of *Artemia franciscana* (Crustacea), *Hexartha fennica* (Rotifera), and *Fabrea salina* (Ciliophora). *Marine Biology* 142:895–904.

Payne, M.F. and Rippingale, R.J. 2001. Effects of salinity, cold storage and enrichment on the calanoid copepod *Gladioferens imparipes*. *Aquaculture* 201:251–262.

Pepin, P. and Penney, R.W. 1997. Patterns of prey size and taxonomic composition in larval fish: are there general size-dependent models? *Journal of Fish Biology* 51 (Supplement A):84–100.

Rippingale, R.J. and Payne, M.F. 2001. Intensive cultivation of a calanoid copepod for live food in fish culture. On-line version, http://www.aims.gov.au/pages/research/hatchery-feeds/pdf/frdc-final-report-96398.pdf.

Sargent, J.R., McEvoy, L.A., and Bell, J.G. 1997. Requirements, presentation and sources of polyunsaturated fatty acids in marine fish larval feeds. *Aquaculture* 155:117–127.

Støttrup, J.G. 2000. The elusive copepods: their production and suitability in marine aquaculture. *Aquaculture Research* 31:703–711.

Støttrup, J.G., Bell, J.G., and Sargent, J.R. 1999. The fate of lipids during development and cold-storage of eggs in the laboratory-reared calanoid copepod, *Acartia tonsa* Dana, and in response to different algal diets. *Aquaculture* 176:257–269.

Støttrup, J.G. and McEvoy, L.A. 2003. *Live Feeds in Marine Aquaculture*. Ames, Iowa, USA: Iowa State Press. 336 pp.

Uye, S. 1980. Development of neretic copepods *Acartia clausi* and *A. steuri*. I. Some environmental factors affecting egg development and the nature of resting eggs. *Bulletin of the Plankton Society of Japan* 27:1–9.

Uye, S. 1985. Resting egg production as a life history strategy of marine planktonic copepods. *Bulletin of Marine Science* 37:440–449.

Watson, N. and Smallman, B. 1971. The role of photoperiod and temperature in the induction and termination of arrested development in two species of freshwater cyclopoid copepods. *Canadian Journal of Zoology* 49:855–862.

Williams-Howze, J. 1997. Dormancy in the free-living copepod orders Cyclopoida, Calanoida, and Harpacticoida. *Oceanography and Marine Biology: Annual Review* 35:257–321.

Zillioux, E. and Gonzalez, J. 1972. Egg dormancy in the neretic calanoid copepod and its implications to overwintering in boreal waters. *Archives of Limnology. European Marine Biology Symposium* 5:217–230.

2

The Potential to Mass-Culture Harpacticoid Copepods for Use as Food for Larval Fish

John W. Fleeger

ABSTRACT

Harpacticoid copepods are a promising alternative food resource for larval and juvenile fishes that require live feeds in mariculture. Harpacticoids, fed alone or as a supplement, have been shown to promote faster growth than rotifers, brine shrimp, or both together. Research suggests that this is most likely related to small body size from first nauplius to adult, which favors ingestion by gape-limited fishes, and the observation that harpacticoids are naturally rich in essential fatty acids needed for marine fish growth. In addition, cultures of harpacticoids have the potential to be productive enough to be of value in mariculture because harpacticoids (a) have high reproductive output and short generation times; (b) are able to withstand variable conditions of temperature, salinity, and waste accumulation; and (c) have flexible food requirements regarding nutritional content and particle shape and size that facilitate the use of formulated feeds, monospecific algae, or both. As a result harpacticoid densities may exceed 100,000 individuals per liter in mass cultures. Finally, harpacticoid biodiversity is high both within and among habitats, and there are many species available for mass culture. Characteristics associated with a primarily benthic lifestyle, however, render many harpacticoid species inappropriate for large-scale mariculture applications. Most have been cultured only in the presence of substrates, complicating the harvest of early developmental stages because few harpacticoid nauplii swim. Furthermore very few species encyst or undergo diapause, requiring that high yields from mass cultures be synchronized with the birth or arrival of larval fish. Escape responses and behavioral attributes have been little studied. Many adult and juvenile harpacticoids are relatively poor swimmers, suggesting a reduced escape response. Alternatively, harpacticoids are commonly protected by long, heavy caudal spines that may aid resistance to ingestion. Currently, the search continues for harpacticoids with favorable features for mass culture that are also high-quality food for mariculture applications. Additional species with a small body size, good swimming abilities in all developmental stages, and ability to be cultured in the absence of substrate are needed. Species in the genera *Tisbe* and *Nitokra* display many of these features and should be the focus of additional research to determine their utility in fish mariculture.

INTRODUCTION

Many marine fishes produce small pelagic eggs. Larvae hatched from small eggs require a source of live food very soon after the onset of exogenous feeding. Larvae of these species raised in mariculture are usually fed the rotifer *Brachionus plicatilis* followed by the brine shrimp *Artemia.* The early feeding regime in the rearing process often limits survival and growth, and thus, the success and/or cost of fish production. Several lines of reasoning suggest that harpacticoid copepods have potential as an alternative food resource in larval fish mariculture that might replace or supplement rotifers, brine shrimp, or both. The purpose of this review is to summarize current understanding of this potential. Mass-culture techniques and population growth characteristics of harpacticoid species known to have potential for mariculture applications are considered below. To complement the recent review of Cutts (2002), features of the biology of harpacticoids that may be construed as positive and negative regarding their use in larviculture are discussed, and suggestions for future avenues of research to foster harpacticoid use in larviculture are highlighted.

There are over 3,000 described species in the Order Harpacticoida, one of 10 orders in the subclass Copepoda of the crustacean class Maxillopoda (Huys and Boxshall 1991). The number of recently confirmed cryptic species (e.g., Rocha-Olivares et al. 2001) and the high number of relatively poorly explored habitats (e.g., tropical mangrove forests, the deep sea) suggest that the true number of harpacticoids worldwide is much higher.

Adult harpacticoid copepods are typically small in body length (~1 mm) and width (~350 μm) as well as mass (~3 μg dry mass). Most harpacticoids are free living, although ectoparasitic and commensal harpacticoids are common on a variety of animals. For example, many species inhabit the protected gill chambers of decapod crustaceans. About 1,000 harpacticoid species live exclusively in freshwater habitats.

The diversity of habitats and lifestyles exploited by marine harpacticoids is great. Although a few harpacticoids are planktonic, the majority are benthic. Harpacticoids are especially abundant in muddy sediments and on foliose macroalgae (Hicks and Coull 1983). Harpacticoids are typically the second most abundant taxon, after nematodes, in the sedimentary meiobenthos. Many species are associated with the epiphytes of seagrasses and macrophytes (Hall and Bell 1988; Rutledge and Fleeger 1993). They live interstitially in sand, in mud as burrowers and tube dwellers (Chandler and Fleeger 1984), and on surfaces such as at the sediment–water interface or on microalgal-covered hard substrates (Danovaro and Fraschetti 2002; Atilla et al. 2003). Although many harpacticoids are passively eroded into the water column when currents exceed some critical erosion velocity (Palmer 1988), some species actively emerge from the sediment by swimming (Armonies 1988; Walters and Bell 1994; Suderman and Thistle 1998; Thistle 2003), potentially increasing encounter rates with predators (Gregg and Fleeger 1997).

Harpacticoids are natural prey to many larval and juvenile fishes (Gee 1989; Coull 1990). Comparisons between the faunal compositions found in fish gut contents with samples from the environment suggest that juvenile fishes may actively select harpacticoids (Feller and Coull 1995). Strong selection may be related to the shallow depth profile of many harpacticoids in sediments (Fleeger et al. 1995a) and the higher frequency of emergence by harpacticoids compared with other meiofaunal groups (McCall and Fleeger 1995). Mean abundances of harpacticoids in fish gut contents often exceed 1,000 harpacticoids per juvenile or adult fish (McCall and Fleeger 1993).

Harpacticoid copepods are an important food resource for many species of marine fish (Coull 1990). Many fishes undergo an ontogenetic shift from a diet comprised largely of harpacticoids to larger-bodied prey when they reach a standard length of about 35 mm (McCall and Fleeger 1995). Harpacticoid copepods, however, may serve as prey for the entire life of some marine fishes (Tipton and Bell 1988; Toepfer and Fleeger 1995) or until a larger size is reached (McCall and Fleeger 1993; Feller and Coull 1995). Species of flatfish, gobies, salmonids, sciaenids, and blennies are sometimes considered to be obligatory harpacticoid feeders, at least for a portion of their lives (Coull 1990; McCall and Fleeger 1995).

HARPACTICOID COPEPODS AND THE GROWTH OF FISH

As has been shown for other copepods, harpacticoids may promote rapid growth, a high reproductive rate, or both, in fishes and invertebrates (Cutts 2002). Volk et al. (1984) showed that food conversion efficiency for juvenile *Oncorhynchus keta* fed the harpacticoid *Tigriopus californicus* was higher than when fed calanoid copepods or amphipods. They attributed differences to the higher caloric content of *T. californicus* compared with amphipods and a poor escape response compared with calanoid copepods. Kreeger et al. (1991) examined the nutritional value of newly hatched *Artemia, Artemia* supplemented with lipid microspheres, and the harpacticoid copepod *T. californicus*. They found that mysid shrimp survival, growth, and the proportion of females brooding offspring were not significantly different among dietary treatment groups, but that the number of viable offspring produced by mysids was improved by adding lipid microspheres and *T. californicus* to a diet of *Artemia*. They attributed the difference to the observation that both lipid microspheres and *T. californicus* are rich sources of essential fatty acids. Støttrup and Norsker (1997) found that larval turbot (*Scophthalmus maximus*) growth and survival when fed rotifers were enhanced when the harpacticoid *Tisbe holothuriae* was provided with rotifers. Including *T. holothuriae* in a daily dietary regime improved appetite, growth rate, and pigmentation of Dover sole (*Microstomus pacificus*) compared with a diet of *Artemia* alone (Heath and Moore 1997).

FAVORABLE LIVE FEED CHARACTERISTICS OF HARPACTICOIDS

An increased growth rate when harpacticoids are included in the diet of larval fishes may be attributed to many factors, including nutritional quality, favorable body size, and effects on larval appetite. Harpacticoids have superior nutritional qualities compared with brine shrimp and rotifers that may stimulate fish growth (Cutts 2002). Highly unsaturated fatty acids (HUFA), especially n-3 fatty acids, are typically represented in harpacticoids in high concentrations, compared with other live feeds used in mariculture. Table 2.1 shows values of selected phospholipid fatty acids (PLFAs) from *Amphiascoides atopus* grown in a laboratory mass culture (the system described by Sun and Fleeger 1995) compared with 1-day old *Artemia* nauplii (analysis conducted by Microbial Insights, Inc.). Note that *A. atopus* is rich in the fatty acids, 22:6*n*3 and 22:5*n*3, essential to marine fish growth, while these fatty acids are absent in *Artemia*. Sun and Fleeger (1995) report that 28% of all phospholipids in *A. atopus* are essential n-3 fatty acids. Other authors have found similarly high amounts of n-3 fatty acids in harpacticoids (Watanabe et al. 1983; Norsker and Støttrup 1994; Nanton and Castell 1999, Rhodes and Boyd, this volume). Compared with brine shrimp and rotifers, harpacticoids appear to have uniformly favorable ratios and amounts of essential fish oils.

At least some harpacticoids are able to synthesize longer-chained HUFA (Norsker and Støttrup 1994), thereby increasing levels of n-3 fatty acids that are essential to fishes. Nanton and Castell (1998) suggest that harpacticoids have enzymes (i.e., Δ-5, Δ-6 desaturase and elongase) that are necessary for the conversion of shorter chain n-3 polyunsaturated fatty acids to the longer-chained essential fatty acids docosahexaenoic acid (DHA) and eicosapentaenoic acid (EPA). In the absence of this enzymatic capability, animals may require a diet that matches their DHA and EPA requirements. Because harpacticoids can synthesize n-3 HUFA, feeds without enriched levels of marine fish oil can be used to produce a favorable DHA/EPA composition. Thus, harpacticoids may have few specific nutritional needs in culture and may be able to synthesize long-chained HUFAs regardless of their diet (e.g., harpacticoids show favorable nutritional qualities when fed algal monocultures and formulated feeds). Fukusho et al. (1980), however, reported that harpacticoids fed n-yeast resulted in increased survival and growth of the larvae of mud dab (*Limanda yokohamae*) compared with harpacticoids fed baker's yeast. In addition, Cutts (2002) reports that synthesis of HUFA may be a rate-limiting step for harpacticoid growth, and dietary supplement of fatty

Table 2.1. Selected fatty acids (expressed in pmol) from a phospholipid fatty acid analysis conducted on *Amphiascoides atopus* and day-old *Artemia*

Species	Dry wt/g tissue	18:4*n*3	20:5*n*3	20:4*n*3	20:2*n*3	22:6*n*3	22:5*n*3
A. atopus	1.41	0.25	6.3	0.05	0.08	10.8	0.95
Artemia	1.92	7.9	4.0	0.07	0.02	0	0

Note: Values represent means of two replicate analyses.

acids to harpacticoids in culture may increase productivity and naupliar survival. A priority for future research should be to design diets that optimize growth and nutritional value of harpacticoid copepods in culture (see Carli et al. 1995).

The body size of harpacticoids may be highly conducive to capture by larval and juvenile fishes. Body length among harpacticoid species varies from about 40–80 μm in the first nauplius stage to about 300–1,500 μm in adults. Body widths are usually about one-fourth to one-third of the body length in adult and juvenile harpacticoids; nauplii are typically about as wide as long (although width exceeds length in many species [Chandler and Fleeger 1984]). Given the small gape of many larval and juvenile fishes, this range is favorable for consumption. Body width of prey may be more important than length to larval fish, because fish capture prey head-on. It is possible that the smallest larval fishes may feed on naupliar stages and that larval and juvenile fishes may consume larger harpacticoids with growth. If harpacticoids are more easily consumed because they are smaller, their body size alone may contribute to the observed better fish growth compared with brine shrimp.

GROWTH IN LABORATORY CULTURE

Chandler (1986) discussed the potential for the culture of harpacticoid copepods in sedimentary systems, primarily for the purposes of ecological, physiological, and toxicological research. Culture systems with or without sediment capable of producing hundreds to thousands of individuals required for such studies are common (examples include Battaglia 1970; Strawbridge et al. 1992; Ingole 1994; Chandler et al. 1997; Lotufo and Fleeger 1997; Kovatch et al. 1999), but most species so cultured have not been grown under conditions expected to produce the numbers of copepods necessary for mariculture applications. In addition, some harpacticoids (e.g., *Tisbe* spp., *Amphiascoides* spp.) are common fouling organisms in fish or macroinvertebrate culture systems and are nearly ubiquitous in laboratory culture systems. These observations suggest that many harpacticoid species have the potential for mass culture. Given the great biodiversity of harpacticoids in understudied habitats, many unexamined candidate species for mass culture probably exist.

Some harpacticoid species, however, have proven to be very successful in mass culture. *Tisbe* spp. (Kahan et al. 1982; Zhang and Uhlig 1993; Støttrup and Norsker 1997; Nanton and Castell 1998), *Tigriopus* spp. (Fukusho 1980; Volk et al. 1984), *Schizopera elatensis* (Kahan et al. 1982), *Euterpina acutifrons* (Kraul 1990), *A. atopus* (Sun and Fleeger 1995), and *Nitokra lacustris* (Rhodes 2003) have been cultured at densities that have the potential to be scaled to the needs of mariculture. Of these, only *E. acutifrons* is a holo-planktonic, suspension-feeding copepod, and most research has centered on species in the genera *Nitokra*, *Tisbe*, *Tigriopus*, and *Amphiascoides*.

Various approaches have been used to mass culture harpacticoid copepods (Støttrup 2000). Kahan et al. (1982) grew three species in a mesh basket suspended in a larger tank in which larval fish were maintained. Fukusho (1980) cultured *Tigriopus japonicus* in 10 × 15 m outdoor tanks, to a water depth of 1.4 m, with a species of rotifer (see also Tseng and Hsu 1984). Sun and Fleeger (1995) used trays with a combined basal area of 4 m^2 and layered with pebbles to mass culture *A. atopus*. Støttrup and Norsker (1997) conducted batch culture in flat trays and continuous culture in a self-contained 150-L conical reactor (with continuous delivery of phytoplankton), in which plastic balls provided a substratum for *T. holothuriae*. Rhodes (2003) used seawater without substrate in large(> 10 L) tanks or trays to culture *N. lacustris*. Certainly, culturing methods and yields can be optimized in future research, especially given recent advances in filtering capacity and the management of waste products. Although harpacticoids grown in mass culture have been fed many substances, including shrimp pellets, a lettuce/*Mytilus* paste, vegetable juices, yeast, and formulated feeds, including commercially produced fish flakes, the most common feed is natural or cultured algae (usually in monoculture).

Among these species, densities ranging from 1 to 115/mL have been achieved under conditions of mass culture (Støttrup 2000; Rhodes 2003). Various means have been designed to measure yield or production of harpacticoids. Fukusho (1980) harvested 2–3 kg, wet weight, of *T. japonicus* over regular intervals for 89 days, while Sun and Fleeger (1995) estimated that 2–8 million adult and juvenile *A. atopus* and over 5 g dry biomass were removed from cultures on a daily basis for about 4 months. Støttrup and Norsker (1997) measured a daily production of 300,000–500,000 *T. holothuriae* corresponding to 30–100 mg/m^2 in trays and a continuous reactor. No system has been maintained at high levels of production for more than a few months.

Generally, harpacticoid copepod species that grow well in culture tend to be epibenthic in lifestyle and ecological generalists. The adults (and later copepodid stages) of many species successful in culture are capable of actively swimming into the water column. They may do so with stimulation from infrequent but specific environmental conditions that might include the onset of hypoxia (Vopel et al. 1996) or sediment contamination (Lotufo 1997), and possibly the mating cycle (Bell et al. 1988). Some species appear to enter the water column on a regular basis, for example, at the onset of darkness. Most species that have been successful in culture inhabit estuaries or areas that are frequently disturbed. Members of the genera *Tigriopus* and *Amphiascoides* are residents of tide or splash pools, and members of the genus *Tisbe* frequently colonize disturbed muddy sediments (Warwick et al. 1988), although *T. holothuriae* is a commensal species. It is possible that these traits could be exploited to help find additional species. Investigators interested in finding harpacticoids suitable for mass culture may concentrate efforts in stressful and physically varying habitats or examine existing culture systems for fish or invertebrates for harpacticoid species already thriving in a laboratory setting.

FAVORABLE TRAITS FOR LABORATORY CULTURE

Harpacticoid copepod densities in mass culture can exceed 100,000/L, surpassing the success to date with calanoid copepods (Støttrup 2000). There are many reasons for this success. Harpacticoids have few specialized dietary needs, diverse feeding mechanisms, and life-cycle traits that favor rapid growth. These traits, including high reproductive potential, rapid development rate, and a low age of first reproduction, are associated with an opportunistic lifestyle (Cutts 2002). In addition, harpacticoid copepods have broad tolerances to variable environmental conditions, and many species appear to tolerate waste build up in culture. Because most harpacticoids are substrate oriented, population growth may depend on the surface area of solid substrate, and they can therefore be produced in volumetrically dense populations (Cutts 2002).

Although many harpacticoids are capable of suspension feeding (Decho 1988; Suderman and Thistle 1998), most feed primarily by gleaning microflora from surfaces or particles (including detritus) of variable shape, size, and consistency (Hicks and Coull 1983; Lewis et al. 1998). Perhaps as a result, harpacticoids in culture are able to use many different food sources (including algal monocultures and pellet- or flake-style formulated feeds) more readily than other copepods (Norsker and Støttrup 1994). Furthermore, few harpacticoids are predators (Seifried and Durbaum 2000), and adult harpacticoids are unlikely to cannibalize nauplii in culture. Calanoids, as suspension feeders, typically require large volumes of water for suspension feeding, and reach rather low densities, 100–2,000/L, in mass culture (Ogle 1979; Støttrup et al. 1986). Also, adult calanoids may cannibalize nauplii in culture, reducing growth potential (Cutts 2002).

In addition, many harpacticoids show good growth characteristics, including a rapid development rate and relatively high reproductive potential (Cutts 2002). Rhodes (2003) reported that *N. lacustris* has a generation time of 10–12 days, a relatively large brood size, and frequent broods such that 60–140 eggs are produced in the lifetime of a female. In general, population growth in harpacticoid copepods is highly temperature dependent (Bergmans 1984), with most rapid growth for temperate and tropical species $\geq$ 25°C. Harpacticoids also have a relatively high r_{max} (Pinto et al. 2001; Hagiwara et al., 1995). Rhodes (2003) applied population-modeling techniques to optimize population growth in *N. lacustris*.

Harpacticoid copepods are also able to withstand variable conditions while under culture. Many harpacticoids tolerate extreme temperature and salinity conditions (Miliou 1993; Lee and Yan 1994; Powlik et al. 1997). Most harpacticoids, however, are not tolerant of low oxygen conditions (Wetzel et al. 2001), and aeration of cultures is desirable. Moreover harpacticoids are generally not tolerant of contaminants such as metals or hydrocarbons (Coull and Chandler 1992). Harpacticoids, however, are tolerant of waste products such as ammonia that build up in mass culture (Støttrup and Norsker 1997).

LIMITING FACTORS

Relatively few harpacticoid species are known to encyst, produce dormant stages, or undergo diapause. The adults of only one harpacticoid species (*Heteropsyllus nunni*) encyst (Williams-Howze 1996), and there are no reports of diapause eggs. *Heteropsyllus nunni* is a common member of the estuarine harpacticoid assemblage along the North American Atlantic coastline but is seldom found in high abundance; nor has it been successfully mass cultured. Cysts of *H. nunni* are therefore rare and difficult to find. Lonsdale et al. (1993) identified reproductive dormancy in late-stage juvenile females of one harpacticoid species. Dahms (1995) reports an Arctic harpacticoid that enters diapause without encysting. The general lack of dormancy suggests that it is unlikely that a harpacticoid will ever be widely available in an encysted stage that can be stored for on-demand needs in mariculture (as is currently feasible for *Artemia* or calanoid eggs).

Harpacticoids, however, may be capable of retaining good nutritional qualities under conditions of starvation, at least on the order of 5–10 day. Figure 2.1 presents lipid data (for various classes based on a percentage of the total dry weight) for starved *A. atopus* (data collected by thin layer chromatography and flame ion detection after copepods were removed from mass culture and held in flasks containing only seawater). These data show that two lipid classes, phospholipids and triglycerides, remain high through 5 days of starvation, and that all lipid values decreased dramatically between 5 and 10 days of starvation. Because fatty acids essential to fish growth belong to the phospholipid class, and because triglycerides are good energy sources, the nutritional quality of *A. atopus* held in flasks appears to remain high for 5–10 days without addition of food. Harpacticoids may survive in culture with little or no added food for extended times

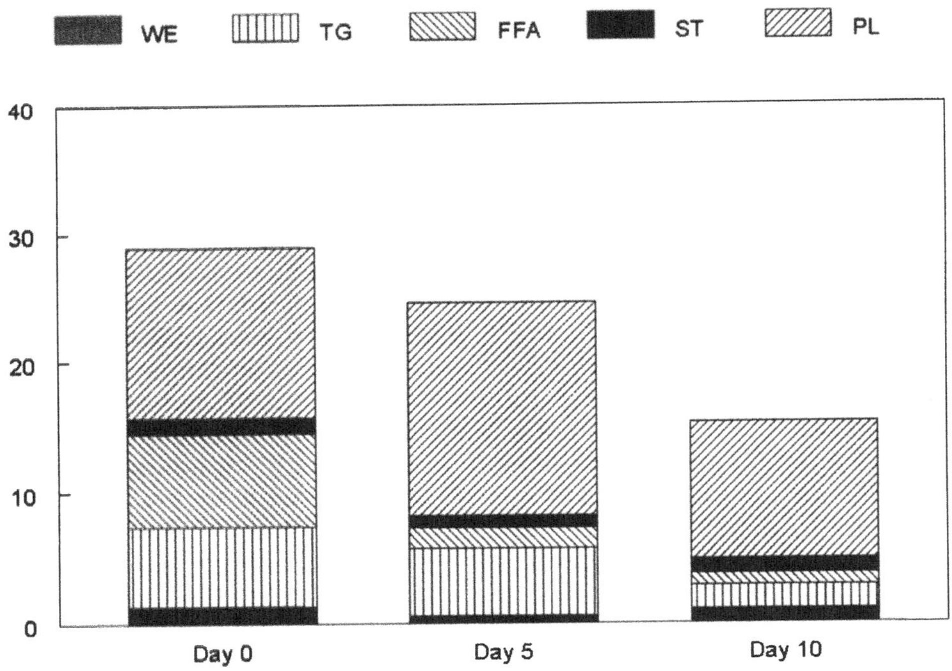

Figure 2.1. Effects of starvation on lipid profiles in *Amphiascoides atopus*. On day 0, lipids comprised 29.0% of total copepod dry weight; on day 5, lipids comprised 24.5% of total copepod dry weight; and on day 10, lipids comprised 15.3% of the total copepod dry weight. WE, wax esters; TG, triglycerides; FFA, free fatty acids; ST, sterols; and PL, phospholipids.

(Weiss et al. 1996), further suggesting that species are broadly tolerant of starvation.

The nutritional value of *Artemia* under starvation conditions appears to decline rapidly, as does its value as a fish feed (Léger et al. 1986). Although additional study of more species is required, harpacticoids may prove to be a good food source for several days after harvest from a mass culture. Harpacticoids cultured in one location could therefore be shipped to a distant site and briefly held with little loss of nutritional value. If this cannot be accomplished, it will be incumbent on harpacticoid culturists to even more finely synchronize the growth of a culture so that production is well correlated with demand (either the birth or arrival of larval fish). Alternatively, copepod cultures will have to be maintained very near to larviculture facilities.

Because most harpacticoid copepods are benthic for at least part of their life cycle, a substratum is required for culture in most species, and the clean harvest of all developmental stages from mass culture presents a serious challenge. Chandler (1986) devised a processed sediment to serve as a culture medium for harpacticoids that is easily sieved and allows good separation between adult harpacticoids and the sediment in which they are cultured. Detritus from the addition of food and fecal pellets and debris, however, are similar in size to larval and juvenile harpacticoids and build up rapidly in culture, making clean separation unlikely and mass culture in the presence of sediment effectively impractical.

Some adult and juvenile harpacticoids, however, have good swimming skills and frequently enter the water column. Filters associated with the circulation of water in culture chambers have been used to capture swimming, mostly postnaupliar, harpacticoids in large numbers with little effort and a minimum of debris (Sun and Fleeger 1995; Støttrup and Norsker 1997). Rhodes (2003) harvested the swimming adults of *N. lacustris* from culture systems by filtering the upper portions of the water column.

Because the nauplii of most harpacticoids are photonegative, with sediment-seeking and sediment-clinging behavior but poor swimming ability, some nauplii have proven difficult to harvest without concurrent collection of large quantities of debris. Even nauplii of harpacticoids grown in containers without sediment may cling to the bottom and aggregate with detritus and fecal and food debris in the absence of a substrate (Rhodes 2003). The nauplii of *T. holothuriae*, however, are an exception in that they are photopositive and swim into the water, facilitating clean capture (Støttrup and Norsker 1997). Although *A. atopus* has proven to be highly productive in mass culture, with small pebbles as a substrate, Sun and Fleeger (1995) were unsuccessful in cleanly capturing the nonswimming nauplii because they could not be quickly and efficiently separated from the detritus of the culture. New approaches might be used to separate nauplii from debris, and several approaches may be worth additional study. Kahan et al. (1982) cultured harpacticoids in a mesh bag with an aperture size that was large enough to allow nauplii to "fall through" the bag and become accessible to larval fish cultured around the bag. Density centrifugation in nontoxic media (such as the sorbitol-colloidal silica mixture used by Schwinghamer, 1981) has not been attempted, but it might be as successful in separating nauplii from large amounts of culture debris as it has been in separating living meiofauna from sediments.

In a related issue, nonswimming harpacticoids (including nauplii) may require suspension in the presence of pelagic larval fishes to facilitate feeding. This might be accomplished with turbulent water circulation that is gentle enough to suspend harpacticoids without interfering with fish-feeding behavior. Støttrup and Norsker (1997), however, found that the swimming nauplii of *T. holothuriae* avoided container walls in feeding trials and were available to larval fish.

AREAS IN NEED OF ADDITIONAL STUDY

The behavior of harpacticoids is not well studied in either sediment or the water column (Palmer 1984; Fleeger et al. 1995b). The few observations that have been made of harpacticoids in water suggest that most leave the sediment to swim under still-water conditions at periodic intervals in short bursts, swim upwards a few centimeters, then drift or swim downward toward the sediment surface (Palmer 1984; Buskey and Fleeger unpublished observations). This suggests that harpacticoid copepods are generally not strong swimmers capable of long distance or sustained movement in the water. Presumably, this modest swimming ability also reduces escape responses associated with swimming and facilitates relatively easy capture by larval fishes, especially for fish that feed near the bottom. Turingan et al. (this volume) videotaped copepods in the presence of larval fish and found that the escape response of *N. lacustris* was reduced compared with the calanoid copepod *Acartia* sp. In the presence of larval fish, *N. lacustris* increased swimming speed from 4 to 17 mm/sec, but *Acartia* increased from 3 to 58 mm/sec. *N. lacustris* was less successful at avoiding capture by fish larvae. Additional research, however, needs to be conducted to quantify swimming speeds and escape responses in harpacticoid copepods. Studies such as that by Buskey et al. (1993) might serve as a good model for work with harpacticoids, and it would be especially interesting to determine if species that are being used in culture (e.g., *T. holothuriae, N. lacustris*) differ in specific behaviors that may favor particular harpacticoid species for specific fisheries. Finally, the influence of turbulence and the presence of predators might change behaviors such as the frequency of swimming events or swimming speeds and should be examined in controlled tests.

Some observations suggest that harpacticoid copepods may have escape responses that do not rely on swimming ability. Many harpacticoids have strong caudal setae that may exceed 20% of the body length. Preliminary observations by Sun (unpublished observations) suggest that *A. atopus* exhibits a vigorous escape response when attacked by larval red drum, *Sciaenops ocellatus*. *A. atopus* appeared to resist predation by repeatedly "stabbing" larval red drum with its caudal setae after being taken into the mouth. This activity by *A. atopus* was repeated often and frequently and appeared to lead to release of a living, unharmed copepod capable of swimming away. Buskey (this volume) observed that the harpacticoid *E. acutifrons* is rejected immediately after being taken into the mouth by small fishes. Other observations suggest that harpacticoids may pass through the digestive tract of small fishes alive. Rhodes (unpublished observations) observed that *T. californicus* broke through the digestive tract and body wall of the sea horse, *Hippocampus reidi,* to escape predation (and thereby harmed or killed its putative predator). Although such observations are usually limited to anecdotal evidence, more study of the behavior of harpacticoids in the presence of predators seems warranted, especially those that are candidates for mass culture. Alternatively, it is possible that ability of harpacticoids to struggle to avoid predation may stimulate appetite in larvae fishes compared with diets with more passive brine shrimp or rotifers. Støttrup and Norsker (1997) and Heath and Moore (1997) both report increased growth and appetite in two different species of larvae fish when harpacticoids were included in dietary regimes.

POTENTIAL SPECIES FOR MARICULTURE

Several harpacticoid species are now firmly established in mass culture with production potential that may be great enough to meet the needs of mariculture. The basic biology of the most studied species (those in the genera *Amphiascoides, Tisbe, Tigriopus,* and *Nitokra* are discussed and considered below) differs in significant ways that may influence the success of these harpacticoids in the "first feeding" of fish larvae. Because harpacticoids that will be successful as larval feeds must be of the right body size and be available to larval fishes, the size of the smallest developmental stages (nauplius and/or copepodids) and their swimming ability are key considerations. Recent reports suggest that body width of prey should be about 80% of gape of the fish larvae (Turingan et al. this volume), and prey should therefore be no wider than about 75 µm to support vigorous growth in many fish larvae (R.J. Shields, University of Stirling, personal communication). Body size and swimming ability among harpacticoids that have been successfully mass cultured varies (Table 2.2). The smallest nauplii are those of *T. holothuriae* and *N. lacustris*, ≤ 75 µm. Their swimming abilities, however, differ. *Tisbe holothuriae* nauplii can swim and have been shown to be available to juvenile fish, while *N. lacustris* swims only in copepodid/adult stages. *Tigriopus japonicus* and *A. atopus* first-stage nauplii are both ≥ 100 µm in width, but *T. japonicus* swims only in late-stage nauplii when it is much larger, and *A. atopus* nauplii do not swim. The smallest copepodids of *N. lacustris,* however, are about the size of rotifers and are able to swim, suggesting that small-bodied harpacticoids may be suitable

Table 2.2. Body lengths and widths of harpacticoids commonly considered for mass culture

	Tisbe[a]		*Tigriopus*[b]		*Amphiascoides*[c]		*Nitokra*[d]	
	L	W	L	W	L	W	L	W
N I	73	71	102	107	35	100	Nr	Nr
N VI	207	140	232	199	104	190	62[d]	78[d]
C I	322	148	350	Nr	325	105	188	88
A♀	773	310	1,003	Nr	925	155	616	173

Note: Measurements in micrometers. L, length; W, width; Nr, not reported; N, nauplius; C, copepodid; A♀, adult female.
[a]From Dahms and Bergmans (1988). *Tisbe gracilis* adult body dimensions are similar to T. *holothuriae*. *T. holothuriae* developmental stages have not been described.
[b]From Ito(1970).
[c]From Lotufo and Fleeger (1995) and Fleeger unpublished.
[d]From Rhodes unpublished. Nauplii data are from an average of 60 individuals from all instars.

to replace or supplement rotifers in larviculture applications. Nauplius and early copepodid body width and behavior have not been considered when choosing species for mariculture applications, but they should be important criteria in selecting additional species for culture.

A. atopus has proven resilient in culture and is capable of a sustained production, with daily yields of several grams of dry biomass of adults and juveniles for several months in cultures of a fairly modest size. *A. atopus* nauplii, however, are relatively large, nonswimming, and are difficult to separate from culture debris. Furthermore, no long-term trials with fish have been conducted to determine if this species is a preferred prey with sustaining nutritional qualities. Perhaps *A. atopus* would be best used in an application in which a copepod of adult body size is needed, for example, in the culture of shrimp or prawns, or in laboratory applications requiring a large mass of copepods (Nieland et al. 1998; Hoyt et al. 2000).

Tigriopus spp. have been the most productive of all cultured harpacticoids in large-tank, water-only systems (Fukusho 1980). Calculations by Tseng and Hsu (1984) suggest the cost of growing sea bream (*Acanthopagrus latus*) is less when fed *T. japonicus* than when fed *Artemia*. Although some research optimizing the diet of *Tigriopus californicus* has been conducted, little is known of *Tigriopus* spp. behavior (see Powlik et al. 1997). Swimming behavior, for example, is initiated only in later-stage nauplii (Powlik et al. 1997). Additional investigation of the utility of *Tigriopus* spp. in mariculture is justified but has been slow to occur outside of Japan. Recent experiences there, however (S. Uye, Hiroshima University, personal communication), suggest that maintenance of *T. japonicus* cultures for mariculture application imposes unrealistic demands in terms of labor costs.

Tisbe spp. have been successfully cultured in several laboratories throughout the world and have been shown to enhance the growth of fishes. *Tisbe holothuriae* cultures are enhanced by the presence of substrates, but all developmental stages swim and can be collected by filtration (Cutts 2002). This species also responds well to exploitation, and high growth rates are found in exploited cultures (Gaudy and Guerin 1982). *T. holothuriae* nauplii have a highly favorable body size, exhibit good swimming characteristics and positive phototaxis and have been shown to be available to larval fishes. Therefore, nauplii are appropriate for the first feeding for larval fishes. Additional research, however, to optimize nutrition and to examine behavior in the presence of fish would be useful. *T. holothuriae* shows excellent promise as a supplement or alternative to both rotifers and brine shrimp in larviculture.

Nitokra lacustris has been most recently recognized as a candidate for mass culture. This species of copepod cultures without substrate, is small in body size throughout its life, has good swimming abilities in all stages after nauplius, shows rapid growth characteristics, and has nutritional flexibility (Rhodes 2003). Although nauplii do not swim, early stage copepodids are < 100 μm at their widest point and are similar to rotifers in body size. Furthermore, *N. lacustris* grows well on formulated feeds, has a favorable fatty acid profile, and demonstrates reduced escape response compared with calanoid copepods

(Rhodes 2003; Turingan et al., this volume). More research with various fishes in feeding trials is justified; *N. lacustris*, however, appears to have promise as an alternative to rotifers in mariculture. Of all the species considered to date, sustained mass culture and utilization in mariculture of *T. holothuriae* and *N. lacustris* appear to be possible, and further research on both species is justified.

FUTURE DIRECTIONS

There is a low probability that a harpacticoid copepod with the ability to produce resting eggs or diapause stages and with potential in mariculture will be found. Harpacticoids reared in mass culture, however, display excellent nutritional qualities and promote the growth of larval fishes. Harpacticoids also achieve high population densities in culture compared with other copepods, and laboratory-reared copepods reduce the chance of spreading diseases in mariculture. Additional research and development, however, are necessary before laboratory-reared harpacticoids will be routinely used to replace or supplement traditional live feeds. Harpacticoid mass cultures must become more efficient and sustainable. In addition, studies that examine copepod behavior, optimize diet, and determine the length of time that harpacticoids can be shipped or stored without losing nutritional quality must be undertaken to determine which species best meet the needs in specific mariculture applications.

Currently, the search continues for additional harpacticoid copepod species with favorable features for mass culture, which are also high-quality live feed for mariculture applications. Finding species with a relatively small body size, but good swimming abilities in all developmental stages and with the ability to grow in the absence of substrate, is the highest priority. Candidate species include those that grow rapidly without a substrate and have relatively small, pelagic nauplii or early-stage copepodids. Members of the families Canuellidae and Longipediidae appear to be excellent candidates. Nauplii in these families are pelagic and exhibit positive phototaxis behavior. Furthermore, some success in culturing has been achieved (Lonsdale and Levinton 1986; Onbe 1984), but no large-scale culturing effort has been attempted for species in these related families. Other promising candidates include species in the genus *Schizopera*. *Schizopera* species are similar in size to *N. lacustris* and have been shown to grow with a high reproductive potential in cultures without substrates (Kahan et al. 1982; Lotufo and Fleeger 1997).

There are undoubtedly many harpacticoid copepod species that may be successfully grown in mass culture and that can serve the needs as a live feed in mariculture. No protocol has been developed to guide investigators in the collection of harpacticoids to help identify species with potential for mass culture. One approach that may prove effective is to collect near-bottom plankton tows, perhaps at dusk, in environments that experience variable temperature, salinity, and frequent disturbance. Individuals from a mixture of species collected at different times, places, or both, could be placed in mass-culture systems. These cultures could be followed through time to determine which species live and reproduce best. Another source of additional harpacticoids is existing culture systems (at government or academic marine laboratories and laboratories with commercial interests in culturing fish or invertebrates) that become fouled by harpacticoids.

ACKNOWLEDGMENTS

I thank Kevin Carman and Crain Garrot for help with lipid analyses and Sidney Marlborough for commenting on a draft of this chapter. Discussions with Adelaide Rhodes have been helpful and are appreciated. Support from Louisiana Sea Grant made this research possible.

REFERENCES

Armonies, W. 1988. Active emergence of meiofauna from intertidal sediment. *Marine Ecology Progress Series* 43:151–159.

Atilla, N., Fleeger, J.W., and Wetzel, M.A. 2003. Abundance and colonization potential of artificial hard-substrate associated meiofauna. *Journal of Experimental Marine Biology and Ecology* 287:273–287.

Battaglia, B. 1970. Cultivation of marine copepods for genetic and evolutionary research. *Helgolaender Wissenschaftliche Meeresuntersuchungen* 20:385–392.

Bell, S.S., Hicks, G.R.F., and Walters, K. 1988. Active swimming in meiobenthic copeopds of seagrass beds: geographic comparisons of abundances and reproductive characteristics. *Marine Biology* 98:351–358.

Bergmans, M. 1984. Life history adaptation to demographic regime in laboratory-cultured *Tisbe furcata* (Copepoda, Harpacticoida). *Evolution* 38:292–299.

Buskey, E.J. 2004. Behavioral characteristics of copepods that affect their suitability as food for larval fishes. This volume.

Buskey, E.J., Coulter, C., and Strom, S. 1993. Locomotory patterns of microzooplankton: potential effects on food selectivity of larval fish. *Bulletin of Marine Science* 53:29–43.

Carli, A., Mariottini, G.L., and Pane, L. 1995. Influence of nutrition on fecundity and survival in *Tigriopus fulvus* Fischer (Copepoda: Harpacticoida). *Aquaculture* 134:113–119.

Chandler, G.T. 1986. High density culture of meiobenthic harpacticoid copepods within a muddy sediment substrate. *Canadian Journal of Fisheries and Aquatic Science* 43:53–59.

Chandler, G.T., Coull, B.C., Schizas, N.V., and Donelan, T.L. 1997. A culture-based assessment of the effects of Chlorpyrifos on multiple meiobenthic copepods using microcosms of intact estuarine sediments. *Environmental Toxicology and Chemistry* 16:2339–2346.

Chandler, G.T. and Fleeger, J.W. 1984. Tube-building by a marine meiobenthic harpacticoid copepod. *Marine Biology* 82:15–19.

Coull, B.C. 1990. Are members of the meiofauna food for higher trophic levels? *Transactions of the American Microscopical Society* 109:233–246.

Coull, B.C. and Chandler, G.T. 1992. Pollution and meiofauna: field, laboratory and mesocosm studies. *Oceanography and Marine Biology Annual Review* 30:191–271.

Cutts, C.J. 2002. Culture of harpacticoid copepods: potential as live feed for rearing marine fish. *Advances in Marine Biology* 44:295–316.

Dahms, H.U. 1995. Dormancy in the Copepoda—an overview. *Hydrobiologia* 306:199–211.

Dahms, H.-U. and Bergmans, M. 1988. Postembryonic development of *Tisbe gracilis* (T. Scott) (Copepoda, Harpacticoida). *Zoologica Scripta* 17:357–370.

Danovaro, R. and Fraschetti, S. 2002. Meiofaunal vertical zonation on hard-bottoms: comparison with soft-bottom meiofauna. *Marine Ecology Progress Series* 230:159–169.

Decho, A.W. 1988. How do harpacticoid grazing rates differ over a tidal cycle? Field verification using chlorophyll-pigment analyses. *Marine Ecology Progress Series* 45:263–270.

Feller, R.J. and Coull, B.C. 1995. Non-selective ingestion of meiobenthos by juvenile spot (*Leiostomus xanthurus*) (Pisces) and their daily ration. *Vie et Milieu* 45:49–60.

Fleeger, J.W., Shirley, T.C., and McCall, J.N. 1995a. Fine-scale vertical profiles of meiofauna in muddy sediments. *Canadian Journal of Zoology* 73:1453–1460.

Fleeger, J.W., Yund, P.O., and Sun, B. 1995b. Active and passive processes associated with initial settlement and post-settlement dispersal of suspended meiobenthic copepods. *Journal of Marine Research* 53:609–645.

Fukusho, K. 1980. Mass production of a copepod *Tigriopus japonicus* in combination with a rotifer *Brachionus plicatilis* fed n-yeast as a food source. *Bulletin of the Japanese Society Scientific Fisheries* 46:625–629.

Fukusho, K., Arakawa, T., and Watanabe, T. 1980. Food value of a copepod, *Tigriopus japonicus*, cultured with w-Yeast for larvae and juveniles of mud dab *Limanda yokohamae*. *Bulletin of the Japanese Society of Scientific Fisheries* 46:499–503.

Gaudy, R. and Guerin, J.P. 1982. Population dynamics of *Tisbe holothuriae* (Copepoda; Harpacticoida) in exploited mass cultures. *Netherlands Journal of Sea Research* 16:208–216.

Gee, J.M. 1989. An ecological and economic review of meiofauna as food for fish. *Zoological Journal of the Linnean Society* 96:243–261.

Gregg, J.C. and Fleeger, J.W. 1997. Importance of emerged and suspended meiofauna to the diet of the darter goby (*Gobionellus boleosoma*). *Journal of Experimental Marine Biology and Ecology* 209:123–142.

Hagiwara, A., Lee, C.S., and Shiraishi, D.J. 1995. Some reproductive characteristics of the broods of the harpacticoid copepod *Tigriopus japonicus* cultured in different salinities. *Fisheries Science* 61:618–622.

Hall, M.O. and Bell, S.S. 1988. Response of small motile epifauna to complexity of epiphytic algae on sea grass blades. *Journal of Marine Research* 46:613–630.

Heath, P.L. and Moore, C.G. 1997. Rearing Dover sole larvae on *Tisbe* and *Artemia* diets. *Aquaculture International* 5:29–39.

Hicks, G.R.F. and Coull, B.C. 1983. The ecology of marine meiobenthic harpacticoid copepods. *Oceanography and Marine Biology Annual Review* 21:67–175.

Hoyt, M., Fleeger, J.W., Siebeling, R., and Feller, R.J. 2000. Serological estimation of prey-protein gut-residence time and quantification of meal size for grass shrimp consuming meiofaunal copepods. *Journal of Experimental Marine Biology and Ecology* 248:105–119.

Huys, R. and Boxshall, G.A. 1991. *Copepod Evolution.* London, U.K.: The Ray Society. 468 pp.

Ingole, B.S. 1994. Influence of salinity on the reproductive potential of a laboratory cultured harpacticoid copepod, *Amphiascoides subdebilis* Willey, 1935. *Zoologica Anzeiger* 232:31–40.

Ito, T. 1970. The biology of a harpacticoid copepod *Tigriopus japonicus* Mori. *Journal Faculty Science Hokkaido University Series VI, Zoology* 17:474–500.

Kahan, D., Uhlig, G., Schwenzer, D., and Horowitz, L. 1982. A simple method for cultivating harpacticoid copepods and offering them to fish larvae. *Aquaculture* 26:303–310.

Kovatch, C.E., Chandler, G.T., and Coull, B.C. 1999. Utility of a full life-cycle copepod bioassay approach for assessment of sediment-associated contaminant mixtures. *Marine Pollution Bulletin* 38:692–701.

Kraul, S. 1990. Production of live prey for marine fish larvae. *Institut Francais de Recherche pour l'Exploitation de la Mer. Brest* 9:595–607.

Kreeger, K.E., Kreeger, D.A., Langdon, C.J., and Lowry, R.R. 1991. The nutritional value of *Artemia* and *Tigriopus californicus* (Baker) for 2 Pacific mysid species, *Metamysidopsis elongata* (Holmes) and *Mysidopsis intii* (Holmquist). *Journal of Experimental Marine Biology and Ecology* 148:147–158.

Lee, W.Y. and Yan, H.Y. 1994. Energy budget of the female harpacticoid *Tisbe carolinensis* (Copepoda, Crustacea). *Comparative Biochemistry Physiology A* 109:817–820.

Léger, P., Bengston, D.A., Simpson, K.L., and Sorgeloos, P. 1986. The use and nutritional value of *Artemia* as a food source. *Oceanography Marine Biology Annual Review* 24:521–623.

Lewis, A.G., Chatters, L., and Raudsepp, M. 1998. Feeding structures and their functions in adult and preadult *Tigriopus californicus* (Copepoda: Harpacticoida). *Journal Marine Biological Association U.K.* 78:451–466.

Lonsdale, D.J. and Levinton, J.S. 1986. Growth rate and reproductive differences in a widespread estuarine harpacticoid copepod (*Scottolana canadensis*). *Marine Biology* 91:231–237.

Lonsdale, D.J., Weissman, P., and Dobbs, F.C. 1993. A reproductive-resting stage in an harpacticoid copepod, and the significance of genetically based differences among populations. *Bulletin of Marine Science* 53:180–193.

Lotufo, G.R. 1997. Toxicity of sediment-associated PAHs to an estuarine copepod: effects on survival, feeding, reproduction and behavior. *Marine Environmental Research* 44:149–166.

Lotufo, G.R. and Fleeger, J.W. 1995. Description of *Amphiascoides atopus,* a new species (Copepoda: Harpacticoida) from a mass culture system. *Proceedings of the Biological Society of Washington* 108:117–124.

Lotufo, G.R. and J. W. Fleeger. 1997. Effects of sediment-associated phenanthrene on survival, development and reproduction of two species of meiobenthic copepods. *Marine Ecology Progress Series* 151:91–102.

McCall, J.N. and Fleeger, J.W. 1993. Recognition and utilization of prey aggregations by juvenile spot (*Leiostomus xanthurus* Lacepede). *Journal of Experimental Marine Biology and Ecology* 174:121–134.

McCall, J.N. and Fleeger, J.W. 1995. Predation by juvenile fish on hyberbenthic meiofauna: a review with data on post-larval *Leiostomus xanthurus. Vie Milieu* 45:61–73.

Miliou, H. 1993. Temperature, salinity and light induced variations on larval survival and sex ratio of *Tisbe holothuriae* Humes (Copepoda: Harpacticoida). *Journal of Experimental Marine Biology and Ecology* 173:95–109.

Nanton, D.A. and Castell, J.D. 1998. The effects of dietary fatty acids on the fatty acid composition of the harpacticoid copepod, *Tisbe* sp., for use as a live food for marine fish larvae. *Aquaculture* 163:251–261.

Nanton, D.A. and Castell, J.D. 1999. The effects of temperature and dietary fatty acids on the fatty acid composition of harpacticoid copepods, for use as a live food for marine fish larvae. *Aquaculture* 175:167–181.

Nieland, D.L., Wilson, C.A., and Fleeger, J.W. 1998. Preliminary evaluation of the use of phosphogypsum for reef substrate in the Gulf of Mexico. *Gulf of Mexico Science* 16:54–63.

Norsker, N.-H. and J.G. Støttrup. 1994. The importance of dietary HUFAs for fecundity and HUFA content in the harpacticoid *Tisbe holothuriae* Humes. *Aquaculture* 125:155–166.

Ogle, J. 1979. Adaptation of brown water culture technique to the mass culture of the copepod *Acartia tonsa. Gulf Research Reports* 6:291–292.

Onbe, T. 1984. The developmental stages of *Longipedia* Americana (Copepoda: Harpacticoida) reared in the laboratory. *Journal of Crustacean Biology* 4:615–631.

Palmer, M.A. 1984. Invertebrate drift: behavioral experiments with intertidal meiobenthos. *Marine Behaviour and Physiology* 10:235–253.

Palmer, M.A. 1988. Dispersal of marine meiofauna: a review and conceptual model explaining passive transport and active emergence with implications for recruitment. *Marine Ecology Progress Series* 48:81–91.

Pinto, C.S.C., Souza-Santos, L.P., and Santos, P.J.P. 2001. Development and population dynamics of *Tisbe biminiensis* (Copepoda: Harpacticoida) reared on different diets. *Aquaculture* 198:253–267.

Powlik, J.J., Lewis, A.G., and Spaeth, M. 1997. Development, body length, and feeding of *Tigriopus californicus* (Copepoda, Harpacticoida) in laboratory and field populations. *Crustaceana* 70:324–343.

Rocha-Olivares, A., Fleeger, J.W., and Foltz, D.W. 2001. Decoupling of molecular and morphological evolution in deep lineages of a meiobenthic harpacticoid copepod. *Molecular Biology and Evolution* 18:1088–1102.

Rhodes, A. 2003. Methods for mass culture for high density batch culture of *Nitokra lacustris*, a marine harpacticoid copepod. In *The Big Fish Bang: Proceedings of the 26th Annual Larval Fish Conference*, edited by H.I. Browman and A.B. Shiftesvik, pp. 449–465. Bergen, Norway, 22–26 July 2002. Bergen, Norway: Institute of Marine Research.

Rhodes, A. and Boyd, L. 2004. Formulated feeds for harpacticoid copepods: implications for population growth and fatty acid composition. This volume.

Rutledge, P.A. and Fleeger, J.W. 1993. Abundance and seasonality of meiofauna, including harpacticoid copepod species, associated with stems of the salt marsh cord grass, *Spartina alterniflora*. *Estuaries* 16:760–768.

Schwinghamer, P. 1981. Extraction of living meiofauna from marine sediments by centrifugation in a silica sol-sorbitol mixture. *Canadian Journal of Fisheries and Aquatic Science* 38:476–478.

Seifried, S. and Durbaum, J. 2000. First clear case of carnivory in marine Copepoda Harpacticoida. *Journal of Natural History* 34:1595–1618.

Støttrup, J.G. 2000. The elusive copepods: their production and suitability in marine aquaculture. *Aquaculture Research* 34:703–711.

Støttrup, J.G. and Norsker, N.H. 1997. Production and use of copepods in marine fish larviculture. *Aquaculture* 155:231–247.

Støttrup, J.G., Richardsen, K., Kirkegaard, E., and Pihl, N.J. 1986. The cultivation of *Acartia tonsa* Dana for use as live food source for marine fish larvae. *Aquaculture* 52:87–96.

Strawbridge, S., Coull, B.C., and Chandler, G.T. 1992. Reproductive output of a meiobenthic copepod exposed to sediment-associated Fenvalerate. *Archives Environmental Contamination Toxicology* 23:295–300.

Suderman, K. and Thistle, D. 1998. Adult female harpacticoid copepods maintain their energy reserves by feeding while suspended during storms. *Marine Ecology Progress Series* 164:245–252.

Sun, B. and Fleeger, J.W. 1995. Sustained mass culture of *Amphiascoides atopus* a marine harpacticoid copepod in a recirculating system. *Aquaculture* 136:313–321.

Thistle, D. 2003. Harpacticoid copepod emergence at a shelf site in summer and winter: implications for hydrodynamic and mating hypotheses. *Marine Ecology Progress Series* 248:177–185.

Tipton, K. and Bell, S.S. 1988. Foraging patterns of two syngnathid fishes: importance of harpacticoid copepods. *Marine Ecology Progress Series* 47:31–43.

Toepfer, C.S. and Fleeger, J.W. 1995. Diet of juvenile fishes *Citharichythys spilopterus, Symphurus plagiusa* and *Gobionellus boleosoma. Bulletin of Marine Science* 56:238–249.

Tseng, W.Y. and Hsu, C.-K. 1984. Studies on the culture of the marine copepod *Tigriopus japonicus* Mori and its value as a food for juveniles of the yellow-fin sea bream *Acanthopagrus latus* (Houttuyn). *Crustaceana* 7:381–389.

Turingan, R.G., Beck, J.L., Krebs, J.M., and Licamele, J.D. 2004. Development of feeding mechanics in marine fish larvae and the swimming behavior of zooplankton prey: Implications for rearing marine fishes. This volume.

Volk, E.C., Wissmar, R.C., Simenstad, C.A., and Eggers, D.M. 1984. Relationship between otolith microstructure and the growth of juvenile chum salmon (*Oncorhynchus keta*) under different prey rations. *Canadian Journal of Fisheries and Aquatic Science* 41:126–133.

Vopel, K., Dehmlow, J., and Arlt, G. 1996. Vertical distribution of *Cletocamptus confluens* (Copepoda, Harpacticoida) in relation to oxygen and sulphide microprofiles of a brackish water sulphuretum. *Marine Ecology Progress Series* 141:129–137.

Walters, K. and Bell, S.S. 1994. Significance of copepod emergence of benthic, pelagic, and phytal linkages in a subtidal seagrass bed. *Marine Ecology Progress Series* 108:237–249.

Warwick, R.M., Carr, M.R., Clarke, K.R., Gee, J.M., and Green, R.H. 1988. A mesocosm experiment of the effects of hydrocarbon and copper pollution of a sublittoral soft-sediment meiobenthic community. *Marine Ecology Progress Series* 46:181–191.

Watanabe, T., Kitajima, C., and Fujita, S. 1983. Nutritional values of live organisms used in Japan for mass propagation of fish: a review. *Aquaculture* 34: 115–143.

Weiss, G.M., G.B. McManus, and H.R. Harvey. 1996. Development and lipid composition of the harpacticoid copepod *Nitocra spinipes* reared on different diets. *Marine Ecology Progress Series* 132:57–61.

Wetzel, M.A., Fleeger, J.W., and Powers, S. 2001. Effects of hypoxia and anoxia on meiofauna: a review with new data from the Gulf of Mexico. In *Coastal Hypoxia; Consequences for Living Resources*, edited by N.N. Rabalais and R.E. Turner, pp. 165–184. Washington, D.C., USA: American Geophysical Union.

Williams-Howze, J. 1996. Biology and morphology of the marine harpacticoid copepod *Heteropsyllus nunni* Coull, during encystment diapause. *Hydrobiologia* 320:179–189.

Zhang, Q. and Uhlig, G. 1993. Effect of density on larval development and female productivity of *Tisbe holothuriae* (Copepoda, Harpacticoida) under laboratory conditions. *Helgolander Meeresuntersuntersuchungen* 47:229–241.

3
Symbiotic Copepods as Live Feed in Marine Finfish Rearing

Ju-shey Ho

ABSTRACT

Copepods of the Family Myicolidae (Poecilostomatoida) are largely symbiotic with marine bivalve mollusks. Seventeen species classified in seven genera are currently known in this family. One of these species, *Pseudomyicola spinosus* (Raffaele and Monticelli), has been reported from more than 50 species of bivalve mollusks occurring in many parts of the tropical, subtropical, and temperate waters of the Pacific, Atlantic, and Indian Oceans. Available information on the biology of *P. spinosus* indicates this species has potential for development into a live feed for finfish larval rearing because it (a) is easy to obtain, (b) has planktonic naupliar larvae, (c) grows well in cultivation, and (d) has high fecundity. To produce a large quantity of naupliar larvae of *P. spinosus*, it is suggested to construct a mussel bed and cultivate the copepod in the water containing the mussel bed by releasing into this setup first copepodid instars of *P. spinosus* reared in the laboratory.

INTRODUCTION

Although species of several genera of calanoids (*Acartia, Centropages, Eurytemora, Gladioferens, Labidocera, Parvocalanus, Pseudodiaptomus, Temora*), harpacticoids (*Tigriopus, Tisbe*), and cyclopoids (*Apocyclops, Oithona*) have been studied for mass production of nauplii to provide live feed for finfish fry in aquaculture, as far as this author is aware no attempt has been made to rear species of poecilostomatoids for the same purpose. The lack of any attempt is largely because the majority of poecilostomatoids are symbiotic, living in association with or parasitic on or in aquatic animals, including many species of fish. The aim of this chapter is to change traditional practices using a species of poecilostomatoid copepod as an example to elucidate the potential of utilizing symbiotic copepods for production of live feed for the rearing of finfish larvae.

SELECTION OF SYMBIOTIC COPEPODS

A species of myicolid copepod, *Pseudomyicola spinosus* (Raffaele and Monticelli), has been selected to illustrate the potential use of symbiotic copepods for the mass production of nauplii. Due to its wide distribution and coupled with the presence of more than one morphotype, the copepod has been called variously *Pseudomyicola ostreae* (Yamaguti 1936) for the specimens found in Japan, *P. glaber* (Pearse 1947) for the specimens found in North Carolina, *P. mirabilis* (Humes 1959) for the specimens found in Madagascar, *P. anomalocardiae* (Narchi 1965) for the specimens found in Brazil, and *Myicola tageli* (Pearse 1947) for the specimens again found in North Carolina. These names have become synonyms for *P. spinosus*.

Pseudomyicola spinosus lives in the mantle cavities of marine bivalve mollusks occurring in coastal intertidal zones, upper subtidal zones, or in both habitats. Based on the available information for the food and feeding of myicolid copepods (Yoshikoshi and Ko 1974), *P. spinosus* seems to feed mostly on the mucus produced by its host. This copepod is selected as a model for mass production because it (a) is easy to obtain,

(b) has planktonic naupliar larvae, (c) grows well in cultivation, and (d) has high fecundity. Each of these four criteria is elaborated below.

AVAILABILITY

Having a constant supply of raw materials is one of the key factors for success in any production industry, including aquaculture. Therefore, the factor to consider first and foremost for live feed rearing among symbiotic copepods is how easily the organism can be obtained.

P. spinosus is the most widely distributed myicolid copepod living in association with more than 50 species of bivalve mollusks, such as clams, oysters, and mussels (Ho 1992; Kim 1998). This symbiotic copepod has been reported in North America (Wilson 1935; Pearse 1947; Ho 1980; Tang 2001), Mexico (Caceres-Martinez et al. 1996), the Pacific coast of Panama (Humes, 1984), the Caribbean (Humes 1968), Brazil (Narchi 1965; Humes 1968), Europe (Raffaele and Monticelli 1885; Stock 1959, 1960; Kleeton 1964; Laubier and Reyes 1964; Dethlefsen 1972), the Black Sea (Porumb and Andriescu 1964), Senegal (Humes 1968), Madagascar (Humes 1959, 1968), Australia (Pregenzer 1983), New Zealand (Dinamani and Gordon 1974), Korea (Suh and Choi 1990; Kim 1998), and Japan (Yamaguti 1936; Ho 1980; Do et al. 1984; Do and Kajihara 1986). Although it has yet to be reported in a number of areas, including India, the west coast of South America, and countries in Southeast Asia, based on currently known patterns of its distribution and host specificity, the apparent absence of *P. spinosus* in these areas is often attributed to a lack of published surveys on symbiotic copepods in these regions of the world.

Studies that report on the occurrence of *P. spinosus* in various other parts of the world show a high prevalence of symbiosis as follows: Dethlefsen (1972) reported a prevalence of 72.5% from the German North Sea coast; Do and Kajihara (1986) reported 85.6% from Matsushima in Tokyo Bay, Japan; Caceres-Martinez et al. (1996) found 100% from Baja California in Mexico; and Tang (2001) observed 82% from lower Newport Bay in California, USA. The maximum number of copepods present in a single host is also large. Caceres-Martinez et al. (1996) reported finding 59 copepods associated with edible mussels (*Mytilus* spp.) from Baja California, Mexico, while Tang (2001) found 35 in bay mussels (*M. galloprovincialis*) from southern California in the United States.

Several methods have been used to collect *P. spinosus* from bivalves. The best method for collecting live *P. spinosus* from the mantle cavities of marine bivalve mollusks is as outlined below.

Collection Methods

1. Scrub and wash the shells to clean all epizoons and epiphytes.
2. Open one by one the cleaned bivalves in a glass Petri dish (about 10 cm in diameter or larger, depending on the size of the bivalve host).
3. Wash the opened bivalves one by one in a 1,000-mL or 2,000-mL beaker half filled with seawater. Wash only the same species of bivalves in a given beaker. Do not discard the seawater used for "washing" the mollusks.
4. After opening and washing all bivalve samples of a given species, pour the wash water retained in the beaker through a plankton net.
5. Examine the debris collected in the plankton net in a Petri dish (with a little seawater, that is, enough to cover the bottom of the dish) under a dissection microscope.
6. Remove the copepods carefully with a pipette, and place them in a separate dish with filtered seawater.
7. Examine also the Petri dish in which the bivalves were opened. Remove copepods with a pipette.

Identification of Copepod Species

The following steps are recommended for identifying the species collected.

1. Preserve in 70% alcohol one specimen of each species collected in a dish.
2. Transfer the preserved specimen after a couple of hours to a depression slide filled with lactic acid. Leave the specimen in lactic acid for 2 or 3 hours until it becomes semitransparent.
3. Cover the "cleared" copepod in a depression slide with a coverslip and examine the preparation under a compound microscope. The copepod can be rolled to an ideal position for examination by gently pushing the edge of the coverslip. To make an accurate identification of the symbiotic copepod, it is necessary to

dissect apart the minute appendages and examine them under high magnification ($\geq \times 1{,}000$). Sometimes, a special hanging-drop method, as described by Humes and Gooding (1964), is required.

Planktonic Nauplii

Symbiotic copepods are mostly sedentary and reproduce with a reduced number of naupliar stages. *P. spinosus*, however, is unusual among symbiotic copepods in having a life cycle comprising 12 stages in two phases, with seven stages in the first phase and five stages in the second phase. While the first phase is free living and planktonic, the second phase is symbiotic and sedentary. The planktonic phase consists of six naupliar stages and one copepodid stage, copepodid I. The first copepodid stage is an infective stage, which actively seeks its host and dies if it fails to locate a bivalve in 48–72 hours.

The planktonic stages of *P. spinosus* are about the same size as those of calanoid species that have been used in experiments for live feed production, with lengths of the naupliar larvae ranging from 270 to 290 μm (Nakamura et al. 1979). Therefore, they are ideal for serving as live feed for finfish fry. Although copepodid I is an infective stage, it is host specific and infects only bivalve mollusks. Consequently, it can serve as live feed for finfish fry.

Cultivation

Because *P. spinosus* is an obligatory symbiont, a bivalve host is necessary for cultivation of its adults to obtain eggs for the production of naupliar larvae. Kajihara et al. (1980) developed an interesting rearing method to circumvent this need. They used a small Petri dish (3 cm in diameter and 2 cm in depth) containing a strip of mussel gill to rear the infective stages of *P. spinosus*.

For rearing, the egg sacs were first removed from the adult female copepods living in the mantle cavities of the marine bivalves. Egg sacs were kept in a large Petri dish (15 cm in diameter and 1.5 cm in depth) with filtered seawater. When the larvae grew to the first copepodid stage, they were transferred individually to a "surrogate host," that is, a strip of (5 mm wide) mussel gill placed in a small Petri dish containing 30 ppt filtered seawater. The filtered seawater used in the rearing and used with the strip of mussel gill in the Petri dish was changed every 2 to 3 days. The temperature was kept at 20°C, and each Petri dish was covered with a dark cap throughout the rearing (Kajihara et al. 1980).

In 1983, the author employed the rearing method devised by Kajihara et al. (1980) to test the rearing of *P. spinosus* from the eggs in 10 sacs obtained from ovigerous females collected from mussels (*M. galloprovincialis* Lamarck) attached to boat floats in Long Beach Marina, Long Beach, California. At 18°C in 30 ppt seawater, the male reached the adult stage in about 32 days from hatching, and the female in about 36 days. About a week after mating, the female started to produce eggs.

High Fecundity

By employing the same rearing method, Kajihara and Nakamura (1985) used 20 virgin females to study the life span and fecundity of *P. spinosus*. They produced 18 of the 20 experimental female copepods in the laboratory from fertilized eggs (laboratory females). The other two (wild females) were cultivated from specimens in their fifth copepodid stage taken from mussel mantle cavities. The major difference between the two types was that the wild females were handled less during their larval development. Each female in the experiment was kept individually in a surrogate host after mating.

Kajihara and Nakamura (1985) reported the following results of the experiment on *P. spinosus*: (a) The life span and the number of ovipositions varied with the individual; (b) the number of eggs laid at one oviposition decreased with the increasing age of the female; and (c) the number of eggs at each oviposition was smaller in laboratory females than in wild females. An interesting phenomenon was revealed in their experiment. Although the life span of the 20 female *P. spinosus* in the experiment varied from 46 to 849 days (2.3 years) and the number of ovipositions varied from 2 to 106, the interval between ovipositions showed less variation, ranging from 6 to 8 days in most cases. In other words, the mated females of *P. spinosus* produced eggs about every week, with an average of 12.3 (7–19) eggs at each oviposition. All eggs were viable and hatched into nauplii.

If fecundity is expressed in terms of the number of eggs produced at each oviposition, the fecundity of *P. spinosus* was low in the study by Kajihara and Nakamura (1985). It ranged from 3.2 to 18.9 eggs for the laboratory females and 13.2 to 15.1 eggs for the wild females. Fecundity of females collected from mussels in nature, however, is generally reported to be higher. For instance, Tang (2001) reported that *P. spinosus* breeds continuously throughout the year in southern California, with the peak breeding events occurring in spring producing on an average 33 ± 10.9 eggs per ovigerous female. On the other hand, Kajihara and Nakamura (1985) recorded that the natural females collected from mussels in Yokosuka Harbor in Tokyo Bay carried on average 25.4 ± 8.8 eggs in February and 42.8 ± 9.3 eggs in April.

CONCLUSION

It is probable that production of a smaller number of eggs by female *P. spinosus* in the study reported by Kajihara and Nakamura(1985) was due to the use of a surrogate host, which required the breeding copepods to be handled every 2 to 3 days, thus creating stress in the experimental animals. In addition, the nutritional value of a strip of trimmed, nonliving mussel gill must be substantially lower than the whole gill lamella of a living mussel. To produce a large quantity of planktonic stages of *P. spinosus* for live feed, therefore, the natural host, and not a surrogate host, should be used for cultivation. Accordingly, producing a mussel bed is the first requirement in the cultivation of *P. spinosus* for live feed.

To mass produce naupliar larvae of *P. spinosus*, it is suggested to first rear instars of the copepodid I stage in Petri dishes with filtered seawater and then transfer them to a tank (or an aquarium) containing a mussel bed. Since it takes from 28.6 to 38.6 days for the female *P. spinosus* to grow to the adult stage, another 5.8 to 8.2 days to produce the first batch of eggs, and 6.8 days for the eggs to hatch, the nauplii in this culture system start to appear in the water in about 41 days after introduction of the instars of copepodid I into the tank.

Additional experiments based on the above-described studies are needed because the following details are not fully understood: (a) the best and most effective way to construct a mussel bed, (b) the maximum capacity of symbiont carrying on the mussel host, and (c) the frequency for introducing the first copepodid instars to the rearing tank. When the answers are found, cultivation of *P. spinosus,* or some other species of myicolid copepod, may be comparable to the rearing of free-living calanoids, harpacticoids, or cyclopoids as live food for fish larvae in aquaculture.

REFERENCES

Caceres-Martinez, J., Vasquez-Yeomans, R., and Suarez Morales, E. 1996. Two parasitic copepods, *Pseudomyicola spinosus* and *Modiolicola gracilis,* associated with edible mussels, *Mytilus galloprovincialis* and *Mytilus californianus*, from Baja California, NW Mexico. *Journal of Shellfish Research* 15:667–672.

Dethlefsen, V. 1972. Zur Parasitologie der Miesmuschel (*Mytilus edulis* L., 1758). *Berichte der Deutschen Wissenshaftlichen Kommission für Meeresforschung* 22:344–371.

Dinamani, P. and Gordon, D.B. 1974. On the habits and nature of association of the copepod *Pseudomyicola spinosus* with the rock oyster *Crassostrea glomerata* in New Zealand. *Journal of Invertebrate Pathology* 24:305–310.

Do, T.T. and Kajihara, T. 1986. Studies on parasitic copepod fauna and biology of *Pseudomyicola spinosus,* associated with blue mussel, *Mytilus edulis galloprovincialis* in Japan. *Bulletin of the Ocean Research Institute, University of Tokyo* 23:1–63.

Do, T.T., Kajihara. T., and Ho, J.S. 1984. The life history of *Pseudomyicola spinosus* (Raffaele & Monticelli, 1885) from the blue mussel, *Mytilus edulis galloprovincialis* in Tokyo Bay, Japan, with notes on the production of atypical male. *Bulletin of the Ocean Research Institute, University of Tokyo* 17:1–65.

Ho, J.S. 1980. Origin and dispersal of *Mytilus edulis* in Japan deduced from its present status of copepod parasitism. *Publications of the Seto Marine Biological Laboratory* 25:293–313.

Ho, J.S. 1992. Phylogenetic analysis of the Myicolidae, a family of poecilostome copepods chiefly parasitic in marine bivalve mollusks. *Acta Zoologica Taiwanica* 3:67–77.

Humes, A.G. 1959. Copépodes parasites de mollusques à Madagascar. *Mémoires de l'Institut Scientifique de Madagascar, 1958, série F* 2:285–342.

Humes, A.G. 1968. The cyclopoid copepod *Pseudomyicola spinosus* (Raffaele & Monticelli) from marine pelecypods, chiefly in Bermuda and West Indies. *Beaufortia* 14:203–226.

Humes, A.G. 1984. *Ostrincola* and *Pseudomyicola* (Crustacea: Copepoda; Poecilostomatoida) associated with marine bivalve mollusks on the Pacific coast of Panama. *Proceedings of the Biological Society of Washington* 97:589–600.

Humes, A.G. and Gooding, R.U. 1964. A method for studying the external anatomy of copepods. *Crustaceana* 6:238–240.

Kajihara, T., Hayashi, M., and Nakamura, K. 1980. Rearing method of a parasitic copepod, *Pseudomyicola spinosus* (Cyclopoida), in laboratory. *Bulletin of Plankton Society of Japan* 27:123–124.

Kajihara, T. and Nakamura, K. 1985. Lifespan and oviposition of the parasitic copepod *Pseudomyicola spinosus* under rearing conditions. *Marine Biology* 87:55–60.

Kim, I.H. 1998. *Illustrated Encyclopedia of Fauna and Flora of Korea. Volume 38. Cirripedia, Symbiotic Copepoda, Pycnogonida.* Seoul, Korea: Ministry of Education. 1038 pp.

Kleeton, G. 1964. Sur la presence de *Pseudomyicola spinosus* (Raff. & Mont.) (Crustacea, Copepoda) dans l'Atlantique, avec une note sur la synonymie de *P. spinosus* et *P. glaber* Pearse. *Beaufortia* 11:171–177.

Laubier, L. and Reyes, D. 1964. Sub-spécietion chez un copépode parasite *Pseudomyicola spinosus* (Raff. & Mont.) et description de deux sous-espèces nouvelles. *Vie Milieu, supplement* 17 (Volume Jubilaire dédié à Georges Petit):41–46.

Nakamura, K., Kajihara, T., and Oka, M. 1979. Development of a cyclopoid copepod, *Pseudomyicola ostreae* Yamaguti, associated with a marine mussel, *Mytilus edulis galloprovincialis* Lamarck: external structure of nauplius in six stages. *Proceedings of the Japanese Society of Systematic Zoology* 17:38–44.

Narchi, W. 1965. A new species of *Pseudomyicola* Yamaguti, 1936 (Copepoda, cyclopoida). *Anais da Academia Brasileira de Ciências* 37:359–361.

Pearse, A.S. 1947. Parasitic copepods from Beaufort, North Carolina. *Journal Elsha Mitchell Scientific Society* 63:1–16.

Porumb, F. and Andriescu, I. 1964. Asupra prezentei a douå copepoda in cavitatea paleală a midiei (*Mytilus galloprovincialis* L.) din apele Romînesti ale Mårii Negre. *Analele Ştiintifice ale Universitătii "Al. I. Cuza" Din Iaşi* 10:93–100.

Pregenzer, C. 1983. Survey of metazoan symbionts of *Mytilus edulis* (Mollusca: Pelecypoda) in southern Australia. *Australian Journal of Marine and Freshwater Research* 34:387–396.

Raffaele, F. and Monticelli, F.S. 1885. Descrizione di un nouva *Lichomolgus* parassita del *Mytilus galloprovinicalis* Lk. *Atti della R. Accademia dei Lincei,* (4) (*Memorie della Classe di Scienze Fisiche, Matematiche e Naturali*) 1:302–307.

Stock, J.H. 1959. Copepoda associated with Neapolitan Mollusca. *Pubblicazioni della Stazione Zoologica di Napoli* 31:43–58.

Stock, J.H. 1960. Sur quelques copepods associés aux invertebres des côtes du Roussilion. *Crustaceana* 1:218–257.

Suh, H.L. and Choi, S.U. 1990. Two copepods (Crustacea) parasitic on the blue mussel, *Mytilus galloprovincialis,* from the Yongsan River estuary in Korea. *Bulletin of Korean Fisheries Society* 23:137–140.

Tang, D. 2001. Population dynamics of *Pseudomyicola spinosus,* a parasitic copepod of the bay mussel (*Mytilus galloprovincialis*) in Lower Newport Bay, California. M.Sc.diss., California State University, Long Beach, California, USA.

Wilson, C.B. 1935. Parasitic copepods from the Pacific Coast. *American Midland Naturalist* 16:776–797.

Yamaguti, S. 1936. Parasitic copepods from mollusks of Japan, I. *Japanese Journal of Zoology* 7:113–127.

Yoshikoshi, K. and Ko, Y. 1974. Food and feeding of three species of cyclopoid copepods associated with marine pelecypods (preliminary note). *Bulletin of the Faculty of Fisheries, Nagasaki University* 38:109–115.

4
Birth Control Effects of Diatoms on Copepod Reproduction: Implications for Aquaculture Studies

Adrianna Ianora

ABSTRACT

Copepods are probably the most numerous multicellular organisms on Earth. With over 11,500 species, they are more numerous than insects on land, even though insects are far more diverse (Humes 1994). The success of these small crustaceans seems to largely depend on their high reproductive rates and fast development times. This chapter reviews some of the literature on the factors controlling egg production and development of a few dominant species of marine planktonic copepods, with particular emphasis on food-quality effects, since these have recently been shown to be extremely important in determining high survivorship of copepods. In particular, the chapter focuses on diatom–copepod interactions and on the production by diatoms of plant metabolites that strongly modify reproductive and developmental processes in these crustaceans, beyond the classical primary metabolites such as proteins, fatty acids, and vitamins. These metabolites include a family of unsaturated aldehydes that are produced as an activated chemical defense against predatory animals, potentially sabotaging future generations of copepods by inducing abortions, birth defects, and poor development rates. This biological model is new for the marine environment, since most of the known negative plant–animal interactions are generally related to poisoning processes, or feeding deterrence, but never to reproductive failure. The production by diatoms of teratogens that induce structural malformations in the offspring exposed to them during gestation should be assessed prior to using certain diatom species in aquaculture as live feed for the mass cultivation of copepods.

INTRODUCTION

Copepods, like other crustaceans, have separate sexes, with females and males that differ due to sexually dimorphic characteristics that develop in the later copepodid stages. Females are usually larger than males, more long-lived, and with modified first genital segments. Males are usually smaller in size, often short-lived, and often have greatly modified fifth legs and antennules that function in mating.

The reproductive systems of females and males are described in detail by Mauchline (1998). Briefly, in females there is a median ovary dorsal to the gut, extending from the midregion of the cephalosome and last thoracic segment. Paired oviducts originate laterally and extend on either side of the gut to the urosome, opening into the genital atrium. Lateral pouches, termed spermathecae, in which the spermatozoa are stored, open into the genital atrium, fertilizing the eggs prior to extrusion through the genital pores. Males have a single testis located medially in the midregion of the cephalosome, from which the vas deferens emerges and leads to the seminal vesicle. This in turn leads to the spermatophore sac, which contains a single spermatophore filled with spermatozoa. At copulation, the spermatophore is extruded through the genital opening and is attached to the female genital segment.

Little is known about the quantity of spermatophores produced during the lifetime of a male copepod. Production rates probably depend on whether a species requires remating for the continued production of viable clutches. For example, species of *Calanus* seem to mate only once (e.g., *C. finmarchicus* [Marshall and Orr 1955]; *C. pacificus* [Runge 1984]), in which case, spermatophore production rates may be lower compared with other species, such as *Temora stylifera*, which require multiple remating (Ianora et al. 1989), with about one spermatophore being produced every 2 days for up to 50 days (Ianora and Poulet 1993).

Most species of copepods broadcast their eggs freely into the surrounding water, but others carry them attached to the genital segment in egg masses until the nauplii are hatched. Species that broadcast their eggs are more prolific than those that do not (Kiørboe and Sabatini 1995), even though egg mortality in broadcast spawners is higher as well, as in most other invertebrates (Giangrande et al. 1994). According to Giangrande and coworkers, however, the final product in terms of recruitment rate may be the same for the two strategies because of the different mortality rates occurring during the two modes of development.

The eggs of most species of copepods develop directly to hatching, but some species of copepods have been shown to produce two egg types, one subitaneous and the other a diapausing egg that is produced to overcome periods of environmental adversity. Diapausing eggs can remain in a state of arrested development for up to 40 years (Marcus et al. 1994), but in most cases, they remain dormant for less than 1 year and serve to seed the copepod population the following spring when adverse conditions are terminated (Santella and Ianora 1990). It is believed that day length and low temperatures are the main environmental factors triggering diapause egg production in most copepod species (Grice and Marcus 1981), while a temperature shock seems to be an essential prerequisite for the hatching of these eggs (Romano et al. 1996).

Copepods tend to be seasonal breeders at high latitudes but continuous breeders in subtemperate and tropical waters. It has been suggested that breeding at high latitudes is timed to take advantage of phytoplankton blooms (e.g., Marshall and Orr 1930), but when food conditions are more stable, breeding tends to be continuous (Ianora 1998). At high latitudes there can be several generations within a single breeding season, with overlapping cohorts resulting from eggs spawned by individual females from different generations. Generation times can either be short, with an adult life span of 2 weeks to 2 months in most cases, or much longer (up to 2 years), with arrested development (diapause) occurring during one or more stages of the life cycle. On the other hand, continuous breeding is typical of subtemperate and tropical copepods. In this case, generation times tend to be short, with multiple overlapping cohorts, but with periods of greater or lower breeding intensity, depending on the copepod species (Ianora 1998).

At present, there is great interest in the use of copepods as live food organisms in the larval rearing of finfish species. There are several reasons for this: copepods have very high reproductive rates and fast development times (Ianora 1998; Mauchline 1998) and have a high nutritional content (Evjemo et al. 2003). If, however, copepods are to substitute traditional live feeds, new methods need to be developed for the mass cultivation of these crustaceans to match those currently used for culturing *Artemia* and rotifers. In the following sections, the literature on what is known about the factors affecting egg production rates, hatching viability, and postembryonic development of copepods is reviewed. Recent findings indicating that certain diatom diets have antiproliferative effects on copepod reproduction by reducing hatching viability, with serious consequences for the mass culturing of these organisms for aquaculture purposes, are also discussed. It is also possible that these antimitotic compounds may accumulate in copepod tissues and be transferred to higher trophic levels such as larval fish, as in the case of several dinoflagellate toxins (Tester et al. 2000). Thus food characteristics other than classical biochemical constituents such as proteins and fatty acids need to be assessed prior to using certain diatom diets to culture copepods.

COPEPOD EGG PRODUCTION RATES

Potential egg production rates vary widely among copepod species, with species that carry their eggs being less fecund than those that broadcast their eggs into the sea. The most fecund copepod species recorded thus far is *Centropages typicus*, with egg production rates as high as 230 eggs per female per day (Dagg 1978; Mauchline 1998). One of the least fecund is *Valdiviella insignis*, a bathypelagic copepod producing only two eggs per brood (Mauchline 1998). Kiørboe and Sabatini (1995) calculate that broadcast spawners produce on average 7.5 times as many eggs as sac spawners, with overall average maximum egg production rates of 40 ± 23 and 5.3 ± 2.2 eggs per female per day at 15°C. Broadcast spawners, however, have weight-specific fecundities that are on average 2.5 times the weight-specific fecundity of sac spawners, since carried eggs are about three times larger than freely spawned eggs.

Rates of egg production by a female are influenced by a number of factors, including temperature, which generally promotes an increase in the number of eggs produced per female per day until a saturation level, beyond which a further increase in temperature induces a decrease in egg production rates (Ianora 1998). At high temperatures, there is also a reduction in female longevity so that even if egg production rates increase, the reproductive output of females declines. For example, as shown by Uye (1981), at low temperatures (2.5°C), reproductive output of *Acartia clausi* was low (652 eggs during the life span of the females), but longevity was high (80 days). Highest reproductive output (1,281 eggs) occurred at 17°C when female longevity was intermediate (30 days). At high temperatures (25°C), reproductive output was again low (560 eggs), with female longevity dropping to 20 days.

Laboratory studies have also clearly demonstrated a positive correlation between fecundity and food quantity up to a saturation level beyond which the reproductive rate remains unchanged (e.g., Kiørboe et al. 1985), with saturation levels of about 400–500 μg carbon/L, corresponding to about 10^2–10^3 cells/mL. Different copepod species, however, reach maximum egg production at differing food concentrations, with smaller species generally requiring less food than larger ones. According to Runge (1988), for example, large species such as *C. pacificus* reach maximum or near-maximum rates at food concentrations that are four times higher than small copepods such as *A. clausi*, *Paracalanus parvus*, and *Pseudocalanus* sp. Small species would therefore be favored in areas with low and fluctuating food sources compared with larger species.

There is a time lag between food ingested and its being converted into production of eggs, which varies among species. Tester and Turner (1990) calculated a time lag of 9.5 hours for *Acartia tonsa*, 16.5 hours for *Centropages velificatus*, 65.5 hours for *Labidocera aestiva*, 89 hours for *C. typicus*, and 91 hours for *Anomalocera ornata*, indicating considerable interspecific variability in the lag period for conversion of ingested food to egg production.

The effects of food limitation have been well documented for several copepod species. Marshall and Orr (1955) observed that egg production in *C. finmarchicus* was always greater in fed than in starved females and that after 1 week, few of the starved females were laying eggs. When starved females were fed, egg laying reinitiated almost immediately, but egg production was only about two-thirds that of females fed all the time. Durbin et al. (1992) also found that egg production rates for *Acartia hudsonica* females maintained at low food levels (66 cells/mL of the diatom *Thalassiosira weissflogii*) declined to zero by day 7 compared with females maintained in continuous high food treatment (5×10^3 cells/mL). When females were switched from low to high food treatment, egg production reinitiated, but was, once again, only about two-thirds of that of females maintained in continuous high food treatment.

Food limitation in the wild can also induce dramatic effects on copepod egg production rates. Laabir et al. (1998), for example, observed that egg production rates for female *Calanus helgolandicus* were extremely variable throughout the year in the English Channel. Fecundity was highest in late spring (20–30 eggs/female/day), somewhat lower in summer and autumn (about 15 eggs/female/day), and extremely low (1–5 eggs/female/day) in winter. When, however, females with low spawning activity in winter were brought to the laboratory and fed, egg production rates resumed to levels observed in spring, indicating that low egg production rates were due to low ambient food conditions.

High egg production rates are considered correlated to phytoplankton blooms even though this correlation is often difficult to demonstrate, especially when chlorophyll *a* is used as an index of phytoplankton availability. For example, Ianora et al. (1992) and Ianora and Poulet (1993) did not find a correlation between egg production rates of *C. typicus* and *T. stylifera* in the Gulf of Naples, Italy, and ambient chlorophyll *a*, whereas Laabir et al. (1998) found that egg production of *C. helgolandicus* off Roscoff, France, was significantly correlated with chlorophyll *a*. Frost (1985) also found that the phytoplankton bloom in Dabob Bay, Washington, USA, induced increased egg production in *C. pacificus*, but not in the co-occurring *Pseudocalanus* sp. Egg production, therefore, may be affected by food availability in some species but not in others. The lack of a clear relation between copepod egg production rates and phytoplankton biomass indicates that chlorophyll *a* by itself is not necessarily a good indicator of food availability. Chlorophyll content may not necessarily be constant for a given species, and other characteristics of the food (nutritional content, presence of toxins) may be more important than chlorophyll content to promote higher copepod egg production rates.

Estimates of total fecundity during the female's life span differ depending on the species. Values are often greater than 1,000 when animals have been cultured for more than 1 month under conditions of plentiful food. In the Ecophysiology Laboratory in Naples, the author and her coworkers have mainly cultured two copepod species, *C. helgolandicus* and *T. stylifera*, for various lengths of time, to test the effect of different algal diets on copepod reproduction and longevity. Table 4.1 is a synopsis of these results for total egg production, summed over a period of 15 days or longer (up to 81 days), with different diatom, dinoflagellate, and flagellate diets. The results show that *C. helgolandicus* and *T. stylifera* respond differently to different diets and that the same food can often induce different effects in the two species. For example, diatoms generally induced higher total egg production rates over a period of 15 days in *C. helgolandicus* than dinoflagellates, while the opposite was true for *T. stylifera*. Among the diatoms tested, however, some induced higher egg production rates than others. With diets of *Thalassiosira rotula*, *T. weissflogii*, *Skeletonema costatum*, *Lauderia annulata*, and *Cylindrotheca closterium*, egg production rates for 15 days in *C. helgolandicus* were about twice (from 338–562 eggs/female) those with the other six diatoms tested (from 128 to 218 eggs/female). High total egg production rates were also obtained with the dinoflagellate *Akashiwo sanguinea* (395 eggs/female), but in general, dinoflagellates resulted in lower egg production rates than diatoms in *C. helgolandicus*. By contrast, dinoflagellates were an optimal food for egg production in *T. stylifera*. When egg production rates were summed for 15 days, production rates were higher with four of the five dinoflagellates tested (from 667 to 815 eggs/female) than all of the seven diatoms tested (from 132 to 650 eggs/female). Only one diatom, *T. rotula*, promoted egg production rates (650 eggs/female) similar to those resulting from dinoflagellate diets. The two flagellates, *Isochrysis galbana* and *Rhodomonas baltica*, did not promote high egg production rates, with values similar to some of the diatoms. None of the diets promoted maximum egg production rates similar to those reported in the wild for these species (about 110 eggs/female/day for *T. stylifera* [Ianora and Poulet 1993]; about 60 eggs/female/day for *C. helgolandicus* [Irigoien et al. 2000]), indicating that the algal diets in the laboratory were not as good as natural food.

Interestingly, the same diet often induced different egg-production rates in the two copepods. For example, *S. costatum* promoted very high egg production in *C. helgolandicus* (508 eggs/female) but was the worst diatom tested for *T. stylifera* (132 eggs/female). In other cases the opposite was true, with the same algae inducing high egg production in both species, as in the case of the diatom *T. rotula*. When *C. helgolandicus* and *T. stylifera* were cultured for longer than 15 days to test the effects of diet on longevity, egg production rates generally dropped to less than 50%, indicating that fecundity became progressively lower during the second half of the lifetime of the female. Of the two copepods, *T. stylifera* was potentially more fecund, reaching egg-production rates that were about 1.5 times greater than that of *C. helgolandicus*, confirming field studies on maximum egg production rates for these species. *Temora stylifera*, however, had a shorter life span (maximum 52 days) than *C. helgolandicus* (maximum 81 days), indicating that the potential reproductive output of this species during its lifetime may be lower.

The experimental protocol for maintaining copepods in the laboratory in these experiments

Table 4.1. Copepod egg production rates on different algal diets

Copepod tested		Algae tested	Total eggs after 15 days at 20°C	Number of eggs/ female/day	Total eggs after >16 days at 20°C	Number of eggs/ female/day
Calanus helgolandicus	Diatoms	*Thalassiosira rotula*	502	33.5	1,139 (81 d)	14.1
		Thalassiosira weissflogii	555	37		
		Skeletonema costatum	508	33.9		
		Chaetoceros socialis	213	14.2		
		Chaetoceros curvisetus	128	8.5		
		Lauderia annulata	562	37.5		
		Ditylum brightwelii	158	10.5		
		Pseudonitzschia delicatissima	195	13		
		Cylindrotheca closterium	338	22.5		
		Phaeodactylum tricornutum	219	14.6		
		Asterionellopsis glacialis	190	12.6		
	Dinoflagellates	*Lingulodinium polyedrum*	297	19.8	543 (80 d)	6.8
		Prorocentrum micans	154	10.3	290 (77 d)	3.7
		Prorocentrum minimum	274	18.3		
		Akashiwo sanguinea	395	26.3	586 (73 d)	8
Temora stylifera	Diatoms	*Thalassirosira rotula*	650	43.3	1,286 (40 d)	32.2
		Skeletonema costatum	132	8.8		
		Skeletonema pseudocostatum	180	12		
		Chaetoceros socialis	429	28.6		
		Chaetoceros curvisetus	454	30.3	58 (27 d)	21.5
		Cylindrotheca closterium	160	10.7		
		Phaeodactylum tricornutum	343	22.9	379 (22 d)	17.2
	Dinoflagellates	*Lingulodinium polyedrum*	815	54.3	1,416 (38 d)	37.3
		Prorocentrum micans	667	44.5	1,056 (42 d)	25.1
		Prorocentrum minimum	744	49	1,774 (47 d)	37.7
		Akashiwo sanguinea	768	51.2	905 (21 d)	43.1
		Pentapharsodinium tyrrhenicum	32	21.3		
	Flagellates	*Isochrysis galbana*	444	29	975 (52 d)	18.8
		Rhodomonas baltica	304	20.3	398 (28 d)	14.2

Note: d, day(s).

was the following. Copepods were collected in the Gulf of Naples from 0 to 50 m depth, and transferred to the laboratory. There, *C. helgolandicus* females were sorted and incubated individually in crystallizing dishes containing 100 mL ambient bay water for 24 hours to determine in situ egg production and egg viability. *Temora stylifera* males and females were incubated under the same conditions but as individual pairs since this species requires remating for the continued production of viable clutches (Ianora et al. 1989). After 24 hours, *C. helgolandicus* females and *T. stylifera* couples were transferred to new containers with 100 mL of 0.45 μm filtered seawater enriched with phytoplankton cultures having cell concentrations adjusted for each diet to obtain similar carbon concentrations, equivalent to a mean daily carbon input of 100 μg. Such concentrations were close to or in excess of daily food requirements reported in the literature (e.g., Frost 1985). Phytoplankton cultures were grown in K medium (Keller et al. 1987) for dinoflagellates and flagellates, and f/2 medium (Guillard and Ryther 1962) for diatoms, at 20°C and on a 12-hour light/12-hour dark cycle. Each day, copepods were transferred to new containers with fresh algal medium, and a daily tally was kept of egg production and egg-hatching success. To determine hatching success, eggs were fixed with alcohol 48 hours after spawning, and hatched nauplii were counted after they had settled to container bottoms. Experiments were run until death of the females in a controlled temperature chamber at 20°C and a photoperiod of 12 hours light/12 hours dark.

COPEPOD EGG HATCHING VIABILITY

Until the late 1980s it was generally assumed that most of the eggs spawned by copepods developed normally to hatching, leading to the production of living nauplii. This view has changed since the early reports of very high egg mortality in the field, in the order of 90% for some species (Peterson and Kimmerer 1994; Kiørboe et al. 1988; Laabir et al. 1998; Miralto et al. 1999; 2003). It was believed that high mortality was due to an absence of remating (Parrish and Wilson 1978) or to cannibalism (Liang and Uye 1996), but there is now strong experimental evidence that the main causes are physiological, with eggs unable to develop to hatching due to the toxic properties of food ingested by adult females. In particular, numerous laboratory studies have shown that when copepods are fed certain diatoms, the eggs produced either fail to develop to hatching or hatch into malformed nauplii (reviewed by Ianora et al. 2003).

Ianora and Poulet (1993) initially showed that although egg production rates in the copepod *T. stylifera* were similar when females were fed monoalgal cultures of either the diatom *T. rotula* or the dinoflagellate *Prorocentrum minimum*, there was a fourfold to sixfold difference in hatching success. Poulet et al. (1994) proposed that low hatching success was not due to a nutritional deficiency, but to the presence of antimitotic compounds in diatoms that blocked copepod embryogenesis. A series of successive papers showed that several diatom species induced similar effects in other copepod species (Ianora et al. 1995; Uye 1996; Starr et al. 1999) and that this effect was diatom density dependent, with stronger effects at higher diatom cell concentrations and longer periods during which diatoms were fed upon (Chaudron et al. 1996; Starr et al. 1999). It was demonstrated that reduced hatching success was due to the production of metabolites by the diatom cells themselves and not to bacteria associated with diatom cultures (Ianora et al. 1996); blockage was also not due to anoxia (Miralto et al. 1995). It was shown that some copepods such as *C. helgolandicus* and *C. pacificus* could recover within 3–4 days from the negative effects of a diatom food when females were switched to a non-diatom diet (Laabir et al. 1995; Uye 1996), but there is also evidence that not all copepod species are capable of recovering (Ianora et al. 2003). Mixed diets dilute but do not delete the negative effects of a diatom diet. For example, Turner et al. (2001) showed that hatching success in *T. stylifera* was reduced by approximately half, and it took about twice as long for total blockage of egg development to occur when females were offered a mixture of the diatom *T. rotula* and the dinoflagellate *P. minimum*. The same authors suggested that low hatching success was due to the toxic properties of the food rather than to a nutritional deficiency, since if essential nutrients were lacking, they probably could have been supplanted by the other food item provided *ad libitum*.

Miralto et al. (1999) were the first to isolate the molecules responsible for the antimitotic activity of diatoms, mainly the unsaturated aldehydes 2-*trans*-4-*trans*-decadienal, 2-*trans*-4-*trans*-7-*cis*-decatrienal, and 2-*trans*-4-*cis*-7-*cis*-decatrienal. These authors showed that diatom-derived aldehydes not only arrested the hatching of copepod eggs and cleavage of sea urchin embryos, but also inhibited the proliferation of human carcinoma cells. The same authors also reported that hatching success in the copepod *A. clausi* was greatly modified during two major diatom blooms in the North Adriatic Sea when egg viability plummeted to 12% and 24%, respectively, of the total number of eggs produced, compared with 90% during postbloom conditions. Miralto et al. (1999) suggested that diatom-derived aldehydes were the probable agents of reproductive failure in copepods when diatoms were the major food source.

Pohnert (2000) has shown that the unsaturated aldehydes decadienal and decatrienal are derived from eicosanoic (C_{20}) fatty acids seconds after mechanical damage to diatom cells due to feeding by copepods, similar to the wound reaction in higher plants (Rosahl 1996). By way of a cascade of enzymatic events initiated by phospholipids, fatty acids are liberated from membrane storage sites and transformed into lipid hydroperoxides through the action of lipoxygenases. These are then transformed within seconds into unsaturated aldehydes by the action of lyases (Pohnert 2002; Pohnert and Boland 2002). Diatom unsaturated aldehydes are therefore released only upon breakage of cells, as an activated defense against potential predators, and are not secreted from intact cells.

Pohnert et al. (2002) and d'Ippolito et al. (2002b; 2003) have now shown that diatom species can produce unsaturated aldehydes other than those originally described by Miralto et al. (1999), depending on the fatty acid precursors, and that not all diatom species are capable of producing such compounds. For example, Pohnert et al. (2002) have reported that the *T. rotula* strain from the Gulf of Naples produced large quantities of the unsaturated aldehydes, decadienal and decatrienal, while the strain isolated from coastal California waters did not produce these compounds but only the saturated aldehyde tridecanal and other homologues. Of the two strains, only the first reduced hatching success in *C. helgolanicus*.

The biological activity of diatom aldehydes seems to be due to their characteristic α, β, γ, δ unsaturated structure and not merely to the reactive aldehydic group. For example, Pohnert et al. (2002) have shown that saturated diatom aldehydes such as tridecanal have no effects on cleavage of sea urchin eggs exposed to saturating concentrations of these molecules. Tosti et al. (2003) have shown that small molecules carrying only an aldehydic group, namely formaldehyde and acetaldehyde, have no effects on fertilization currents in ascidian embryos, as opposed to the diatom unsaturated aldehyde, decadienal. Therefore, diatom species capable of producing unsaturated aldehydes with an α, β, γ, δ chemical conformation should be better defended against potential predators than those that only produce saturated aldehydes. In theory, this should provide a selective survival advantage in species carrying this capacity, since such "protected" species would experience an explosive increase because of their protection from grazers. Copepod recruitment would therefore be expected to be greater during postbloom conditions rather than in concomitance to diatom blooms, as already reported by Miralto et al. (1999, 2003).

Elucidation of the chemistry of diatom defense is growing at a rapid pace, and it is now evident that such compounds may be responsible for the antiproliferative effects observed in copepods fed diatom diets. An example of reduced hatching success when female *C. helgolandicus* are fed diatom as opposed to dinoflagellate diets is shown in Fig. 4.1. As already shown in Table 4.1, diatoms (*T. rotula*, *S. costatum*, *L. annulata*) generally induced higher egg production rates than dinoflagellates (*P. minimum*, *Lingulodinium polyedrum*, *A. sanguinea*) in this copepod species. In all cases, even though fecundity was higher with diatom diets, hatching success was greatly reduced with time, with differing effects depending on the diatom species. Of the three species, *S. costatum* induced the strongest inhibition, with hatching success dropping to 0% after 9 days of feeding on this alga. The other two diatoms *T. rotula* and *L. annulata* induced very similar responses, with hatching success diminishing steadily to 5–10% after 15 days of feeding. Methods regarding these experiments have already been described in the section on copepod egg production rates.

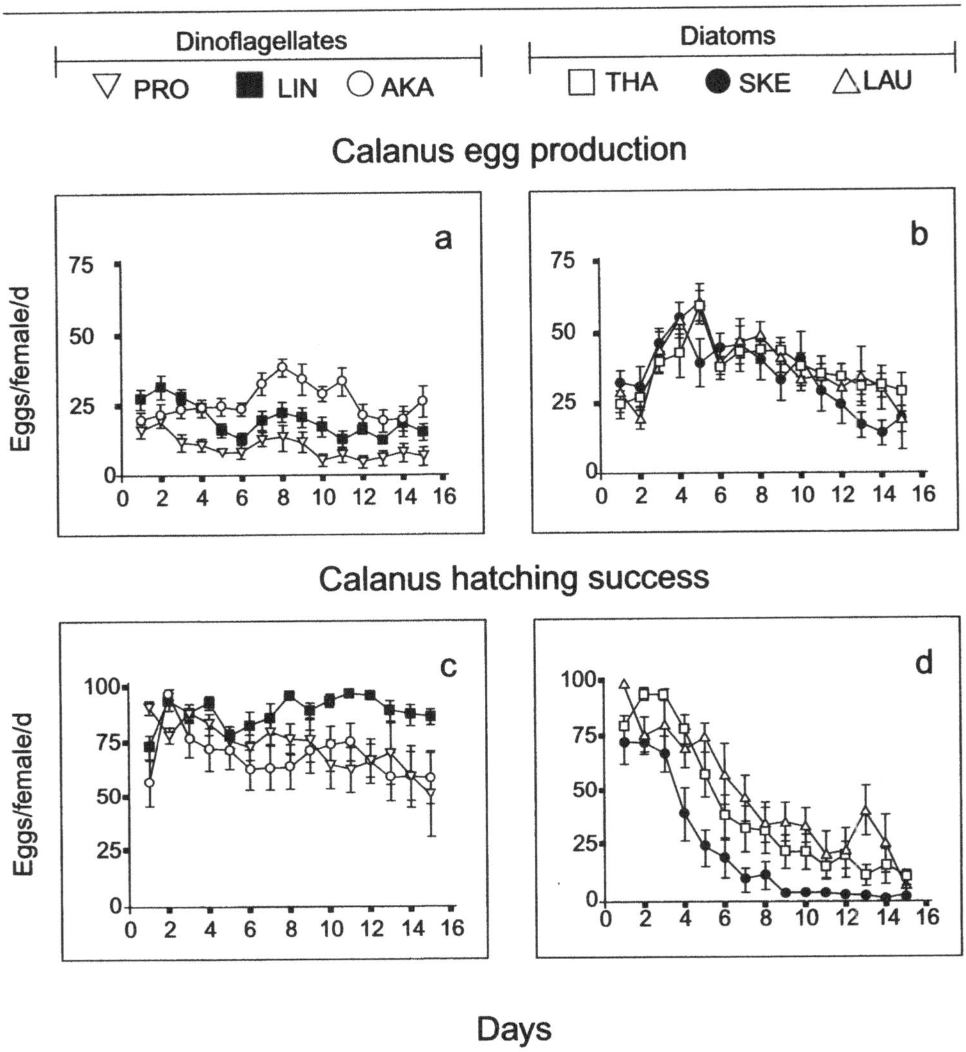

Figure 4.1. Laboratory experiments showing (A, B) daily egg-production rates and (C, D) percentage of egg viability for the copepod *Calanus helgolandicus* fed different dinoflagellate (PRO, *Prorocentrum minimum*; LIN, *Lingulodinium polyedrum*; AKA, *Akashiwo sanguinea*) and diatom (THA, *Thalassiosira rotula*; SKE, *Skeletonema costatum*; LAU, *Lauderia annulata*) diets. Data for LIN and AKA modified from Ianora et al. (1999). All other data are unpublished results. Experimental protocols to measure egg production and hatching success are the same for all copepod–diatom combinations and are described in the text. Data are means ± S.E. for 15 replicates.

Copepods, however, can respond differently to the same diatom species. As shown in Fig. 4.2, the diatom *T. rotula* induced different reproductive responses in the copepods *A. clausi*, *C. typicus*, *C. helgolandicus*, and *T. stylifera*. Egg production rates were relatively high and stable in *T. stylifera* and *C. helgolandicus*, lower but nonetheless stable in *A. clausi*, but decreased rapidly with time in *C. typicus*. The same copepods also responded differently to these diatoms in terms of hatching success. While *T. rotula* induced dramatic effects in *C. typicus*, the effect was somewhat milder in *T. stylifera*, and there was no effect in *A. clausi* and *C. helgolandicus*, denoting species-specific responses to the toxic metabolites in diatoms. This suggests that some copepod species may have evolved counterdefenses to diatom aldehydes in the continuing "coevolutionary arms race" between plant defense and animal response. For example, some copepod species may

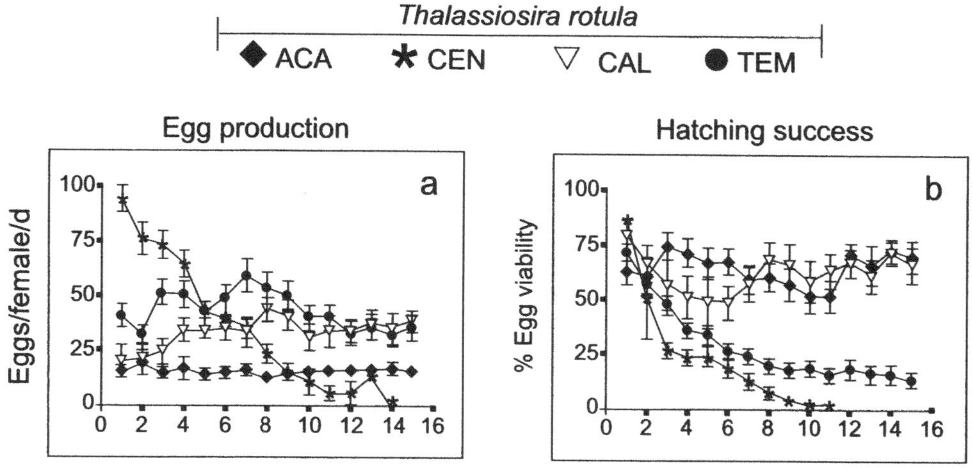

Figure 4.2. Laboratory experiments showing (A) egg production and (B) hatching success for the copepods *Acartia clausi* (ACA), *Centropages typicus* (CEN), *Calanus helgolandicus* (CAL), and *Temora stylifera* (TEM), which were fed the diatom *Thalassiosira rotula*. Data for *A. clausi* are redrawn from Ianora et al. (1996), *C. typicus* from Miralto et al. (1995), *C. helgolandicus* from Ianora et al. (2003), and *T. stylifera* from Turner et al. (2001). Data are means ± S.E.

be capable of detoxifying diatom aldehydes through the production of enzymes such as aldehyde dehydrogenases or glutathion reductases or through other mechanisms, thereby reducing or deleting the negative effects of diatom toxins. Such copepods would therefore be favored with respect to others in that they would be able to exploit diatom blooms without incurring high offspring mortality.

The capability of producing detoxifying enzymes may explain past conflicting results in which diatoms at times did not induce hatching inhibition in copepods. For example, Ban et al. (1997) and Starr et al. (1999) reported that the diatom *S. costatum* isolated from the Pacific Coast of the United States did not reduce hatching success in *C. finmarchicus*, while the same strain, when tested on *C. helgolandicus*, induced similar negative effects on hatching success as the *S. costatum* strain from the Adriatic Sea (Ianora et al. 2003). Further studies are warranted at this point to better clarify the differing effects of diatoms in different copepod species.

COPEPOD POSTEMBRYONIC DEVELOPMENT

Copepods have complex life cycles, with 13 life-cycle stages (egg, six naupliar [NI–NVI], five copepodid [CI–CV], adult). According to Landry (1983), the prefeeding stages NI and NII are relatively short since the newly born nauplius still relies on maternal yolk reserves, while the first feeding stages NII or NIII, depending on the species, are relatively long, and longer than other juvenile stages, possibly due to the time necessary to recover from the weight lost during the nonfeeding stage. The same author also reports that the CV stage is longer than the other juvenile stages, probably due to physiological changes in preparation for maturity, and that males generally develop faster than females.

It has been suggested that some species may show what is termed *isochronal development*, in which all life-cycle stages are of equal duration. Miller et al. (1977) initially developed this rule for species of the genus *Acartia*, but since then, this type of development has also been reported for the genus *Pseudocalanus* (Corkett and McLaren 1978). On the other hand, Landry (1983) has argued that isochronal development probably applies only to a limited number of species, since most species have at least one or more stages with prolonged development.

A number of studies have determined the generation times of temperate copepods and the abiotic and biotic factors affecting development (reviewed by Mauchline 1998). Briefly, it has been argued that temperature is the most important factor determining development times (Huntley and Lopez 1992). Klein Breteler and Schogt (1994) have argued that food is limiting in many periods of the year and that food shortage will therefore also be important in affecting copepod development rate, body length, and weight. For example, at superabundant food levels and high temperatures (20°C), Klein Breteler and Schogt (1994) found that *A. clausi* reached adulthood in 15 days, while at low food concentrations and low temperatures (5°C), it took 85 days. When temperature was maintained constantly high (20°C), nauplii reached adulthood in 40 days at low food concentrations, 20 days at intermediate levels, and 15 days at highest food concentrations. Alternatively, when food concentrations were always superabundant, nauplii reached adulthood after 80 days at 5°C, 40 days at 10°C, and 15 days at 15°C and 20°C.

Copepods have been reared to adulthood on a variety of algal diets, but it is not clear how food quality affects generation times. Paffenhöfer and Harris (1979) wrote a comprehensive review on cultivation techniques used to maintain and rear various holozooplankton groups, including copepods. Their text (see their Table II) provides a detailed synopsis of studies in which copepods were successfully reared on mixed algal diets (Zillioux and Wilson 1966; Mullin and Brooks 1967; Katona 1970; Landry 1975; McLaren 1976), where diatom diets were generally mixed with dinoflagellate (e.g., species of *Gymnodinium* and *Gonyaulax*) diets, flagellate (e.g., *I. galbana*) diets, or both. The problem with mixed diets, however, is that it is difficult to discern the differing effects of, for example, diatom and other algal diets and, therefore, the effects of diatom-derived aldehydes on postembryonic development. Table 4.2 shows only those studies, listed by Paffenhöfer and Harris (1979), in which copepods have been successfully reared to adulthood on a monalgal diatom diet, as well as more recent studies in which copepods have been reared to adulthood on a sole diatom food. Table 4.2 includes only 12 studies, which is surprising, given that diatoms are believed to support good development rates and low larval mortality in copepods. In most studies, however, diatom diets generally resulted in poorer developmental performance compared with nondiatom foods, since they lengthened the generation time and increased mortality rates. For example, Paffenhöfer (1970) found that *C. helgolandicus* grew faster with the dinoflagellate *Gymnodinium splendens* (18 days) compared with the diatoms *S. costatum* (36 days) and *Lauderia borealis* (24 days) and that mortality was lower with *G. splendens* (2.3%) compared with *S. costatum* (33.9%) and *L. borealis* (13.5%). Smith and Lane (1985) also found very slow generation times when *C. typicus* was fed the diatom *T. weissflogii* (40 days), but these authors did not report mortality rates. Koski et al. (1998) found that *Pseudocalanus elongatus* grew well on the diatom *T. weissflogii*, but development and growth rates were slower compared with the flagellate

Table 4.2. Studies on generation times of copepods

Species	Food organism	Generation time (egg to adult)	Temperature (°C)	Source
Euterpina acutifrons	PHA	10–25 d	16–23	Bernard (1963)
Calanus helgolandicus	LAU, SKE, CHA	18–54 d	15	Paffenhöfer (1970)
Calanus helgolandicus	THAF	22–44 d	12–17	Mullin and Brooks (1970)
Calanus helgolandicus	LAU		15	Paffenhöfer (1976)
Calanus helgolandicus	SKE, THA	14–18 d	20	Ianora et al. (2003)
Rhincalanus nasutus	THAF, DIT	22–53 d	10–15	Mullin and Brooks (1970)
Pseudocalanus elongatus	THA	26–32 d	12	Paffenhöfer and Harris (1976)
Pseudocalanus elongatus	TWEI	25 d	15	Koski et al. (1998)
Temora longicornis	THA	24–33 d	12	Harris and Paffenhöfer (1976)
Temora stylifera	THA	20 d	20	Carotenuto et al. (2002)
Centropages typicus	THA	40 d	10–15	Smith and Lane (1985)

Note: PHA, *Phaeodactylum tricornutum*; LAU, *Lauderia borealis*; SKE, *Skeletonema costatum*; CHA, *Chaetoceros curvisetus*; THAF, *Thalassiosira fluviatilis*; THA, *Thalassiosira rotula*; DIT, *Ditylum brightwelii*; TWEI, *Thalassiosira weissflogii*; d, day(s).

Rhodomonas spp. and the dinoflagellate *Gymnodinium simplex*.

There are also several studies that report that copepods were unable to grow to adulthood on a monoalgal diatom diet. For example, Mullin and Brooks (1970) found that *C. helgolandicus* were unable to grow with the diatom *Ditylum brightwellii* compared with another copepod, *Rhincalanus nasutus*, which was successfully reared to adulthood on this diet. Carotenuto et al. (2002) were able to rear *T. stylifera* to adulthood only on the diatom *T. rotula*, but not on the diatoms *S. costatum* and *Phaeodactylum tricornutum*. Ianora et al. (2003) were able to rear *C. helgolandicus* on both *T. rotula* and *S. costatum*, but not *P. tricornutum*. Hirche (1980) found that all nauplii of the copepod *Calanoides carinatus* died soon after passing the naupliar phase when fed the diatom *P. tricornutum*, compared with the dinoflagellate *G. simplex*. In this case, nauplii reached adulthood after 20 days, with a final survival of 13%. Peterson (1986) found that even when the diatom *T. weissflogii* was mixed with the flagellate *I. galbana*, *Calanus marshallae* was able to grow in only one experiment, with a generation time of 36 days and 19% final survivorship; in all other replicates larvae did not grow beyond the copepodid stage V.

On the other hand, there are at least four studies reporting that diatoms are a good food source for copepod development. Harris and Paffenhöfer (1976) were able to successfully rear *Temora longicornis* on the diatom *T. rotula*, with a generation time of 24 days and final mortality of 37%. Paffenhöfer and Harris (1976) reported similar generation times (29 days) and final mortality (25%) when *P. elongatus* was reared on the same diatom. Paffenhöfer (1976) reported that *C. helgolandicus* grew well on a diet of *L. borealis*, with a final mortality of less than 10%. Payne and Rippingale (2000) found that of four microalgae tested on the estuarine copepod *Gladioferens imparipes*, *I. galbana* promoted the fastest development, followed by *Chaetoceros muelleri* and then by *Dunaliella tertiolecta* and *Nannochloropsis oculata*. In terms of survival (in percentage), however, *I. galbana* promoted highest survival (90%), followed by *D. tertiolecta* (60%) and *C. muelleri* (45%); *N. oculata* resulted in little or no survival.

These conflicting results seem to denote species-specific responses to diatom diets, with some diatom species (e.g., *T. rotula*) that are capable of promoting successful development compared with others (e.g., *S. costatum*) that promote poor development in some copepods (*T. stylifera*) but not others (*C. helgolandicus*). In general, however, diatoms are often a suboptimal food for larval growth, but why? Most studies have assumed that this may be due to a nutritional deficiency, such as an inadequate mineral/biochemical composition of the phytoplankton (Koski et al. 1998). Brown and Jeffrey (1995), however, examined the chemical composition of diatoms and found that this algal class is rich in high-quality proteins (31%–38% dry weight) and lipids (18%–20% dry weight). Similar results were also found by Brown et al. (1989), who found that diatoms were generally rich in proteins (30%–460%) and lipids (10%–20%) and induced successful growth in crustaceans. Carotenuto et al. (2002) were the first to propose that unsuccessful development with some diatom diets was not due to a nutritional deficiency, but rather to the presence of unsaturated aldehydes that have already been shown to induce developmental failure of embryos. These authors reared *T. stylifera* to adulthood with only one (*T. rotula*) of three (*S. costatum* and *P. tricornutum*) diatoms tested. When females were preconditioned for 24 hours with the control dinoflagellate diet *P. minimum*, and nauplii were then reared on the same diet as the mother, survivorship improved from 34% in nauplii from nonconditioned mothers to 80% in nauplii from preconditioned females. Survivorship improved from 7% in nauplii from nonconditioned mothers to 34% in nauplii from preconditioned mothers when larvae were reared on a diet of *T. rotula*. With the two other diatoms tested, *S. costatum* and *P. tricornutum*, larvae still failed to develop to adulthood even after females were preconditioned with the dinoflagellate *P. minimum*. Carotenuto et al. (2002) reported that birth defects in hatched nauplii occurred only when females were fed with diatom diets and not when nauplii were generated from wild females with high initial hatching success. In this case, nauplii fed on a diatom diet developed without showing any anatomical deformities, even though they were unable to complete their life cycle and died for unknown reasons. The authors concluded that diatoms, therefore, not only induced insidious abortive effects during embryogenesis but also antigrowth effects on later larval stages.

Deformed copepod nauplii were first reported by Poulet et al. (1995) in *C. helgolandicus* and have since been found in several other copepods (Uye 1996; Starr et at. 1999; Ban et al. 2000; Carotenuto et al. 2002; Ianora et al. 2003). These nauplii are produced when female copepods are fed diatom diets in laboratory studies, but they have also been recorded from wild females in concurrence with natural diatom blooms (e.g., Ban et al. 2000). A scanning electron micrograph of a deformed nauplius of *C. helgolandicus* NI is shown in Fig. 4.3. The body is highly irregular, and the swimming and feeding appendages are reduced to stubs; the head and tail regions of the body are difficult to distinguish. These nauplii die soon after birth because they are unable to swim or feed properly.

Ianora et al. (2004) found that the degree of teratogenesis (i.e., malformation) increased with increasing exposure of females to a diatom diet. After 9 days, female *C. helgolandicus* fed the diatom *S. costatum* spawned only abnormal nauplii that showed a series of birth defects, including malformed bodies with aberrant or reduced number of swimming and feeding appendages. Caldwell et al. (2002) reported similar adverse effects of diatom-derived aldehydes on polychaetes and echinoderms, with anatomical malformations such as an incomplete cilary band in trochophore larvae and stunted asymmetrical arms in echinopluteus larvae.

By definition, teratogens are substances that induce structural malformations in the offspring of organisms exposed to them during gestation. Such compounds were unknown for marine plants until the recent discovery of diatom-derived aldehydes, even though they are well known for higher terrestrial plants. Classic examples of teratogenic compounds are nicotine, the principal chemical present in tobacco, and caffeine, present in many plants processed for use as beverages and foods. Other common teratogens are the cytochalasins and solanines present in the potato, *Solanum tuberosum*, and the coumarins in parsley, *Petroselinum crispum*. All of these teratogens have been shown to induce malformations during gestation in laboratory animals (Scharden and York 1995). The discovery that diatoms are capable of producing such compounds as an activated defense against their predators, thereby crippling future generations of herbivores, has far-reaching implications on our understanding of the factors governing recruitment rates of filter-feeding copepods at sea.

IMPLICATIONS FOR THE MASS CULTIVATION OF COPEPODS IN AQUACULTURE

Microalgae are widely used in aquaculture as live feeds for mollusks, crustaceans, and some fish larvae, as well as for zooplankton that, in turn, are used as feed in mariculture food chains. For microalgae to be nutritionally complete diets, however, they must not only contain a balanced mixture of essential nutrients, but also be devoid of toxic substances that interfere with growth. At present, there is much concern regarding use of live feeds with high concentrations of *n*-3 highly unsaturated fatty acids (*n*-3 HUFA) such as eicosapentanoeic (20:5*n*-3) and docosahexaenoic (22:6*n*-3) acids (Brown et al. 1997), since these are reported as supporting good larval growth (e.g., Harrison 1990). It has been shown, however, that *n*-3 HUFAs are highly unstable and can become toxic at high concentrations (McEvoy et al. 1995; Jüttner 2001). For example, Jüttner (2001) found that 20:5*n*-3 induced 100% mortality in the freshwater crustacean *Thamnocephalus platyrus* at a concentration of 35.7 μM. The use of this fatty acid as a nutrient supplement should therefore be closely monitored. In fact, studies on the effects of *n*-3 HUFAs in the diets of Japanese flounder *Paralichthys olivaceus* have demonstrated that although egg production was highest in fish fed the highest level of *n*-3 HUFA, egg quality parameters such as percentage of buoyant eggs, hatching rate, and percentage of normal larvae were significantly higher in the group fed the lowest *n*-3 HUFA diet (Furuita et al. 2002). Diatoms are rich sources of *n*-3 and other HUFAs (Brown et al. 1997), but this may not necessarily imply that all diatoms promote high hatching success and larval growth. For example, *T. rotula* is extremely rich in C_{20} fatty acids (d'Ippolito et al. 2002a; Pohnert et al. 2002), yet this diatom promotes poor copepod embryonic and postembryonic development (Fig. 4.1 and Table 4.2).

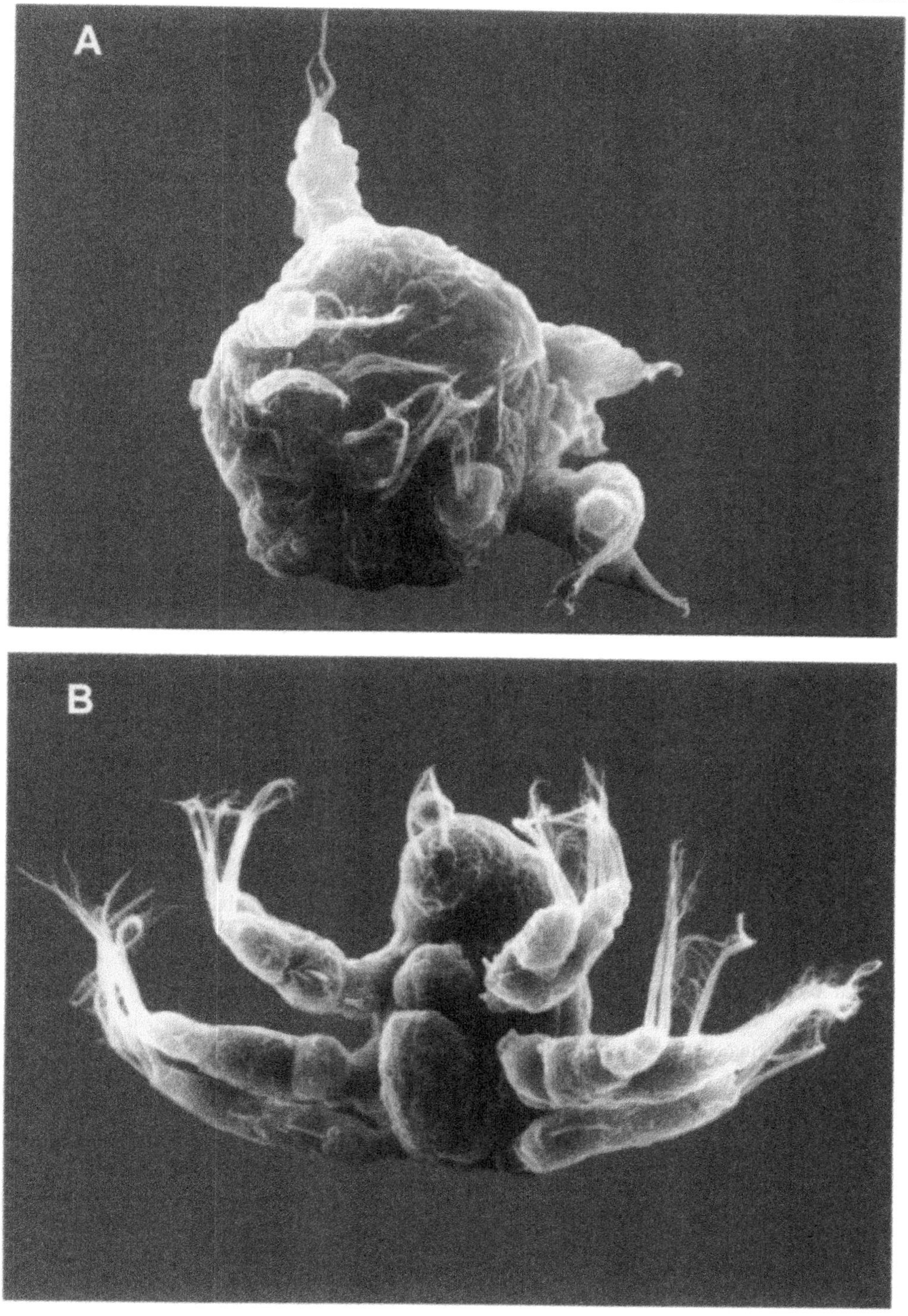

Figure 4.3. Scanning electron images of (A) teratogenic malformed NI nauplius of the copepod *Calanus helgolandicus* generated from females fed the diatom *Skeletonema costatum* for 3–4 days and (B) normal NI nauplius generated from *C. helgolandicus* females fed the control dinoflagellate diet *Prorocentrum minimum* for the same length of time.

Diatoms are also potential sources of teratogens that interfere with the normal development and growth of grazers. Such compounds may have little or no effect on adult females, but can sabotage the offspring by inducing abortions, birth defects, and poor development rates (Ianora et al. 2003). The nutritional value of a mariculture feed should therefore be evaluated not only for its content of primary components such as proteins, fatty acids, and vitamins, but also evaluated to see if any microalgae used are capable of producing teratogens such as those produced by diatoms. Not all diatoms synthesize these compounds (Pohnert et al. 2002), and animals may differ in their capacity to detoxify them (Ianora et al. 2003), but the presence of these compounds may pose serious problems for rearing copepods as food for fish larvae. Most mariculture diets already use a mixture of microalgae, including diatoms, for the mass cultivation of copepods (e.g., Reitan et al. 1997; Nanton and Castell 1998; Hernéndez Molejón and Alvarez-Lajonchère 2003). Thus, possible deleterious effects of diatoms would, in this case, be diluted but possibly not abolished. New methods for the identification and quantification of α, β, γ, δ unsaturated aldehydes have recently been developed and can now be employed to routinely screen for the presence of such compounds (d'Ippolito et al. 2002a).

It is also interesting to consider the maternal effects of diet on copepod embryonic and postembryonic development. Carotenuto et al. (2002) showed that by preconditioning females with a good diet, such as the dinoflagellate *P. minimum* or the autotrophic flagellate *I. galbana*, percentage of survivorship of nauplii to adulthood increased by about 60% when nauplii were fed the same diet as the mother. These authors also showed that potential teratogens were only transferred through maternal diets since nauplii from egg batches with high initial hatching success developed normally to adulthood on a diatom diet. Such nauplii showed no anatomical deformities, compared with nauplii generated from mothers that had fed on a diatom diet and that had been generated from batches of eggs with low initial hatching success. This suggests that maternal diets are as important, if not more important, than larval diets and that future aquaculture efforts should consider providing good food for both mothers and their offspring. In the long run, this would maximize the growth potential of larval copepods by improving offspring fitness and percentage of survivorship.

Another potential problem when rearing copepods using diatom diets is the possibility that these compounds are accumulated in copepod tissues and are transferred to higher trophic levels, as demonstrated for saxitoxins and other phycotoxins (e.g., Tester et al. 2000), with subsequent fish kills and poisoning of humans (Turner and Tester 1997). It is yet unknown whether diatom-reactive aldehydes are transferred to and sequestered in copepod oocytes in an unmodified form or whether structural modifications occur when aldehydes are exposed to the digestive system in copepods. Work currently under way demonstrates that a range of aldehydes and other hydrocarbons are present in the oocytes of polychaete worms (Caldwell 2004). The same may also be true for copepods. Further studies of the mechanisms and interactions discussed above should be given high priority in future research on the culturing of copepods as potential mariculture feed for finfish or other fish larvae.

ACKNOWLEDGMENTS

I wish to thank Flora Palumbo for computer graphics and Giuseppe Gargiulo for scanning electron micrographs.

REFERENCES

Ban, S., Burns, C., Castel, J., Chaudron, Y., Christou, E., Escribano, R., Fonda Umani, S., Gasparini, S., Guerrero Ruiz, F., Hoffmeyer, M., Ianora, A., Kang, H., Laabir, M., Lacoste, A., Miralto, A., Ning, X., Poulet, S., Rodriguez, V., Runge, J., Shi, J., Starr, M., Uye, S., and Wang, J. 1997. The paradox of diatom–copepod interactions. *Marine Ecology Progress Series* 157:287–293.

Ban, S., Lee, H.-W., Shinada, A., and Toda, T. 2000. *In situ* egg production and hatching success of the marine copepod *Pseudocalanus newmani* in Funka Bay and adjacent waters off southwestern Hokkaido, Japan: association to diatom bloom. *Journal of Plankton Research* 22:907–922.

Bernard, M. 1963. Le cycle vital en laboratoire d'un copépode pélagique de Méditerranée *Euterpina acutifrons* Claus. *Pelagos* 1:35–48.

Brown, M.R. and Jeffrey, S.W. 1995. The amino acid and gross composition of marine diatoms poten-

tially useful for mariculture. *Journal of Applied Phycology* 7: 521–527.
Brown, M.R., Jeffrey, S.W., and Garland, C.D. 1989. Nutritional aspect of microalgae used for mariculture: a literature review. *CSIRO Marine Laboratory Reports* 205:1–44.
Brown, M.R., Jeffrey, S.W., Volkman, J.K., and Dunstan, G.A. 1997. Nutritional properties of microalgae for mariculture. *Aquaculture* 154:315–331.
Caldwell, G.S. 2004. Diatom mediated disruption of invertebrate reproduction and development. Ph.D. thesis, University of Newcastle-upon-Tyne, U.K.
Caldwell, G.S., Olive, P.J.W., and Bentley, M.G. 2002. Inhibition of embryonic development and fertilization in broadcast spawning marine invertebrates by water soluble diatom extracts and the diatom toxin, 2,trans,4,trans decadienal. *Aquatic Toxicology* 60:123–137.
Carotenuto, Y., Ianora, A., Buttino, I., Romano, G., and Miralto, A. 2002. Is postembryonic development in the copepod *Temora stylifera* negatively affected by diatom diets? *Journal of Experimental Marine Biology and Ecology* 276:49–66.
Chaudron, Y., Poulet, S.A., Laabir, M., Ianora, A., and Miralto, A. 1996. Is hatching success diatom density-dependent? *Marine Ecology Progress Series* 144:185–193.
Corkett, C.J. and McLaren, I.A. 1978. The biology of *Pseudocalanus. Advances in Marine Biology* 15:1–231.
Dagg, M. 1978. Estimated, in situ, rates of egg production for the copepod *Centropages typicus* (Krøyer) in the New York bight. *Journal of Experimental Marine Biology and Ecology* 34:183–196.
d'Ippolito, G., Iadicicco, O., Romano, G., and Fontana, A. 2002a. Detection of short-chain aldehydes in marine organisms: the diatom *Thalassiosira rotula. Tetrahedron Letters* 43:6137–6140.
d'Ippolito, G., Romano, G., Iadiccio, O., Miralto, A., Ianora, A., Cimino, G., and Fontana, A. 2002b. New birth-control aldehydes from the marine diatom *Skeletonema costatum. Tetrahedron Letters* 43:6133–6136.
d'Ippolito, G., Romano, G., Caruso, T., Spinella, A., Cimino, G., and Fontana, A. 2003. Production of octadienal in the marine diatom *Skeletonema costatum. Organic Letters* 5:885–887.
Durbin, E.G., Durbin, A.G., and Campbell, R.G. 1992. Body size and egg production in the marine copepod *Acartia hudsonica* during a winter–spring diatom bloom in Narragansett Bay. *Limnology and Oceanography* 37:342–360.
Evjemo, J.O., Reitan, K.I., and Yngvar, O. 2003. Copepods as live food organisms in the larval rearing of halibut larvae (*Hippoglossus hippoglossus* L.) with special reference on the nutritional value. *Aquaculture* 227:191–210.
Frost, B.W. 1985. Food limitation of the planktonic marine copepods *Calanus pacificus* and *Pseudocalanus* sp. in a temperate fjord. *Archiv für Hydrobiologie (Beih Ergeb Limnol)* 21:1–13.
Furuita, H., Tanaka, H., Yamamoto, Y., Suzuki, N., and Takeuchi, T. 2002. Effects of high levels of n-3 HUFA in broodstock diet on egg quality and egg fatty acid composition of Japanese flounder, *Paralichthys olivaceus. Aquaculture* 210:323–333.
Giangrande, A., Geraci, S., and Belmonte, G. 1994. Life-cycle and life-history diversity in marine invertebrates and the implications in community dynamics. *Oceanography and Marine Biology Annual Review* 32:305–333.
Grice, G.D. and Marcus, N.H. 1981. Dormant eggs of marine copepods. *Oceanography and Marine Biology Annual Review* 19:125–140.
Guillard, R.R.R. and Ryther, J.H. 1962. Studies of marine planktonic diatoms. *Journal of Microbiology* 8:229–239.
Harris, R.P. and Paffenhöfer, G.-A. 1976. Feeding, growth and reproduction of the marine planktonic copepod *Temora longicornis* Müller. *Journal of the Marine Biological Association of the United Kingdom* 56:875–888.
Harrison, K.E. 1990. The role of nutrition in maturation, reproduction and embryonic development of decapod crustaceans. *Journal of Shellfish Research* 9:1–28.
Hernédez Molejón, O.G., and Alvarez-Lajonchère, L. 2003. Culture experiments with *Oithona oculata* Farran, 1913 (Copepoda: Cyclopoida), and its advantages as food for marine fish larvae. *Aquaculture* 219:471–483.
Hirche, H.-J. 1980. The cultivation of *Calanoides carinatus* Krøyer (Copepoda: Calanoida) under different temperature and food conditions with a description of the eggs and nauplii. *Journal of the Marine Biological Association of the United Kingdom* 60:15–125.
Humes, A.G. 1994. How many copepods? *Hydrobiologia* 292/293:1–7.
Huntley, M.E. and Lopez, M.D.G. 1992. Temperature-dependent production of marine copepods: a global synthesis. *American Naturalist* 140:201–242.
Ianora, A. 1998. Copepod life history traits in subtemperate regions. *Journal of Marine Systems* 15:337–349.
Ianora, A., Mazzocchi, M.G., and Grottoli, R. 1992. Seasonal fluctuations in fecundity and hatching success in the planktonic copepod *Centropages typicus. Journal of Plankton Research* 14:1483–1494.

Ianora, A., Miralto, A., Buttino, I., Poulet, S.A., and Romano, G. 1999. First evidence of some dinoflagellates reducing male copepod fertilization capacity. *Limnology and Oceanography* 44:147–153.

Ianora, A., Miralto, A., Poulet, S.A., Carotenuto, Y., Buttino, I., Romano, G., Casotti, R., Pohnert, G., Wichard, T., Colucci D'Amato, L., Terrazzano, G., and Smetacek, V. 2004. Aldehdye suppression of copepod recruitment in blooms of a ubiquitous planktonic diatom. *Nature* 429:403–407.

Ianora, A. and Poulet, S.A. 1993. Egg viability in the copepod *Temora stylifera. Limnology and Oceanography* 38:1615–1626.

Ianora, A., Poulet, S.A., and Miralto, A. 1995. A comparative study of the inhibitory effect of diatoms on the reproductive biology of the copepod *Temora stylifera. Marine Biology* 121:533–539.

Ianora, A., Poulet, S.A., and Miralto, A. 2003. The effects of diatoms on copepod reproduction: a review. *Phycologia* 42:351–363.

Ianora, A., Poulet, S.A., Miralto, A., and Grottoli, R. 1996. The diatom *Thalassiosira rotula* affects reproductive success in the copepod *Acartia clausi. Marine Biology* 125:279–286.

Ianora, A., Scotto di Carlo, B., and Mascellaro, P. 1989. Reproductive biology of the planktonic copepod *Temora stylifera. Marine Biology* 101:187–194.

Irigoien, X., Harris, R.P., Head, R.N., and Harbour, D. 2000. The influence of diatom abundance on the egg production rate of *Calanus helgolandicus* in the English Channel. *Limnology and Oceanography* 45:1433–1439.

Jüttner, F. 2001. Liberation of 5,8,11,14,17-eicosapentaenoic acid and other polyunsaturated fatty acids from lipids as a grazer defense reaction in epilithic diatom biofilms. *Journal of Phycology* 37:744–755.

Katona, S.K. 1970. Growth characteristics of the copepods *Eurytemora affinis* and *E. herdmani* in laboratory cultures. *Heloländer wissenschaftliche Meeresuntersuchungen* 20:373–384.

Keller, M.D., Selvin, R.C., Claus, W., and Guillard, R.R.L. 1987. Media for the culture of oceanic ultraphytoplankton. *Journal of Phycology* 23: 633–638.

Kiørboe, T., Mohlenberg, F., and Hamburger, K. 1985. Bioenergetics of the planktonic copepod *Acartia tonsa*: relation between feeding, egg production and respiration, and composition of specific dynamic action. *Marine Ecology Progress Series* 26:85–97.

Kiørboe, T., Munk, P., Richardson, K., Christensen, V., and Paulen, H. 1988. Plankton dynamics and herring larval growth, drift and survival in a frontal area. *Marine Ecology Progress Series* 44:205–219.

Kiørboe, T. and Sabatini, M. 1995. Scaling of fecundity, growth and development in marine planktonic copepods. *Marine Ecology Progress Series* 120:285–298.

Klein Breteler, W.C.M. and Schogt, N. 1994. Development of *Acartia clausi* (Copepoda, Calanoida) cultured at different conditions of temperature and food. *Hydrobiologia* 292/293:469–479.

Koski, M., Klein Breteler, W., and Schogt, N. 1998. Effect of food quality on rate of growth and development of the pelagic copepod *Pseudocalanus elongatus* (Copepoda, Calanoida). *Marine Ecology Progress Series* 176:169–187.

Laabir, M., Poulet, S.A., Harris, R.P., Pond, D.W., Cueff, A., Head, R.N., and Ianora, A. 1998. Comparative study of the reproduction of *Calanus helgolandicus* in well-mixed and seasonally stratified coastal waters of the western English Channel. *Journal of Plankton Research* 20:407–421.

Laabir, M., Poulet, S.A., Ianora, A., Miralto, A., and Cueff, A. 1995. Reproductive response of *Calamnus helgolandicus*. II. *In situ* inhibition of embryonic development. *Marine Ecology Progress Series* 129:97–105.

Landry, M. 1983. The development of marine calanoid copepods with comment on the isochronal rule. *Limnology and Oceanography* 28:614–624.

Landry, M.R. 1975. The relationship between temperature and the development of life stages of the marine copepod *Acartia clausi* Giesbr. *Limnology and Oceanography* 20:254–257.

Liang, D and Uye, S. 1996. Population dynamics and production of the planktonic copepods in a eutrophic inlet of the Inland Sea of Japan. II. *Acartia omorii. Marine Biology* 125:109–117.

Marshall, S.M. and Orr, A.P. 1930. A study of the spring diatom increase in Loch Striven. *Journal of the Marine Biological Association United Kingdom* 16:853–878.

Marshall, S.M. and Orr, A.P. 1955. *The Biology of a Marine Copepod, Calanus finmarchicus (Gunnerus)*. London, U.K.: Oliver and Boyd. 188 pp.

Marcus, N.H., Lutz, R., Burnett, W., and Cable, P. 1994. Age, viability, and vertical distribution of zooplankton resting eggs from an anoxic basin: evidence of an egg bank. *Limnology and Oceanography* 39:154–158.

Mauchline, J. 1998. The biology of calanoid copepods. *Advances in Marine Biology* 33:1–710.

McEvoy, L.A., Navarro, J.C., Bell, J.A., and Sargent, J.R. 1995. Autoxidation of oil emulsions during the *Artemia* enrichment process. *Aquaculture* 134:101–112.

McLaren, I.A. 1976. Inheritance of demographic and production parameters in the marine copepod *Eurytemora herdmani. Biological Bulletin* 151:200–213.

Miller, C., Johnson, J., and Heinle, D. 1977. Growth rules in the marine copepod genus *Acartia. Limnology and Oceanography* 22:326–334.

Miralto, A., Barone, G., Romano, G., Poulet, S.A., Ianora, A., Russo, G.L., Buttino, I., Mazzarella, G., Laabir, M., Cabrini, M., and Giacobbe, M.G. 1999. The insidious effect of diatoms on copepod reproduction. *Nature* 402:173–176.

Miralto, A., Guglielmo, L., Zagami, G., Buttino, I., Granata, A., and Ianora, A. 2003. Inhibition of population growth in the copepods *Acartia clausi* and *Calanus helgolandicus* during diatom blooms. *Marine Ecology Progress Series* 254:253–268.

Miralto, A., Ianora, A., and Poulet, S.A. 1995. Food type induces different reproductive responses in the copepod *Centropages typicus*. *Journal of Plankton Research* 17:1521–1534.

Mullin, M.M. and Brooks, E.R. 1967. Laboratory culture, growth rate and feeding behaviour of a planktonic marine copepod. *Limnology and Oceanography* 12:657–666.

Mullin, M.M. and Brooks, E.R. 1970. Growth and metabolism of two planktonic, marine copepods as influenced by temperature and type of food. In *Marine Food Chains*, edited by J.H. Steele, pp. 74–95. Edinburgh, U.K.: Oliver and Boyd.

Nanton, D.A. and Castell, J.D. 1998. The effects of dietary fatty acids on the fatty acid composition of the harpacticoid copepod, *Tisbe* sp., for use as a live food for marine fish larvae. *Aquaculture* 163:251–261.

Paffenhöfer, G.-A. 1970. Cultivation of *Calanus helgolandicus* under controlled conditions. *Helgoländer wissenschaftliche Meeresuntersuchungen* 20:346–359.

Paffenhöfer, G.-A. 1976. Feeding, growth, and food conversion of the marine planktonic copepod *Calanus helgolandicus*. *Limnology and Oceanography* 21:39–50.

Paffenhöfer, G.-A. and Harris, R.P. 1976. Feeding, growth and reproduction of the marine planktonic copepod *Pseudocalanus elongatus* Boeck. *Journal of the Marine Biological Association of the United Kingdom* 56:327–344.

Paffenhöfer, G.-A. and Harris, R.P. 1979. Laboratory culture of marine holozooplankton and its contribution to studies of marine planktonic food webs. *Advances in Marine Biology* 16:211–308.

Parrish, K.K. and Wilson, D.F. 1978. Fecundity studies on *Acartia tonsa* (Copepoda: Calanoida) in standardized culture. *Marine Biology* 46:65–81.

Payne, M.F. and Rippingale, R.J. 2000. Evaluation of diets for culture of the calanoid copepod *Gladioferens imparipes*. *Aquaculture* 187:85–96.

Peterson, W.T. 1986. Development, growth, and survivorship of the copepod *Calanus marshallae* in the laboratory. *Marine Ecology Progress Series* 29:61–72.

Peterson, W.T. and Kimmerer, W.J. 1994. Processes controlling recruitment of the marine calanoid copepod *Temora longicornis* in Long Island Sound: egg production, egg mortality, and cohort survival rates. *Limnology and Oceanography* 39:1594–1605.

Pohnert, G. 2000. Wound-activated chemical defense in unicellular planktonic algae. *Angewandte Chemie International Edition* 39:4352–4354.

Pohnert, G. 2002. Phospholipase A2 activity triggers the wound-activated chemical defense in the diatom *Thalassiosira rotula*. *Plant Physiology* 129:103–111.

Pohnert, G. and Boland, W. 2002. The oxylipin chemistry of attraction and defense in brown algae and diatoms. *Natural Products Reports* 19:108–122.

Pohnert, G., Lumineau, O., Cueff, A., Adolph, S., Lange, M., Cordevant, C., and Poulet, S. 2002. Are volatile aldehydes from diatoms the only line of chemical defence against copepods? *Marine Ecology Progress Series* 245:33–45.

Poulet, S.A., Ianora, A., Miralto, A., and Meijer, L. 1994. Do diatoms arrest embryonic development in copepods? *Marine Ecology Progress Series* 111:79–86.

Poulet, S.A., Laabir, M., Ianora, A., and Miralto, A. 1995. Reproductive response of *Calanus helgolandicus*: I. Abnormal embryonic and naupliar development. *Marine Ecology Progress Series* 216:151–165.

Reitan, K.I., Rainuzzo, J.R., Øle, G., and Olsen, Y. 1997. A review of the nutritional effects of algae in marine fish larvae. *Aquaculture* 155:207–221.

Romano, G., Ianora, A., and Miralto, A. 1996. Respiratory physiology in summer diapause embryos of the neustonic copepod *Anomalocera patersoni*. *Marine Biology* 127:229–234.

Rosahl, S. 1996. Lipoxygenases in plants — their role in development and stress response. *Zeitschrift für Naturforschung* 51c: 123–138.

Runge, J.A. 1984. Egg production of the marine, planktonic copepod, *Calanus pacificus*. *Limnology and Oceanography* 25:134–145.

Runge, J.A. 1988. Should we expect a relationship between primary production and fisheries? The role of copepod dynamics as a filter of trophic variability. *Hydrobiologia* 167/168:61–71.

Santella, L. and Ianora, A. 1990. Subitaneous and diapause eggs in Mediterranean populations of *Pontella mediterranea* (Copepoda: Calanoida): a morphological study. *Marine Biology* 105:83–90.

Scharden, J.L. and York, R.G. 1995. Teratogenic alkaloids in foods. In *The Toxic Action of Marine and Terrestrial Alkaloids*, edited by M.S. Blum, pp. 281–327. Fort Collins, Colorado, USA: Alaken.

Smith, S.L. and Lane, P.V.Z. 1985. Laboratory studies of the marine copepod *Centropages typicus*: egg production and development rates. *Marine Biology* 85:153–162.

Starr, M., Runge, J.A., and Therriault, J.-C. 1999. Effects of diatom diets on the reproduction of the planktonic copepod *Calanus finmarchicus*. *Sarsia* 84:379–389.

Tester, P.A. and Turner, J.T. 1990. How long does it take copepods to make eggs? *Journal of Experimental Marine Biology and Ecology* 141:169–182.

Tester, P.A., Turner, J.T., and Shea, D.A. 2000. Vectorial transport of toxins from the dinoflagellate *Gymnodinium breve* in through copepods to fish. *Journal of Plankton Research* 22:47–61.

Tosti, E., Romano, G., Buttino, I., Cuomo, A., Ianora, A., and Miralto, A. 2003. Bioactive aldehydes from diatoms block the fertilisation current in ascidian oocytes. *Molecular Reproduction and Development* 66:72–80.

Turner, J.T., Ianora, A., Miralto, A., Laabir, M., and Esposito, F. 2001. Decoupling of copepod grazing rates, fecundity and egg-hatching success on mixed and alternating diatom and dinoflagellate diets. *Marine Ecology Progress Series* 220:187–199.

Turner, J.T. and Tester, P.A. 1997. Toxic marine phytoplankton, zooplankton grazers, and pelagic food webs. *Limnology and Oceanography* 42:1203–1214.

Uye, S. 1981. Fecundity studies of neritic calanoid copepods *Acartia clausi* Giesbrecht and *A. steueri* Smirnov: a simple empirical model of daily egg production. *Journal of Experimental Marine Biology and Ecology* 50:255–271.

Uye, S. 1996. Induction of reproductive failure in the planktonic copepod *Calanus pacificus* by diatoms. *Marine Ecology Progress Series* 133:89–97.

Zillioux, E.J. and Wilson, D.F. 1966. Culture of a planktonic calanoid copepod through multiple generations. *Science* 151:996–998.

5
Maximizing the Nutritional Values of Copepods in Aquaculture: Managed versus Balanced Nutrition

G. S. Kleppel, Sarah E. Hazzard, and Carol A. Burkart

ABSTRACT

Numerous fish species feed on copepods at some stage in their life cycles, mostly as larvae and juveniles. Copepods, therefore, are a logical choice as food for fishes under cultivation, particularly during early life. To be most useful, the objectives of copepod cultivation should be to (a) maximize the number of eggs produced and hatching, (b) optimize the nutritional value of the nauplii for the fish, and (c) match the rate of copepod development to the mouth size of the feeder. Although much about the nutrition of copepods is understood, the balance of nutritional constituents that optimizes fitness in nature may be different from those that produce the most and best food for fishes under cultivation. In nature, fit copepods maximize production of viable eggs that, upon hatching, grow rapidly to the reproductive stage. This strategy is not completely consistent with the goals of aquaculture, where the preference might be to prolong the stage-length(s) most desirable as food. When the copepod is no longer in a food-optimal stage, the goal becomes maximizing the rate of development to recruitment. Copepod egg production and hatching, though clearly influenced by species-specific physiologies and behaviors, are intimately associated with food biochemistry, involving both the nutritional and sometimes the teratogenic attributes of the microplanktonic. Detection and capture of food influences the ability of the diet to support growth and production of copepods, and the presence of large amounts of protein, lipid, and carbohydrate in the diet does not guarantee optimal production of food-useable copepods. It may be efficacious to adjust the diet to maximize the time the copepod spends in a food-usable size class as well as its nutritional quality when at that size.

INTRODUCTION

It is well known that crustaceans are important in the diets of fishes and that copepods are among the principal foods of several economically valuable fish species (Lebour 1921; 1924; Hardy 1924; Marshall et al. 1939; Marshall and Orr 1955; Lasker 1975; Kleppel et al. 1980). Copepods were present in the guts of more than 80% of the fish species studied by Turner (1984) during some part of the life cycle. While some species, for example, *Corycaeus anglicus*, are predators of fish larvae (Brewer et al. 1984), the overwhelming majority of planktonic copepod species are thought to be high-quality food sources for fishes. Thus, they are particularly well suited for use as food in fish culture. In this chapter, the authors briefly review some of the pertinent literature on copepod nutrition and consider how the nutrition of copepods in culture might be managed to optimize their value as food for the fishes in aquaculture, particularly during early development stages. The authors focus on the Calanoida, though much

Corresponding author: G.S. Kleppel, Department of Biological Sciences, University at Albany, SUNY, 1400 Washington Avenue, Albany, New York 12222, U.S.A.; Phone.\: (518) 442-4338; Fax: (518) 442-4767; E-mail: gkleppel@csc.albany.edu.

of what is presented will be applicable to the Harpacticoida and Cyclopoida as well.

GOALS OF COPEPOD CULTURE IN AQUACULTURE

Copepods can be obtained for aquaculture systems by capturing wild organisms or by cultivation. While wild harvesting is normally less costly and labor intensive than culturing, it offers less control over nutritional quality and composition. Nonetheless, in certain harbors and bays, virtual monocultures of copepods can be routinely collected by size-selective screening. The animals can be concentrated (e.g., many Acartidae are positively phototactic) and either fed directly or frozen for later use.

The focus here, however, is on the cultivation of copepods for aquaculture operations. The authors do not intend to provide the details of culturing, but rather to describe aspects of copepod nutrition that are germane to their culture as food for fish cultivation and suggest a strategy for managing the nutrition of the copepods to optimize their use in this respect. In nature, success is measured by fitness. The fittest species produces the greatest number of offspring that reproduce. The contribution of nutrition to fitness is to provide the dietary constituents that convey physical advantage (e.g., size, musculature), minimize the time length of ontogeny, and maximize the number and health of offspring. These are not necessarily the measure of success for a population when it is to be used as food in aquaculture. Here it is suggested that nutrition might become a tool to manage the development and biochemistry of the population to optimize its quality for its predators.

Among the criteria for selecting a species for cultivation are the following: (a) relative ease of captive rearing, (b) hardiness and high growth rate, and (c) a relatively high, continuous egg production rate and high egg viability.

Several calanoid copepod taxa have been cultured successfully. These include (but are not limited to), *Acartia* spp. (Kiørboe 1989; Jónasdóttir 1992; Kleppel et al. 1998a), *Calanus* spp. (Paffenhöfer 1971; Peterson 1988; Nejstgaard et al. 1997), *Eucalanus* spp. (Paffenhöfer and Van Sant 1985), *Temora* spp. (Jónasdóttir 1992), *Centropages* spp. (Smith and Lane 1985; 1987), and *Paracalanus* spp. (Paffenhöfer 1984). The choice of taxon depends on geography, size of the predator (fish), and whether culturing will take place indoors or outside. For example, for reasons that are not clear, *Calanus finmarchicus* tends to be easier to rear in outdoor systems than indoors. Acartidae and Temoridae are often available in bays and lagoons close to shore. They frequently occur in large numbers (some > 50 animals/L; unpublished data), and they are relatively hardy and amenable to many kinds of culturing protocols. Furthermore, Acartidae reproduce rapidly and relatively continuously, the average fecundity of *Acartia tonsa* being 25 eggs per female per day (Kleppel et al. 1998b). Under nutritionally replete conditions, at optimal temperature (which varies by population), egg production rates in excess of 100 per female per day can be achieved for many calanoid species, though not sustained (Marshall and Orr 1955; Peterson 1988; McManus and Foster 1998). Many of the widely cultivated Calanidae are coastal and require more effort to capture. There are some near-coastal regions (particularly off islands) and fjords where *Calanus* spp., an important food of cod (*Gadus morhua*) larvae (Marshall and Orr 1955), can be captured. *Calanus* spp., *Eucalanus* spp., and *Temora* spp. are widely cultivated for research. They are a bit more difficult than *Acartia* spp. and require a bit more expertise. Furthermore, temperature constraints may limit the value of high-latitude and temperate *Calanus* species as a food in intensive aquaculture systems (Marcus and Alatalo 1989; Støttrup 2000). Dagg (1977) observed that egg production by nearshore species, such as *Acartia* spp., is sensitive to food availability, showing a rapid response (within 8 hours) to patchiness. Egg production by coastal genera, such as *Calanus* and *Centropages*, is less sensitive to variations in food concentration. Species of these genera, however, are more likely than those of *Acartia* to produce nonviable eggs (i.e., do not hatch within 48 hours) as opposed to "resting eggs" (Marcus 1990). Nonhatching eggs may be produced when adequate amounts of food are not available, or alternatively, as a result of certain qualitative attributes of the food supply (Poulet et al. 1994; Burkart and Kleppel 1998; Ianora et al. 1995; Jónasdóttir and Kiørboe 1996). *Acartia tonsa* in Florida Bay exhibited low egg-production rates associated with food quality (Burkart 1998; also see, Kleppel et al. 1998b; Kleppel and Burkart 1995) and possibly the presence of toxic contaminants in the system (G. Scott personal communication, Dec. 1999). In most cases, however, egg-hatching success was greater than 80% (Fig. 5.1). Ianora et al. (1995) and Turner et al. (2001), on the other hand, have shown that prolonged exposure to diatoms may have a teratogenous effect (i.e., eggs are produced but do not hatch). This effect is thought to result from the enzymatic conversion of specific fatty acids to aldehyde derivatives that interfere with normal embryonic development (Pohnert et al. 2002). Observations of the production and hatching of eggs by *A. tonsa* in Florida Bay (Burkart 1998) support the notion that diatoms tend, for reasons likely to be linked to biochemistry, to be poor foods (Fig. 5.2). It is

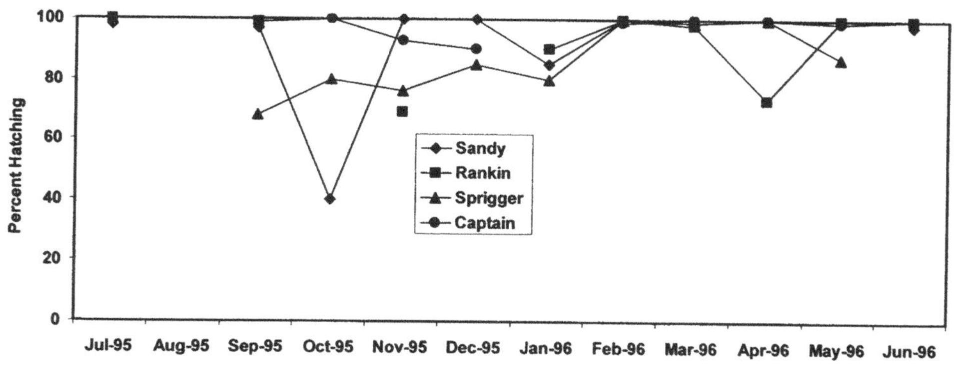

Figure 5.1. Hatching success of the eggs of *Acartia tonsa* at four locations in Florida Bay, July 1995 to June 1996. (Source of data: Burkart 2000).

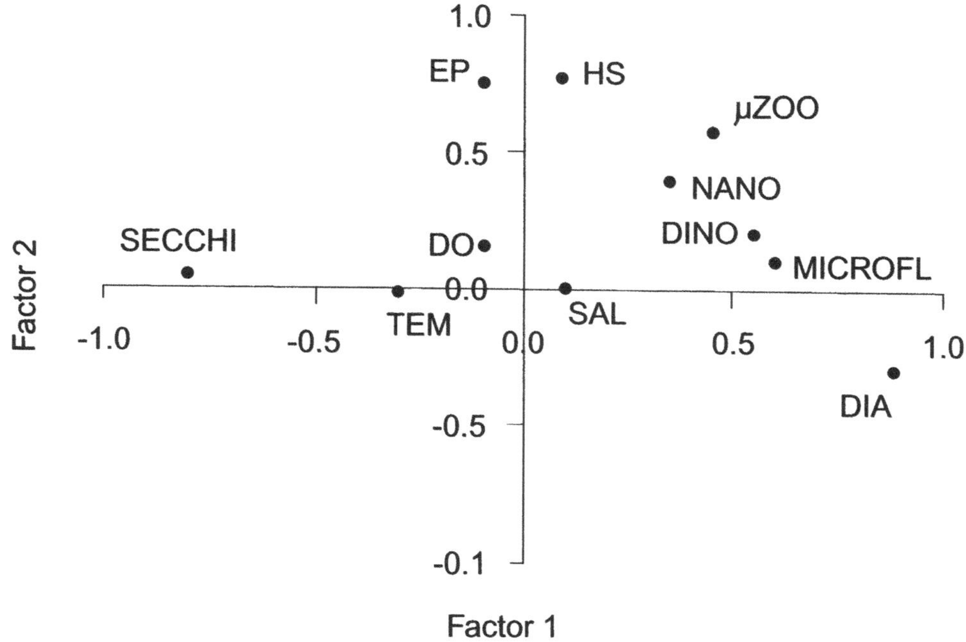

Figure 5.2. Plot of the first two factors in a principal components analysis showing the distribution of variables associated with the egg production (EP) and hatching success (HS) of the calanoid copepod *Acartia tonsa* in Florida Bay. Egg production and hatching success are most closely associated with the availability of specific microplankton taxa, notably, dinoflagellates (DINO) and microzooplankton (µZOO; principally ciliates). Diatoms (DIA) are inversely associated with egg production and hatching success. Other variables for which associations are weak are nanoplankton (NANO; < 5 µm protists), microflagellates (MICROFL), and a suite of environmental variables: transparency, temperature (TEM), dissolved oxygen (DO), and salinity (SAL). After Burkart (2000).

as yet unclear whether aldehyde production is widespread among diatom taxa, whether they are produced in other algal classes, and how aldehyde production is triggered (Pohnert et al. 2002).

It is important to consider the time between ingestion and production when trying to manipulate and manage the nutritional composition of the copepod that is being reared. As a general rule, species that occur nearshore tend to convert dietary nutrients to new biomass (somatic growth or eggs) relatively rapidly, but not to store large quantities of lipid. Tester and Turner (1990) reported that *A. tonsa* converted food to eggs within 24 hours. Coastal and oceanic species, such as many Calanidae, retain a larger portion of their food as storage products, and while more resistant to starvation, they may or may not use recently ingested nutrients for the production of eggs. This adds to the complexity of managing the nutrition of the copepod, but as a general rule, if "bloom" conditions are maintained in the culture, egg production will likely reflect recent feeding; smaller amounts of lipid will be sequestered for storage.

The specific nutritional objectives for copepod cultivation in aquaculture are to (a) maximize the availability of the desired copepod life-cycle stage for the longest time possible; (b) optimize the nutritional value of the copepod for the fish; (c) maximize the growth rate of the copepod to reproductive adult after it develops beyond its optimal size as food; and (d) maximize egg production and hatching success. The specific approach for addressing these objectives will depend on the copepod species being cultivated, the physical plant of the aquaculture system, and the nature of the product.

COPEPOD NUTRITION: A BRIEF REVIEW

Much of what is known about copepod nutrition is based on information about other crustaceans. A comprehensive review of even the limited literature on copepods is not possible here. Instead, the authors provide a summary of what they consider key aspects of copepod nutrition and refer the reader to sources of more detailed information (for lipids see Sargent and Henderson [1986]; for proteins and amino acids see Suyama et al. [1965]; Cowey and Corner [1963]; for carbohydrates see Raymont et al. [1969]; Head [1992]). Calanoid copepods are thought to require 10 amino acids (Table 5.1) and are unable to produce several fatty acids (Table 5.2) required for embryo development. Roman (1991) observed that prior to copepodid stage IV (i.e., CIV), most of the carbon sequestered from the diet by *A. tonsa* was used for protein synthesis. The male continued sequestering carbon for protein synthesis through CVI, though feeding declined. In copepods that ulti-

Table 5.1. Essential amino acids (as percenage of dry weight) of copepods present in *Oxyrrhis marina*, a heterotrophic dinoflagellate; *Debariomyces hanseni*, a marine yeast; *Thalassiosira weissflogii*, a centric diatom; *Calanus pacificus*, a calanoid copepod; and menhaden, a fish

Amino acid[a]	*O. marina*[b]	*D. hanseni*[b]	*T. weissflogii*[c]	*C. pacificus*[c]	Menhaden[d]
Arginine	1.5	5.0	4.4	5.2	3.6
Histidine	2.0	4.0	1.8	1.8	1.4
Isoleucine	1.4	1.6	4.7	4.1	2.8
Leucine	1.5	2.3	7.3	6.9	4.6
Lysine	0.5	1.8	9.7	8.8	4.7
Methionine	0.2	2.2	2.2	2.3	1.8
Phenylalanine	1.0	1.9	3.7	2.9	2.4
Threonine	0.9	1.9	4.8	5.6	2.4
Tryptophan	nd	nd	nd	nd	0.7
Valine	1.3	2.0	5.7	5.7	3.3

Note: nd: not detected.
[a]Based on Claybrook (1983.
[b]Kleppel and Burkart (1995).
[c]Cowie and Hedges (1996).
[d]De Silva and Anderson (1995).

Table 5.2. Concentrations of fatty acids in phytoplankton (*Isochrysis galbana, Thalassiosira weissflogii, Rhodomonas baltica*); a heterotrophic dinoflagellate (*Oxyrrhis marina*); mixed microplankton assemblages in Florida Bay during nonbloom and bloom conditions; and in the eggs and adults of the calanoid copepod, *Acartia tonsa*

					FB Microplankton		*Acartia tonsa*	
Fatty acid moiety	*I. galbana*[a] (ng/μg DW)	*T. weissflogii*[a] (ng/μg DW)	*R. baltica*[a] (ng/μg DW)	*O. marina*[b] (ng/μg DW)	Nonbloom[c] (ng/L)	Bloom[c] (ng/L)	Eggs[d] (ng/μg DW)	Females[d] (ng/μg DW)
14:0	na	na	na	2.7	0.8	25.6	3.4	3.5
16:1	1.6	75.9	2.0	1.0	1.1	25.8	10.0	5.5
16:0	5.5	67.3	11.4	1.8	4.1	25.3	14.8	17.5
18:3	2.7	3.5	21.9	0.6	0.3	2.2	5.8	5.3
18:+18.2	8.9	nd	6.6	2.6	1.1	9.4	3.4	3.3
18:0	nd	nd	trace	0.2	nd	1.9	nd	1.0
20:5	0.3	85.3	15.1	0.3	0.5	1.7	0.3	3.0
20:1	na	na	na	0.8	na	na	nd	1.0
20:0	na	na	na	trace	na	0.2	0.6	1.0
22:6	8.8	19.1	99.5	2.6	0.2	1.3	0.3	2.0
22:0	na	na	na	nd	na	na	nd	0.3

Note: Moieties shown in bold type are thought to be key essential fatty acids that are important to reproduction and embryo development. FB, Florida Bay; DW, dry weight; nd, not detected; na, not analyzed.
[a]Støttrup and Jensen (1990).
[b]Kleppel et al. (1998a).
[c]Hazzard and Kleppel (2003).
[d]Hazzard (2001).

mately became females, more carbon was sequestered to lipids during CIV–CVI, with protein sequestration important, but secondary.

Unsaturated fatty acids are key to reproductive success, the 18-, 20-, and 22-carbon chain moieties apparently being extremely important (Støttrup and Jensen 1990; Jónasdóttir et al. 1995; Pond et al. 1996; Hazzard and Kleppel 2003). Some authors have suggested that the key essential fatty acids, that is, the 20- and 22-carbon chains, can be synthesized by chain elongation of other ingested moieties, but this possibility remains controversial. Hazzard (2001) and Hazzard and Kleppel (2003) observed that egg production rates by *A. tonsa* in Florida Bay could be explained largely by the availability of foods containing 18:3 polyunsaturated fatty acids.

Most small, calanoid copepods are omnivorous. They can be selective in their diets (Irigoien et al. 2000) in a manner reasonably consistent with optimal foraging theory (Cowles 1979). In general, diverse diets appear to provide the highest rates of growth and production (Kleppel 1993; Klein Breteler et al. 1990), possibly because they provide the best chance of obtaining the full range of nutritional elements (Roman 1984; Kleppel 1993; Klein Breteler et al. 1999). It is important, in this sense, to ensure that all food organisms are easily captured. For example, Kleppel et al. (1998a) observed that although the yeast *Debariomyces hanseni* contains large amounts of essential amino acids, it is too small to be captured by adult (CVI) *A. tonsa* and is therefore not eaten. It was also observed that while *Isochrysis galbana* is a good source of fatty acids, the cellular fatty acid concentration is low because the cells are small. Thus, the copepod must expend considerable energy to capture enough cells to constitute a ration. On the other hand, the larger (25 μm equivalent spherical diameter, or esd; 465 picograms carbon [pg-C] per cell) heterotrophic dinoflagellate, *Oxyrrhis marina*, feeds efficiently on both the yeast and *I. galbana*. In addition, the nutritional composition of *the O. marina* reflected that of its food (Kleppel et al. 1998a). Thus, copepods responded favorably to the delivery of nutrients in the relatively large "packages" represented by the dinoflagellate. The lesson here is that the nutritional value of a food is necessary but not sufficient to maximize copepod production. The food must be of a size and motility that is easily captured (Paffenhöfer 1984) such that the nutrition delivered per unit of energy expended on capture must be optimized.

OPTIMIZING THE NUTRITIONAL VALUE OF COPEPODS FOR FISHES

The structural biochemistry of an organism reflects the nutritional composition of its food. Thus, protein levels in the diet are important to somatic growth. Diet drives growth by providing structural components for somatic development (e.g., protein for muscle growth) and also by providing the biochemicals for metabolism (e.g., protein for enzymes, lipids and proteins for hormones). Among the aims of nutritional management in aquaculture is to provide food organisms for the copepod that supply nutrients for both needs at the highest level. This, in turn, produces copepods that are nutritious for the fish for which they will be food. "Designer diets" for copepods are becoming increasingly feasible (Harel et al. 2002). While such diets should provide the desired nutrition in as natural a form as possible, vitamins and other nutritional supplements can be provided in microencapsulated form, which have been effective in the culture of mollusks and a few crustaceans (Kreeger and Langdon 1994). Microencapsulated dietary supplements have not, to the knowledge of the authors, been attempted in copepod cultivation for aquaculture, but the approach seems well suited to an aquaculture environment and could provide nutrients and micronutrients that are important both for the copepod and for the fish that ingests it.

MAXIMIZING THE DESIRED COPEPOD LIFE-CYCLE STAGE

Copepods undergo 13 developmental stages. Subitaneous eggs generally hatch within 48 hours. Of the six naupliar stages, nauplius 1 and 2 (NI and NII) usually do not feed. In the subsequent six copepodid (CI–CVI) stages, gender differentiation usually occurs in CV, but only CVI (adult) is reproductive. Details of the copepod life cycle are available in numerous texts (e.g., Parsons et al. 1977) as well as in monographs dealing with individual species (e.g., Marshall and Orr 1955). Although complete development from egg to adult may require as little as 14 days, ontogeny is generally longer (Nicholls 1933). The length of each developmental stage is variable and independent of the length of other stages. Under rigorously controlled environmental conditions, however, the length of each stage may be predictable for a particular species. It is reasonable, therefore, to consider whether the life cycle can be manipulated to increase the amount of time a nauplius or copepodid spends in a particular developmental stage that is desirable as food for a cultivated species in an aquaculture system.

Size and nutritional content determine the desirability of a nauplius or copepodid as food for fishes. The gape of the mouth of a marine fish larva determines the optimal size of its food (Hunter 1981; Cunha and Planas 1995). Among the goals of a nutritional management strategy is to maintain the maximum number of copepods at the optimal size for the target species for the longest period of time (Fig. 5.3). To realize this goal, manipulation of the nutritional environment of the copepod is crucial. Nutritional management should begin at stage NIII. The diet should be "designed" to maximize the growth rate of the copepod to its optimal size (stage) as food. Growth maximization requires that food for the copepod be easy to capture (the energetic cost of catching food affects the net energy gain from the diet) and be rich in protein and lipid. A diet that combines species of *Rhodomonas* spp. and *Isochrysis* spp. would seem to provide desirable protein and lipid (i.e., fatty acid) compositions (Støttrup and Jensen 1990; Klein Breteler et al. 1999), though small changes in culturing conditions can result in dramatic alterations in nutritional composition (Fig. 5.4). When (or slightly before) the copepod reaches a desired stage in its life cycle for use as food, a diatom should be substituted for *Rhodomonas* spp. to lower the protein content of the diet, thereby slowing somatic growth. Metabolic activity can be maintained or enhanced with a diet rich in carbohydrates, as found in many diatoms (Hitchcock 1982), and certain fatty acids found in diatoms may be nutritionally important to fish larvae (Støttrup et al. 1999; Støttrup 2000). It is not known at this point whether aldehydes produced when copepod larvae feed on certain diatoms might be retained and act as teratogens in the adult. If possible, clones that do not produce teratogenic aldehydes (e.g., certain *Thalassiosira rotula*) should be used when diatoms are considered

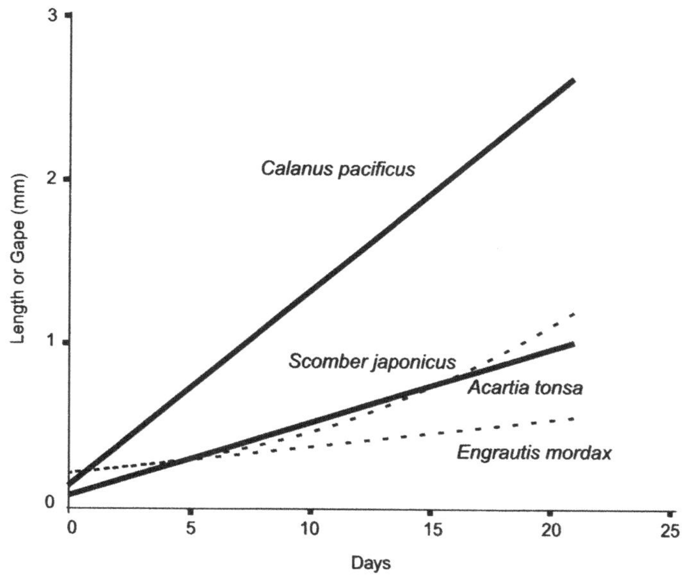

Figure 5.3. Length (mm) versus time (in days) for two copepods, *Calanus pacificus* and *Acartia tonsa*, during an assumed 21-day development period from egg to adult (stage CVI). Length is here assumed to increase linearly over time. The change in mouth gapes (mm) of the larvae of two fishes *Scomber japonicus* and *Engraulis mordax* are plotted over time using data from Hunter (1981). It is seen that *C. pacificus* eggs and larvae up to 1 mm (about 7 days old) will support the development of *S. japonicus* through day 21. *Engraulis mordax* can grow through the same period on *C. pacificus* up to 0.3 mm (about 2 days old). *Acartia tonsa* tracks the growth in mouth gape of *S. japonicas* through the first 21 days, while the increase in the gape of *E. mordax* through day 21 constrains the size of the optimal *A. tonsa* to day 7 (i.e., 0.3 mm length).

for use as food for copepods (Pohnert et al. 2002). Several papers have demonstrated that diet modification affects copepod growth, development, production, and nutritional composition (Koski et al. 1999; Støttrup et al. 1999) as well as the nutritional content and viability of the eggs produced by feeding parents (Burkart 1998; Støttrup et al 1999). Few studies, however, if any, have examined the possibility of using nutritional techniques to constrain copepod development to desirable stages. Diets containing lipids, fatty acids, and fatty acid ratios desirable for fish culture, but with reduced protein, can be produced and should ideally create a lipid rich, slow-growing copepod larva. The lower protein content of these copepod larvae will not produce a structural insufficiency in the fish, because all amino acids should be present. Instead, total protein should be limiting. To remedy this, the ration should simply be increased.

When copepods molt out of their most desirable stage as food, they can be used as food for other stages in the life cycle of the fish or for other organisms that are being cultivated. Alternatively, they can be brought rapidly to CVI and allowed to reproduce. If the goal is the latter, then the nutritional objective becomes expediting somatic growth and fostering the development of viable oocytes in the female. Sequestration of lipid for egg development begins between CIV and CV (Roman 1991). Thus, prior to CIV, the managed diet should optimize protein availability. No later than CIV, foods rich in lipids need to be added to the diet. It is likely that somatic growth can be augmented by the presence of heterotrophic protists in the diet, which facilitate amino acid assimilation.

The ability to provide desired foods in accessible packages (i.e., cell sizes; see for example Paffenhöfer 1984) in excess of ration requirements seems the most reasonable approach to fast-growing the copepod to a reproductively optimized adult. Protein can be supplied by large dinoflagellates such as *Gymnodinium sanguinium* (60 μm esd), an excellent protein source for *Calanus* spp., or *Rhodomonas* spp. (8 μm), *G. sanguineum,* or smaller, naked dinoflagellates. *Ceratium* spp. seem to be avoided by *Calanus* spp. and

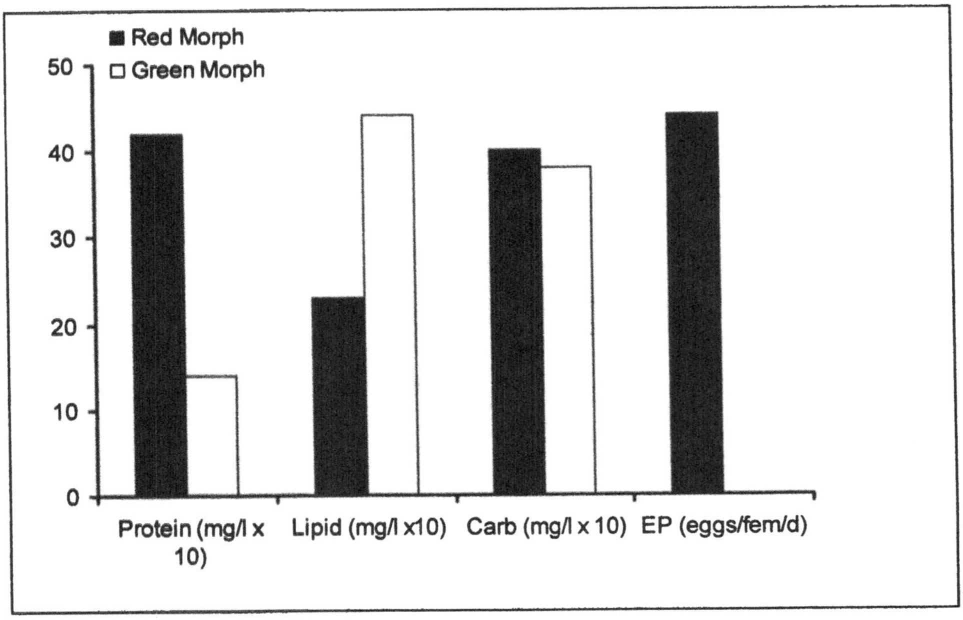

Figure 5.4. Cellular protein, lipid, and carbohydrate contents of two morphotypes of *Rhodomonas lens*. The red morph was cultured in f/2 medium under cool white fluorescent lights (14-hour light/10-hour dark cycle) at 20°C. The green morph experienced the same conditions but with a 3% increase in ultraviolet (UV-ß) irradiation. The small change in the optical environment resulted in a large change in nutritional attributes. Only the red morph supported egg production by *A. tonsa* (Data from Kleppel, unpublished). fem, female; d, day.

Acartia californiensis (Birge 1898; Harvey 1937; Kleppel 1992), and several dinoflagellates produce irritants or toxins (Huntley et al. 1986). Ciliates and heterotrophic dinoflagellates should be part of the diet by stage CII or CIV.

Alternatively, different-size copepod species might be cultivated and provided according to the size requirements of the developing fish larvae. The drawback to this system is the potential cost and difficulty involved in culturing and size fractionating multiple copepod species.

PRODUCTION OF VIABLE EGGS

Egg production and viability are determined only partially in the adult stage. While it is imperative that there be adequate numbers of males present and that temperatures and photoperiod be considered in the culturing protocol, the nutritional biochemistry of the adults is determined during development. As mentioned above, certain species are more likely to produce viable eggs when they do produce eggs (see Burkart and Kleppel 1998). Generally, these species tend not to store food as lipid, and therefore they suffer when food patchiness increases: Fecundity usually declines rapidly (Dagg 1977). Finally, it is unclear to what extent copepod cultures should be supplemented with wild material. The experience of the authors has been that cultures tend to lose their vitality after 8–10 generations (unpublished data). Incorporation of newly captured animals to the culture, while potentially reducing the degree of standardization of the food organism, may improve the quality of the product, for reasons that are not yet understood.

CONCLUSION

Copepods are the natural food of numerous fishes and other organisms of interest to aquaculture. To feed wild copepods or copepods cultured without consideration of how their diet in culture might affect the efficiency of the aquaculture operation or the quality of its product is to miss, the authors believe, much of the potential of copepods in the process. By managing the copepod diet, that is, the nutritional intake and the cost of acquiring nutrients, the authors suggest that nutritionally optimized, size-optimized food for fishes can be produced that meets and standardizes quality in the copepod food product. Although much remains to be learned in this area, research in copepod nutrition is of great interest to the community of copepodologists. At an international workshop on zooplankton nutrition (Kleppel and Hazzard 2002), the need for research and infrastructure to support the understanding of copepod nutrition became apparent. Currently, efforts in Europe in this direction seem to be surpassing those in the United States. Potentially, the ability to manage copepod nutrition for aquaculture will motivate greater attention among funding sources in the United States.

REFERENCES

Birge, E.A. 1898. Plankton studies in Lake Mendota. II. The crustacea of the plankton. *Transactions of the Wisconsin Academy of Science, Arts and Letters* 11:274–448.

Brewer, G.D., Kleppel, G.S., and Dempsey, M. 1984. Apparent predation of ichthyoplankton by zooplankton and fishes in nearshore waters off southern California. *Marine Biology* 80:17–28.

Burkart, C.A. 1998. Variability in copepod hatching success: observations on natural populations and experiments on the role of maternal diet. Ph.D. Thesis. Nova Southeastern University, Davie, Florida, USA.

Burkart, C.A. and Kleppel, G.S. 1998. A new incubation system for the measurement of copepod egg production and egg hatching success in the field. *Journal of Experimental Marine Biology and Ecology* 221:89–97.

Claybrook, D.L. 1983. Nitrogen metabolism. In *The Biology of Crustacea*, Volume 5, *Internal Anatomy and Physiological Regulation*, edited by L.H. Mantel, pp. 163–213. New York, New York, USA: Academic Press.

Cowey, C.B. and Corner, E.D.S. 1963. On the nutrition and metabolism of zooplankton. II. The relationship between the marine copepod *Calanus helgolandicus* and particulate material in Plymouth sea water in terms of amino acid composition. *Journal of the Marine Biological Association, UK* 43:495–511.

Cowie, G.L. and Hedges, J.I. 1996. Digestion and alteration of the biochemical constituents of a diatom (*Thalassiosira weissflogii*) ingested by an herbivorous zooplankton (*Calanus pacificus*). *Limnology and Oceanography* 41:581–594.

Cowles, T.J. 1979. The feeding response of copepods from the Peru upwelling system: Food size selection. *Journal of Marine Research* 13:601–622.

Cunha, I. and Planas, M. 1995. Ingestion rates of turbot larvae (*Scophthalmus maximus*) using different-sized live prey. *ICES Marine Science Symposia* 201:16–20.

Dagg, M. 1977. Some effects of patchy food environments on copepods. *Limnology and Oceanography* 22:99–107.

De Silva, A.A. and Anderson, T.A. 1995. *Fish Nutrition in Aquaculture*. New York, New York, USA: Chapman and Hall. 338 pp.

Hardy, A.C. 1924. The herring in relation to its animate environment. Part I. The food and feeding habits of the herring with special reference to the east coast of England. *Fishery Investigations*, London Ser. II 7. 53 pp.

Harel, M., Koven, W., Lein, I., Bar, Y., Behrens, P., Stubblefield, J., Zohar, Y., and Place, A.R. 2002. Advanced DHA, EPA and ArA enrichment materials for marine aquaculture using single cell heterotrophs. *Aquaculture* 213:347–362.

Harvey, H.W. 1937. Notes on selective feeding by Calanus. *Journal of the Marine Biological Association of the United Kingdom* 22:97–100.

Hazzard, S.E. 2001. Copepod nutrition in Florida Bay: the relationship between the fatty acid composition of the food environment and *Acartia tonsa* egg production. Master's thesis. University of South Carolina, Columbia, South Carolina, USA.

Hazzard, S.E. and Kleppel, G.S. 2003. Egg production of the copepod *Acartia tonsa* in Florida Bay: role of fatty acids in the nutritional composition of the food environment. *Marine Ecology Progress Series* 252:199–206.

Head, E.J.H. 1992. Comparison of chemical composition of particulate material and copepod fecal pellets at stations off the coast of Labrador in the Gulf of St. Lawrence. *Marine Biology* 112:593–600.

Hitchcock, G.L. 1982. A comparative study of the size-dependent organic composition of marine diatoms and dinoflagellates. *Journal of Plankton Research* 4:363–377.

Hunter, J.R. 1981. Feeding ecology and predation of marine fish larvae. In *Marine Fish Larvae. Morphology, Ecology and Relation to Fisheries*, edited by R. Lasker, pp. 33–79. Seattle, Washington, USA: Washington Sea Grant.

Huntley, M.E., Sykes, P., Rohan, S., and Marin, V. 1986. Chemically mediated rejection of dinoflagellate prey by *Calanus pacificus* and *Paracalanus parvus*: mechanisms, occurrence, significance. *Marine Ecology Progress Series* 28:105–120.

Ianora, A., Poulet, S.A., and Miralto, A. 1995. A comparative study of the inhibitory effects of diatoms on the reproductive biology of the copepod *Temora stylifera. Marine Biology* 121:533–539.

Irigoien, X., Head, R.N., Harris, R.P., Cummings, D., Harbour, D., and Meyer-Harms, B. 2000. Feeding selectivity and egg production of *Calanus helgolandicus* in the English Channel. *Limnology and Oceanography* 45:44–54.

Jónasdóttir, S.H. 1992. Chemical composition of food and the reproductive success of the copepods *Acartia tonsa, Acartia hudsonica* and *Temora longicornis*. Ph.D. Thesis. State University of New York, Stony Brook, New York, USA.

Jónasdóttir, S.H., Fields, D., and Pantoja, S. 1995. Copepod egg production in Long Island Sound, USA, as a function of the chemical composition of the seston. *Marine Ecology Progress Series* 119:87–98.

Jónasdóttir, S.H. and Kiørboe, T. 1996. Copepod recruitment and food composition: do diatoms affect hatching success? *Marine Biology* 125:743–750.

Kiørboe, T. 1989. Phytoplankton growth rate and nitrogen content: implications for feeding and fecundity in a herbivorous copepod. *Marine Ecology Progress Series* 5:229–234.

Klein Breteler, W.C.M., Schogt, N., Baas, M., Schouten, S., and Kraay, G.W. 1999. Trophic upgrading of food quality by protozoans enhancing copepod growth: the role of essential lipids. *Marine Biology* 135:191–198.

Klein Breteler, W.C.M., Schogt, N., and Gonzalez, S.R. 1990. On the role of food quality in grazing and development of life stages, and genetic change in body size during cultivation of pelagic copepods. *Journal of Experimental Marine Biology and Ecology* 135:177–189.

Kleppel, G.S. 1992. Environmental regulation of feeding and egg production by *Acartia tonsa* off southern California. *Marine Biology* 112:57–65.

Kleppel, G.S. 1993. On the diets of calanoid copepods. *Marine Ecology Progress Series* 99:183–195.

Kleppel, G.S. and Burkart, C.A. 1995. Egg production and the nutritional environment of *Acartia tonsa*: the role of food quality in copepod nutrition. *ICES Journal of Marine Science* 52:297–304.

Kleppel, G.S., Burkart, C.A., and Houchin, L. 1998a. Nutrition and the regulation of egg production in the calanoid copepod *Acartia tonsa. Limnology and Oceanography* 43:1000–1007.

Kleppel, G.S., Burkart, C.A., Houchin, L., and Tomas, C. 1998b. Egg production of the copepod *Acartia tonsa* in Florida Bay during summer. 1. The roles of food environment and diet. *Estuaries* 21:328–339.

Kleppel, G.S. and Hazzard, S.E. 2002. The significance of zooplankton nutrition in the aquatic sciences. Outcomes of an international workshop on zooplankton nutrition. Columbia, South Carolina, USA: University of South Carolina.

Kleppel, G.S., Word, J.Q., and Roney, J. 1980. Demersal fish feeding in Santa Monica Bay and off Palos Verdes. In Coastal Water Research Project, biennial report for the years 1979–1980, pp. 309–318. Southern California Coastal Water Research Project, Los Angeles, California, USA.

Koski, M., Engstrom, J., and M. Vitasalo, M. 1999. Reproduction and survival of rgw calanoid copepod Eurytemora affinis fed toxic and non-toxic cyanobacteria. *Marine Ecology Progress Series* 186:187–197.

Kreeger, D.A. and Langdon, C.J. 1994. Digestion and assimilation of protein by *Mytilus trossulus* (Bivalvia: Mollusca) fed mixed carbohydrate/protein microcapsules. *Marine Biology* 118:479–488.

Lasker, R. 1975. Field criteria for survival of anchovy larvae: the relation between inshore chlorophyll maximum layers and successful first feeding. *U.S. Fishery Bulletin* 73:453–462.

Lebour, M.V. 1921. The food of young clupeids. *Journal of the Marine Biological Association of the United Kingdom* 12:458–457.

Lebour, M.V. 1924. The food of young herring. *Journal of the Marine Biological Association of the United Kingdom* 13:325–330.

Marcus, N. 1990. Calanoid copepod, cladoceran and rotifer eggs in sea-bottom sediments of northern California coastal waters: identification, occurrence and hatching. *Marine Biology* 105:413–418.

Marcus, N.H. and Alatalo, P. 1989. Conditions for rearing *Calanus finmarchicus* (Gunnerus, 1770) (Copepoda, Calanoida) through multiple generations in the laboratory. *Crustaceana* 57:101–103.

Marshall, S.M., Nicholls, A.G., and Orr, A.P. 1939. On the growth and feeding of young herring in the Clyde. *Journal of the Marine Biological Association of the United Kingdom* 23:427–455.

Marshall, S.M. and Orr, A.P. 1955. *The Biology of a Marine Copepod Calanus finmarchicus (Gunnerus)*. Edinburgh, UK: Oliver and Boyd.

McManus, G.B. and Foster, C.A. 1998. Seasonal and fine-scale spatial variations in egg production and triacylglycerol content of the copepod *Acartia tonsa* in a river-dominated estuary and its coastal plume. *Journal of Plankton Research* 20:767–785.

Nejstgaard, J.C., Gismervik, I., and Solberg, P. 1997. Feeding and reproduction by *Calanus finmarchicus* and microzooplankton grazing during mesocosm blooms of diatoms and the coccolithophore

Emeliania huxleyi. Marine Ecology Progress Series 147:197–217.

Nicholls, A.G. 1933. On the biology of *Calanus finmarchicus*. I: reproduction and seasonal distribution in the Clyde Sea area during 1932. *Journal of the Marine Biological Association of the United Kingdom* 19:83–110.

Paffenhöfer, G.-A. 1971. Grazing and ingestion rates of nauplii, copepodids and adults of the marine planktonic copepod *Calanus helgolandicus. Marine Biology* 11:286–298.

Paffenhöfer, G.-A. 1984. Food ingestion by the marine planktonic copepod *Paracalanus* in relation to abundance and size distribution of food. *Marine Biology* 80:323–333.

Paffenhöfer, G.-A. and Van Sant, K.B. 1985. The feeding response of a marine planktonic copepod to quantity and quality of particles. *Marine Ecology Progress Series* 27:55–65.

Parsons, T.R., Takahashi, M., and Hargrave, B. 1977. *Biological Oceanographic Processes*, 2nd ed. Oxford, UK: Pergamon Press. 343 pp.

Petersen, W.T. 1988. Rates of egg production by the copepod *Calanus marshallae* in the laboratory and in the sea off Oregon, USA. *Marine Ecology Progress Series* 47:229–237.

Pohnert, G., Lumineau, O., Cuelf, A., Adolph, S., Cordevant, C., Lange, M., and Poulet, S. 2002. Are volatile unsaturated aldehydes from diatoms the main line of chemical defence against copepods? *Marine Ecology Progress Series* 245:33–45.

Pond, D., Harris, R., Head, R., and Harbour, D. 1996. Environmental and nutritional factors determining seasonal variability in the fecundity and egg viability of *Calanus helgolandicus* in coastal waters off Plymouth, UK. *Marine Ecology Progress Series* 143:45–63.

Poulet, S.A., Ianora, A., Miralto, A., and Meijer, L. 1994. Do diatoms arrest embryonic development in copepods? *Marine Ecology Progress Series* 111:79–86.

Raymont, J.E.G., Srinivasagam, R.T., and Raymont, J.K.B. 1969. Biochemical studies on marine zooplankton. VII. Observations on certain deep-sea zooplankton. *International Review of Hydrobiology* 54:357–365.

Roman, M.R. 1984. Utilization of detritus by the copepod *Acartia tonsa. Limnology and Oceanography* 29:949–959.

Roman, M.R. 1991. Pathways of carbon incorporation in marine copepods: effects of developmental stage and food quality. *Limnology and Oceanography* 36:796–807.

Sargent, J.R. and Henderson, R.J. 1986. Lipids. In *The Biological Chemistry of Marine Copepods*, edited by E.D.S. Corner and S.C.M. O'Hara, pp. 59–109. Oxford, UK: Clarendon.

Smith, S.L. and Lane, P.V.Z. 1985. Laboratory studies of the marine copepod *Centropages typicus*: egg production and development rates. *Marine Biology* 85:153–162.

Smith, S.L. and Lane, P.V.Z. 1987. On the life history of *Centropages typicus*: response to fall diatom bloom. *Marine Biology* 95:305–314.

Støttrup, J.G. 2000. The elusive copepods: their production and suitability in marine aquaculture. *Aquaculture Research* 31:703–711.

Støttrup, J.G., Bell, J.G., and Sargent, J.R. 1999. The fate of lipids during during development and cold-storage of eggs in the laboratory-reared calanoid copepod, *Acartia tonsa* Dana, and in response to different algal diets. *Aquaculture* 176:257–269.

Støttrup, J.G. and Jensen, J. 1990. Influence of algal diet on feeding and egg production of the calanoid copepod *Acartia tonsa* Dana. *Journal of Experimental Marine Biology and Ecology* 141:87–105.

Suyama, M., Nakajima, K., and Nonaka, J. 1965. Studies on the protein and non-protein nitrogenous constituents of *Euphausia. Bulletin of the Japanese Society of Scientific Fisheries* 31:302–306.

Tester, P.A. and Turner, J.T. 1990. How long does it take copepods to make eggs? *Journal of Experimental Marine Biology and Ecology* 141:169–182.

Turner, J.T. 1984. The feeding ecology of some zooplankters that are important prey items of larval fish. NOAA (National Oceanic and Atmospheric Administration) Technical Report NMFS 7:1–28, Beaufort, North Carolina, USA.

Turner, J.T., Ianora, A., Miralto, A., Laabir, M., and Esposito, F. 2001. Decoupling of copepod grazing rates, fecundity and egg hatching success on mixed and alternating diatom and dinoflagellate diets. *Marine Ecology Progress Series* 220:187–199.

6
Formulated Feeds for Harpacticoid Copepods: Implications for Population Growth and Fatty Acid Composition

Adelaide Rhodes and Leon Boyd

ABSTRACT

This study examines the effect of using a formulated feed, composed mainly of flaxseed oil, on population growth rate and lipid composition of the marine harpacticoid copepod *Nitokra lacustris*. The epibenthic harpacticoid copepod *Nitokra lacustris* is highly adaptable to existing technologies for the culture of live feed for marine species because it matures quickly, produces high numbers of offspring (> 100/female), and has an ideal size range (40–620 μm) for most early-stage marine fish larvae. This copepod species can bioconvert important essential fatty acids for larval marine fish (i.e., docosahexaenoic acid 22:6*n*3, DHA; eicosapentaenoic acid 20:5*n*3, EPA; and arachidonic acid 20:4*n*6; ARA) from linolenic acid (18:3*n*-3) and linoleic acid (18:2*n*-6).

N. lacustris fed on the diet that contained large amounts of linolenic acid (55.0%) and linoleic acid (15.4%) and only trace amounts (< 0.1%) of EPA and DHA had significantly more DHA (8.1%) and EPA (3.0%) than the feed. Additionally, the copepod DHA-to-EPA ratio of 2.7:1 exceeded the target of 2:1 recommended for larval marine fish. The final lipid composition of copepods fed formulated feed did not differ significantly in final EPA and DHA composition from copepods fed live *Tetraselmis* sp. *Tetraselmis* is also a source of linolenic (20.6%) and linoleic (13.0%) acid, as well as EPA (4.5%). Surprisingly, the copepods fed the *Tetraselmis* had a lower amount of EPA than copepods fed the formulated feed. The lower EPA value for the copepods fed *Tetraselmis* resulted in an increase in the DHA-to-EPA ratio from 2.7:1 to 3.9:1. These results indicate that feeding copepods with formulated feeds containing seed oils could be a novel and inexpensive approach to producing high-quality live feeds for marine fish larvae that require EPA and DHA.

INTRODUCTION

The culture of marine finfish depends on having a reliable and nutritious food available to the larvae the first few days of feeding. Many marine species need live feeds for the first few days of exogenous feeding (Watanabe et al. 1978; Watanabe et al. 1983; Næss et al. 1995; Schipp et al. 1999). The most common live feeds in mass production, brine shrimp and rotifers, are not necessarily the natural prey of marine finfish and are often deficient in certain essential fatty acids (Lubzens et al. 1985; Dhert et al. 2001; Sorgeloos et al. 2001; Smith et al. 2002).

Harpacticoid copepods are a good source of the essential fatty acids eicosapentaenoic acid 20:5*n*3 (EPA) and docosahexaenoic acid 22:6*n*3 (DHA)

Corresponding author: Adelaide Rhodes, Essential Live Feeds, 343 Soquel Avenue #197, Santa Cruz, CA 95062, USA; Phone: (831) 477-0166; E-mail: Adelaide@essentiallivefeeds.com.

(Støttrup and Jensen 1990; Norsker and Støttrup 1994; Støttrup 2000). They also constitute a large percentage of the diet of marine finfish in the natural environment (Watanabe et al. 1978; Næss et al. 1995; Aarnio 2000; Prisco et al. 2001). Copepods, however, are more difficult to cultivate at the scale of production involved in hatchery finfish rearing (Schipp et al. 1999; Støttrup 2000; Payne and Rippingale 2001). One factor that increases the time and money involved in rearing copepods is the expense of growing live algae to feed them. The economic and labor costs of producing copepods may be reduced by using a formulated feed instead of cultured algae, especially when marine fish oil does not need to be added as enrichment.

This study examines the value of a formulated feed as a replacement for live feed by examining its effects on the population growth rate and lipid composition of the marine harpacticoid copepod *Nitokra lacustris*. The epibenthic harpacticoid copepod *N. lacustris* is highly adaptable to existing live-feed culture technologies because it has a short generation time (12–14 days), produces large numbers of nauplii (> 100/female/week), and ranges in size from 40 to 620 μm, which is ideal for most early marine fish larvae. The fatty acid profile also contains DHA, EPA, and arachidonic acid 20:4*n*6 (ARA), three important essential fatty acids for larval marine fish.

It has not been established previously whether *N. lacustris*, like other harpacticoid copepods, can bioconvert linolenic acid (18:3*n*3; LNA) and linoleic acid (18:2*n*6; LA) into the highly unsaturated fatty acids (HUFAs) EPA, DHA, and ARA. This study examined whether harpacticoid copepods fed a formulated feed that contained flaxseed oil as the lipid source instead of a direct enrichment of omega-3 fatty acids from fish oil would be able to maintain the population growth rates observed in copepods fed live algae. The formulated feed included flaxseed oil, a terrestrial source of omega-6 and omega-3 fatty acids, instead of a direct enrichment of omega-3 fatty acid from fish oil (Rhodes 2003). Flaxseed contains omega-3 fatty acid and omega-6 fatty acid, which serve as precursors to the more essential omega-3 and omega-6 fatty acids EPA, DHA, and ARA.

METHODS

Cultivation of Copepods

Six *N. lacustris* populations were simultaneously reared in 10-L batch cultures for 31 days. All treatments were started with approximately 2,300 copepods/L, with equal abundance of individuals in each life stage. The average culture temperature was 20°C (± 0.5°C), and the average salinity was 30 ppt (± 0.5 ppt). Three replicate populations were fed formulated feed (Table 6.1), and three populations were fed a continuous supply of the live algae *Tetraselmis suecica*.

The feeding rate on the first day was calculated to be approximately 10 times the requirement of 10,000 adult copepods per liter of similar species. Exact grazing rates are not known for this species and are difficult to determine due to the recycling of fecal pellets and potential ingestion of bacteria and detritus by the nauplii (Decho and Fleeger 1988). The copepods in the formulated feed treatments were fed 1 mL formulated feed per liter (approximately 50,000 cells/mL) on days 0, 10, and 20. Only particles greater than 8 μm in diameter were included when estimating the density of particles using a hemocytometer. *T. suecica* was added to live algae treatments on day 0 (50,000 cells/mL) and allowed to grow for the duration of the experiment. The abundance of *N. lacustris* in

Table 6.1. Formulated feed

Ingredient	Amount	Source
Tomato or vegetable juice	240 mL	V-8 100% Vegetable Juice, Campbell Soup Co., Camden, NJ, USA
Enriched brewer's yeast	10 g	GNC Brand, General Nutrition Centers Inc., Pittsburgh, PA, USA
Liquid vitamin C	1 mL	GNC Brand, General Nutrition Centers Inc., Pittsburgh, PA, USA
Liquid vitamin B complex	2 mL	GNC Brand, General Nutrition Centers Inc., Pittsburgh, PA, USA
Flaxseed oil	5 mL	GNC Brand, General Nutrition Centers Inc., Pittsburgh, PA, USA

Note: The ingredients are combined and mixed in a blender (Osterizer, Sunbeam Product, Inc., Boca Raton, FL, USA). Artificial seawater (30 ppt) is added until the volume of the mixture is 1 L. This volume of feed is blended for 2 minutes, and 1 mL/L of the homogenized mixture is fed to *N. lacustris*.

each population was estimated on days 10, 24, and 31 by stirring the culture and counting the number of copepods per volume of four equal aliquots taken from the middle of the water column using a 1-mL pipet.

Statistical Comparison of Copepod Populations

The instantaneous daily rate of increase, r, was calculated by dividing the difference between the natural logarithm of population counts at the beginning (N_0) and the end (N_t) by the number of days that had passed since the beginning of the experiment (t):

$$r = [\ln(N_t) - \ln(N_o)] * t^{-1} \qquad (6.1)$$

r = instantaneous daily rate of increase
N_0 = starting population size
N_t = population size at time t
t = time since beginning of experiment

The population density and r values given for the population were measured on each sampling day. Population estimates include all life stages. The r values calculated in the experiments assumed that the populations were growing in a constant environment. The finite rate of population growth lambda (λ) for each time interval was calculated from the instantaneous daily rate of increase using equation (6.2). The percentage of daily rate increase or decrease in a population was calculated by subtracting one from lambda.

$$\lambda = e^r \qquad (6.2)$$

r = instantaneous daily rate of increase
λ = population growth rate

Separate tanks under the same conditions were considered replicates. The natural log of final population counts and the r values of treatments run concurrently were compared using a one way analysis of variance (ANOVA) on the three replicates from each treatment.

Fatty Acid Compositional Analyses

After 31 days, all copepods were collected by passing the entire culture through a 35-µm sieve three times. A representative sample of adults was measured for length and width under a stereomicroscope using an optical micrometer. The largest specimens were retained on a 105-µm sieve, rinsed, and frozen (−20°C).

Duplicate lipid extractions and fatty acid compositional analyses were performed on each copepod replicate using ultrasonification of the copepods in a mixture of chloroform-methanol (1:2, v/v) and a modified Christie procedure (1989) as described by Boyd et al (1999). In brief, this involved the addition of 0.005% butylated hydroxyl toluene (BHT) to methanol, followed by washing of crude lipid extracts using 0.88% potassium chloride, drying over anhydrous sodium sulfate, and removal of solvent under a nitrogen flush. After addition of an internal standard (C21:0), fatty acid methyl esters (FAMEs) were prepared using methanolic HCl followed by purification on Fluorosil-packed Pasteur pipet columns using hexane-ethyl ether. Purified FAMEs were dried under nitrogen flush and redissolved in iso-octane for gas liquid chromatograph (GLC) analysis of fatty acids.

The FAMEs were analyzed on a GLC (model 5890A; Hewlett Packard, Avondale, PA, USA) equipped with a flame-ionization detector. A fused silica Omegawax 320 column (30 m × 0.32 mm i.d. with 0.25 mm thickness) was used for separation (Supelco, Bellefonte, PA, USA). The injector and detector temperatures were set at 260°C. Helium was used as the carrier gas at an inlet pressure of 11 psi and a flow rate of 0.75 mL/min. The initial oven temperature was set at 170°C, raised 4°C/min to a final temperature of 240°C, and held for 5 minutes. FAs were identified by comparison of their relative retention times to authentic standards (Nu-Chek Prep, Elysian, MN, USA). The same extraction and analysis procedure was repeated on two samples taken from each feed type.

Individual fatty acid concentrations were expressed as the weight percentage of the total FAMEs identified and were used to determine the differences in fatty acid composition of copepods fed formulated feed versus live algae feed. The normalization technique was used to calculate absolute response factors for all identified fatty acids using standards obtained from Nu-Chek Prep.

Statistical Analysis of Fatty Acid Methyl Esters

The weight percentages of individual fatty acids were arcsin transformed and the treatments were compared for statistical difference using one way analysis of variance (ANOVA) when three replicates were available (copepods treatments versus copepod treatments) and Student's *t* test when only two replicates were available from one set of samples (food versus copepod treatments). Results were considered to be significant if the *p* value was less than 0.05 and marginally significant if the *p* value was greater than 0.05 and less than 0.10.

RESULTS

Fatty Acid Composition

The harpacticoid copepod *N. lacustris* contained significant amounts of DHA, EPA, and ARA, even when the feeds were deficient in these HUFAs (Table 6.2). The harpacticoid copepod *Nitokra lacustris* did not differ in the natural log of the population size (Fig. 6.1) or size of adults (Fig. 6.2) when fed formulated feed instead of live algae. The presence of the HUFAs and the continued reproduction of the animal when fed a diet deficient in HUFAs suggested that *N. lacustris* was able to bioconvert the shorter chain 18:3*n*3 and 18:2*n*6 into EPA, DHA, and ARA to meet its physiological requirements for these fatty acids.

The dominant fatty acids in *T. suecica* were LNA, LA, oleic acid 18:1*n*9, and palmitic acid 16:0. These fatty acids represented more than 60% of the total fatty acids. The fatty acid composition of this strain of *T. suecica* was found to be similar to literature values (Volkman et al. 1989; Zhukova and Aizdaicher 1995; Fábregas et al. 2001). *T. suecica* samples tested also contained ARA and EPA, but did not contain DHA.

The dominant fatty acids in the formulated feed were the same as the live *T. suecica*. The main difference was the much higher proportion of 18:3*n*3 in the formulated feed (55%) in comparison to the live feed (20.6%). The higher amounts of 18:3*n*3 were responsible for the higher overall HUFAs and omega-3 content of the formulated feed in comparison to the live feed. The formulated feed also had higher amounts of 18:2*n*6 (15.4% versus 13.0%) and oleic acid 18:1*n*9 (16.9% versus 10.6%). The formulated feed had a lower proportion of saturated fatty acid than *T. suecica*. The formulated feed contained trace amounts of ARA (0.2%), EPA (0.1%), and DHA (0.1%), probably due to the fortified brewer's yeast. Small amounts of these HUFAs were found in a compositional analysis of the yeast alone (this study). Seventy-five percent of the yeast fatty acid composition consisted of saturated fatty acid (14:0, 16:0, and 18:0) and monounsaturated fatty acid (16:1*n*-13 trans, 18:1*n*9, and 20:1*n*9) (this study).

The five dominant fatty acids in *N. lacustris* fed *T. suecica* and fed the formulated feed were the same: 18:2*n*6, 18:1*n*9, 16:0, 18:0, and 22:6*n*3. Copepods had much less 18:3*n*3 than the feeds provided. In contrast, copepods contained more 18:2*n*-6 than the feeds provided. The amount of the most abundant saturated fatty acid (acids 16:0 and 18:0) increased in both treatments in comparison to the feed. The EPA composition in the copepods fed the live algae was less than half of the composition of the algae. The EPA proportion of the copepods fed the formulated feed was more than 30 times greater than the feed. DHA 22:6*n*3, which was not detected in the *T. suecica* and was found only in trace amounts in the formulated feed (0.1%), was found in significantly higher amounts in the copepods fed *T. suecica* (8.2%) and the copepods fed the formulated feed (8.1%).

Fatty acid composition by category varied little between copepod populations (Table 6.2B and Fig. 6.3A). Results for omega-3 fatty acid without 18:3*n*3 and the omega-6 fatty acid without 18:2*n*6 were analyzed because LNA 18:3*n*3 and LA 18:2*n*6 were such a large component of the omega-3 fatty acid and omega-6 fatty acid, respectively (Table 6.2B and Fig. 6.3A). Without LNA, the proportion of the omega-3 fatty acid increased significantly in copepods fed the formulated feed, while the proportions remained similar to the algae in copepods fed live *T. suecica*. The proportions of omega-6 fatty acid other than 18:2*n*6 increased significantly from the feed levels to the tissue levels for both populations of copepods.

The fatty acid composition of LA (18:2*n*6), LNA (18:3*n*3), ARA (20:4*n*6), EPA (20:5*n*3), and DHA (22:6*n*3) differed between the copepods and the feeds (Fig. 6.3B). The proportion of LA increased significantly from the live algae to the copepods fed the live algae. Copepods in both groups had significantly lower proportions of LNA than the feeds.

Copepods fed the live algae had a marginally higher proportion of ARA and a significantly

Table 6.2. Summary of changes in fatty acid composition of copepods fed *Tetraselmis suecica* and formulated feed

A. Individual fatty acid composition

	T. suecica treatments (% of total fatty acid)		Formulated feed treatments (% of total fatty acid)	
Fatty acid	*T. suecica*	Copepods	Formulated feed	Copepods
13:0	3.0	0.2[a, b]	0.2	1.0[a]
14:0	0.9	0.5	0.5	1.5
16:0	16.1	13.8	6.9	15.1[b]
16:1*n*7	1.7	0.9	—	1.3
16:1*n*13	1.2	1.7	0.8	3.8
16:2*n*4	2.3	1.2	—	1.0
16:4*n*3	2.4	0.1	—	—
18:0	2.9	8.5[b]	4.1	9.2[b]
18:1*n*9	10.6	14.8	16.9	15.5
18:1*n*7	3.4	2.7	0.5	2.9[b]
18:2*n*6	13.0	22.4[b]	15.4	16.8
18:3*n*3	20.6	6.2[a,b]	55.0	3.5[a,b]
20:1*n*9	1.4	1.0	—	1.2
20:2*n*6	0.1	0.5[a]	—	1.4[a]
20:4*n*6	0.6	0.7[a]	0.2	1.0[a,b]
20:4*n*3	1.7	0.6	—	1.1
20:5*n*3	4.5	2.1[b]	0.1	3.0[b]
22:2*n*6	—	1.2	—	3.0
22:4*n*6	1.0	0.4	—	1.1
22:5*n*6	—	2.0	—	1.7
22:5*n*3	0.9	0.7	—	1.0
24:0	—	2.2	—	2.2
22:6*n*3	—	8.2[b]	0.1	8.1[b]
24:1*n*9	0.8	0.8	—	0.8

B. Combined fatty acid composition

Fatty acid category	*T. suecica*	Copepods	Formulated feed	Copepods
Saturated FA	22.9	25.3	11.7	28.9[b]
Monoenic FA	19.0	21.9	18.2	25.6[b]
Dienic FA	15.6	25.2[b]	15.4	22.2[b]
HUFA	31.6	21.1[b]	55.5	20.5[b]
Omega-3	30.0	18.0[b]	55.3	16.7[b]
Omega-6	14.7	27.2[b]	15.6	25.0[b]
Omega-3 without 18:3	9.4	11.7	0.3	13.2[b]
Omega-6 without 18:2*n*6	1.7	4.8[b]	0.2	8.2[a,b]
Ratios	***T. suecica***	**Copepods**	**Formulated feed**	**Copepods**
Omega-3 to 18:3*n*3	0.5	1.9	0.01	3.8
Omega-6 to 18:2*n*6	0.1	0.2	0.01	0.5
Omega-3 to omega-6	2.0	0.7	3.5	0.7
DHA to EPA	No DHA	3.9	1.0	2.7

Note: For copepods, the percentages represent an average of the three replicates in each treatment. Each replicate was tested in duplicate. The food percentages are representative of two independently prepared samples in duplicate. Only those fatty acids that reached > 0.5% in either the feed or copepod compositions are shown.

[a]Significant difference between copepod populations (ANOVA, $p < 0.05$).

[b]Significant difference between copepod populations and their diet (Student t test, $p < 0.05$).

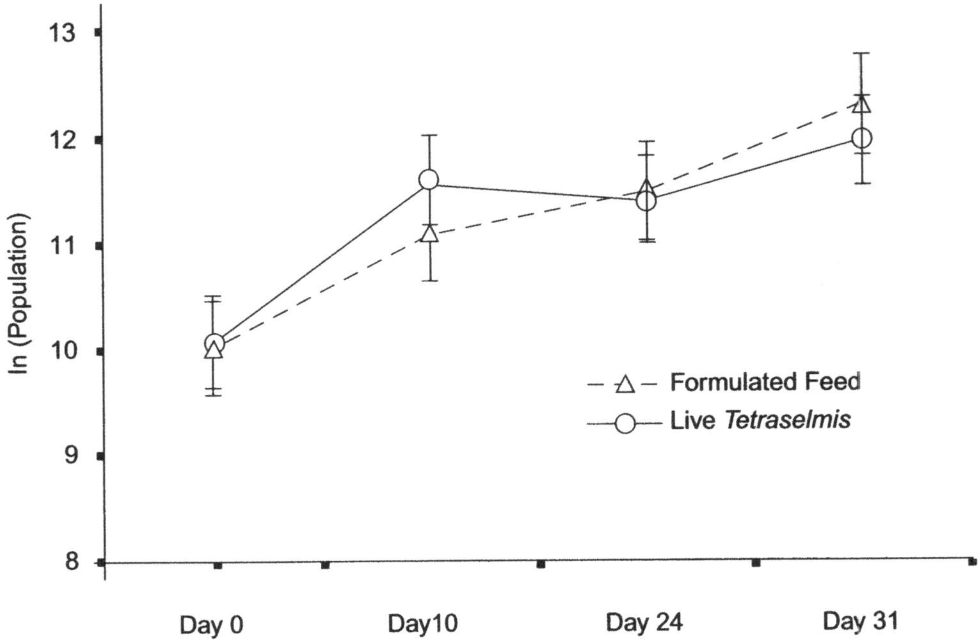

Figure 6.1. Natural log of population estimates made on days 10, 24, and 31 for the copepods fed formulated feed (Δ) and the copepods fed live *Tetraselmis suecica* (○). Values represent an average of three replicates. Error bars represent standard error.

lower proportion of EPA. In contrast, copepods fed formulated feed had significantly higher ARA and EPA. Both groups of copepods had significantly higher levels of DHA than their respective feeds.

Comparisons between copepod populations and their diets yielded more significant differences in individual fatty acid than the comparison between the copepod populations. Three individual fatty acids were found to be significantly higher in copepods fed formulated feed than copepods fed live algae: 13:0, 20:2*n*6, and 20:4*n*6. Copepods fed *T. suecica* had significantly higher amounts of 18:3*n*3.

Some lipids were not detected in feeds, but were detected in *N. lacustris*. In populations fed live *T. suecica*, copepods contained 22:2*n*6, 22:5*n*6, 22:6*n*3, and 24:0, while *T. suecica* did not. In the group fed formulated feed, copepods contained 16:1*n*7, 16:2*n*4, 20:1*n*9, 20:2*n*6, 20:4*n*3, 22:2*n*6, 22:4*n*6, 22:5*n*6, 22:5*n*3, 24:0, and 24:1*n*9, while formulated feed did not.

Copepod compositions were significantly different from their feeds for almost all categories of fatty acids (Table 6.2A). The copepods fed the *T. suecica* had significantly less 13:0, 18:3*n*3, and 20:5*n*3, and significantly more 18:0 and 18:2*n*6 than the *T. suecica*. The copepods fed the formulated feed had significantly less 18:3*n*3 and significantly more 16:0, 18:0, 18:1*n*7, 20:4*n*6, 20:5*n*3, and 22:6*n*3 than the formulated feed.

When the individual fatty acids were combined into general classifications, copepods fed *T. suecica* were significantly different than the feed except for saturated fatty acid, monounsaturated fatty acid, and omega-3 fatty acid other than 18:3*n*3. Copepods fed formulated feed were significantly different in all categories. As in the comparison of individual fatty acid, the composition of fatty acid classes was very similar for the two sets of copepods. Between copepod populations, the only significant difference was that copepods fed formulated feed had a higher amount of omega-6 fatty acid when 18:2*n*6 is not included.

Population Growth and Size Results

No significant difference was detected between treatments for the natural log of the population or the instantaneous daily rates of increase on days

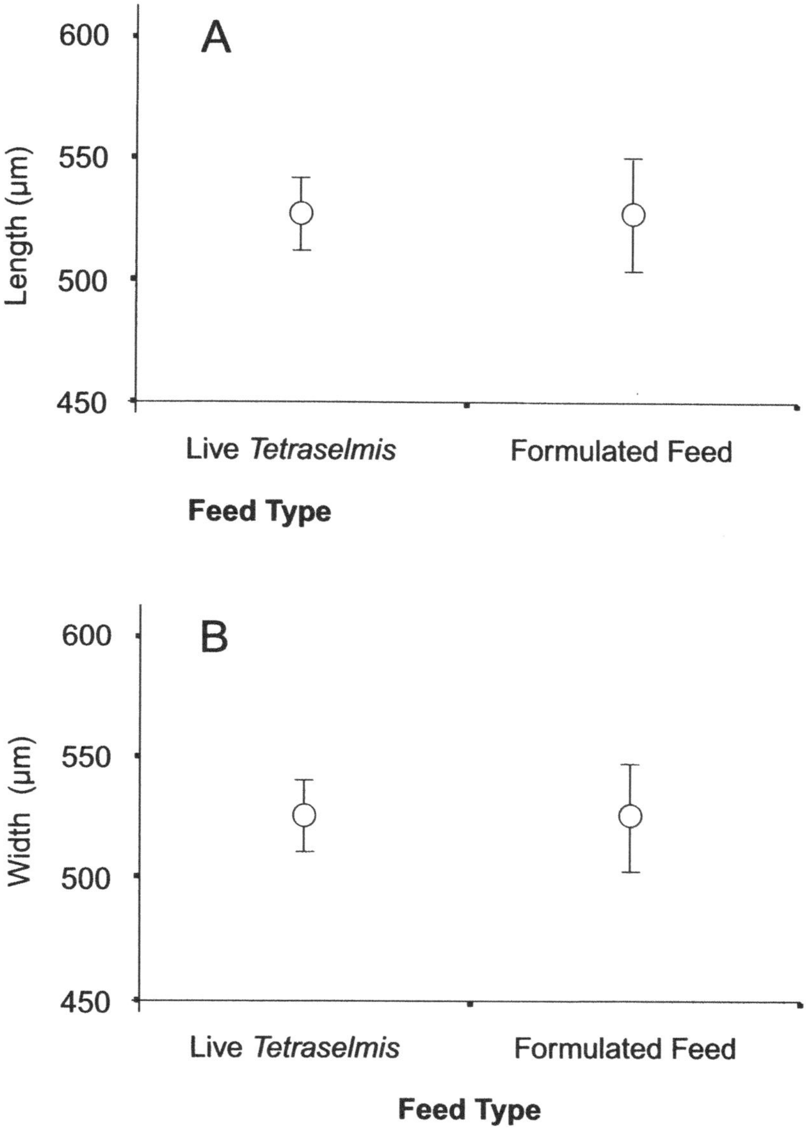

Figure 6.2. (A) Length and (B) width estimates for 30 adult copepods fed live *Tetraselmis suecica* and formulated feed. Error bars represent variance.

10, 24, and 31 (Fig. 6.1, Table 6.3). Copepods fed formulated feed had a final daily population rate of increase of 7.5%, and copepods fed live algae had a final daily population rate of increase of 6.3%. Copepod density at the end of the experiment reached an average value of 21,600 copepods per liter for the formulated feed treatments and 15,400 copepods per liter for the live algae treatments. No significant difference was detected between treatment in the length (527 μm) or width (154 μm) of nonovigerous adult copepods (Fig. 6.2).

DISCUSSION

The harpacticoid copepod *Nitokra lacustris* did not differ in population growth rate, size of adults, or fatty acid composition when fed formulated feed instead of live algae. Other harpacticoid copepod species have demonstrated a similar

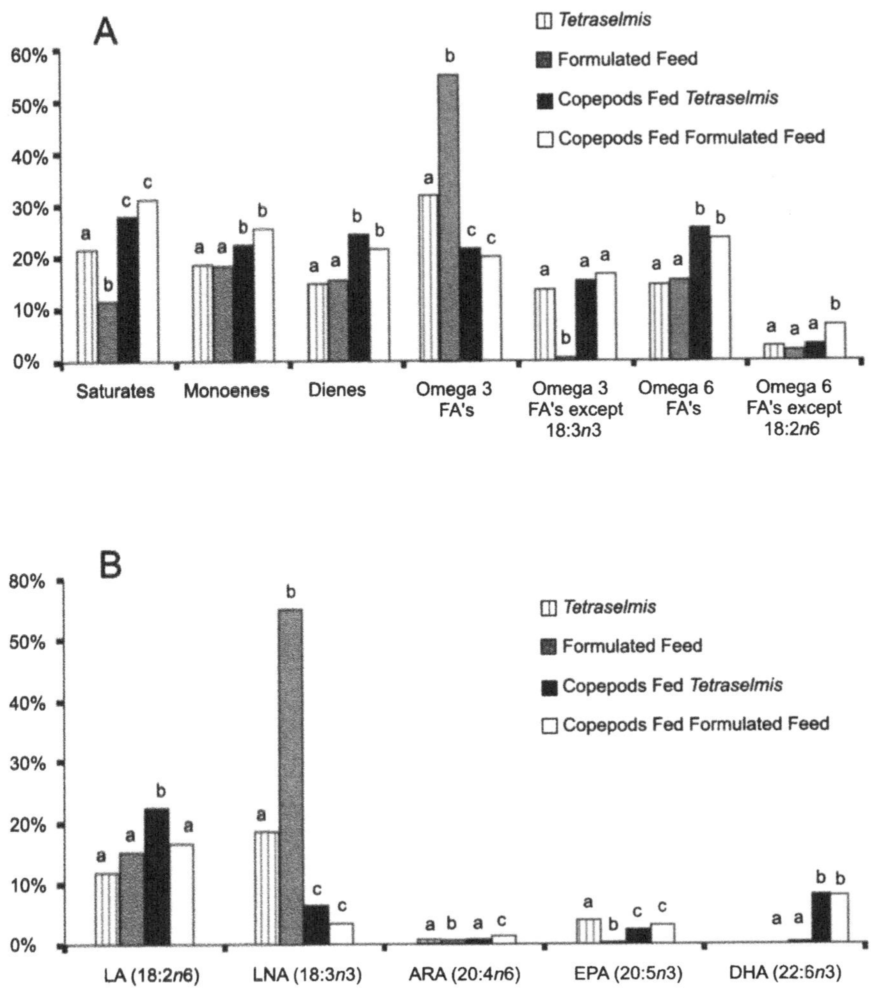

Figure 6.3. Comparison of fatty acid composition of feeds and copepod populations. (A) Saturates, monoenes, dienes, omega-3 fatty acids (FAs), omega-3 fatty acids not including 18:3*n*3, omega-6 fatty acids, and omega-6 fatty acids without 18:2*n*6. (B) LA (18:2*n*6); LNA (18:3*n*6); ARA (20:4*n*6); EPA (20:5*n*3); and DHA (22:6*n*3). Values assigned the same letter (a, b, c) were not significantly different within that category.

adaptability to algal and yeast diets that contain little or no EPA and DHA (Watanabe et al. 1978; Miliou and Moraitou-Apostolopoulou 1991; Norsker and Støttrup 1994; Carli et al. 1995; Nanton and Castell 1998). This suggests that *N. lacustris* and the other harpacticoid copepods contain the Δ-5, Δ-6 desaturase and elongase enzymes necessary for the conversion of the shorter chain LNA (18:3*n*3) to the essential fatty acids EPA (20:5*n*3) and DHA (22:6*n*3) (Norsker and Støttrup 1994; Støttrup 2000). The same enzymes are used in the conversion of LA (18:2*n*6) to ARA (20:4*n*6) (O'Keefe 2002).

Evidence of this bioconversion capability in *N. lacustris* can be found in the difference between the relative fatty acid profiles of the feeds and copepod tissues. In both treatments, copepods fed diets high in 18:3*n*3 (20.6% in *T. suecica* and 55.0% in the formulated feed) had low 18:3*n*3 compositions: 6.2% for copepods fed *T. suecica* and 3.5% for copepods fed formulated feed. Both sets of copepods, however, had a higher ratio of

Table 6.3. Population estimates and instantaneous daily rates of increase

Feed	Day	Population estimate (all developmental stages)	Natural log (Ln) of population estimate	Instantaneous daily rate of increase (r)
Tetraselmis suecica	0	23,317	10.057	—
	10	109,404	11.603	0.1546
	24	90,240	11.410	0.0564
	31	154,228	11.946	0.0609
Formulated feed	0	23,046	10.045	—
	10	66,533	11.105	0.1060
	24	97,655	11.489	0.0602
	31	215,529	12.281	0.0721

Note: Instantaneous daily rates of increase were calculated using equation (6.1). No significant difference was found between treatments on any sampling days for natural log of the population estimate or instantaneous daily rates of increase.

other omega-3 fatty acids to 18:3n3 than did their diets. Copepod populations fed *T. suecica*, which has a 0.5:1 ratio of other omega-3 fatty acid to 18:3n3, had a final ratio of 1.9:1. Likewise, copepods fed formulated feed, which contained more than 99% of its omega-3 fatty acid in the form of 18:3n3, had a final lipid composition ratio of other omega-3 fatty acid to 18:3n3 of 3.8:1.

The concurrent reduction in dietary 18:3n3 fatty acid proportion and increase in the proportion of other omega-3 fatty acid suggested a conversion from 18:3n3 to other omega-3 fatty acids. Adding excessive amounts of 18:3n3, however, as in the formulated diet, did not appear to dramatically increase the overall conversion of omega-3 fatty acid, as reflected by the similarity of the copepod populations' final composition of omega-3 fatty acid other than 18:3n3: 11.7% for copepods fed *T. suecica* and 13.2% for copepods fed formulated feed. Additional 18:3n3 in the diet may have increased the accumulation of EPA. The amount of EPA in copepods fed formulated feed represented a 30-fold increase from the proportion in the feed, while EPA in copepods fed the live *T. suecica* was reduced by more than half from the proportion in the feed. It may also be possible that the excess dietary 18:3n3 in the formulated feed may be channeled into saturated fatty acids and monoenes, which increased more in the case of copepods fed formulated feed than in copepods fed live *T. suecica*. The excess amounts of saturated fatty acid and monoenes may reduce the nutritional value of the copepods for fish that require higher percentages of HUFAs.

The relative proportions of LA (18:2n6) and the other omega-6 fatty acid increased in relation to diet for both treatments. This increase in omega-6, however, was not accompanied by a dramatic reduction in LA, as was seen in the case of LNA. The fivefold increase of ARA (20:4n6) in copepods fed formulated feed, which had very low amounts of this particular omega-6 fatty acid (0.2%), may be evidence for the conversion of LA into other omega-6 fatty acid. The chain of conversion from LA to ARA shares the same enzymes as the conversion pathway from LNA to DHA (O'Keefe 2002). Based on the small increase in the ratio of other omega-6 fatty acid to LA from feed to copepod, this conversion pathway may not be as active as the omega-3 conversion. Because both conversion pathways share the same enzymes, competitive inhibition from the omega-3 conversion may limit the amount of omega-6 conversion (Okuyama et al. 1996).

The effect of formulated feed on population growth and size of *N. lacustris* also indicated that a similar quantity and quality of copepods could be produced in the same amount of time as traditional methods using live algae. The instantaneous daily rates of increase of *N. lacustris* populations fed formulated feed consistently met or exceeded populations fed live *T. suecica*. Supporting this result was the observation that *N. lacustris* cultures in our laboratory have been maintained solely on formulated feed for the last 2 years. The size of the animals, which must be small enough to be captured by the fish larvae, was not altered by the formulated diet.

The ability to convert shorter-chain omega-3 and omega-6 fatty acid into HUFAs has been found in other harpacticoid copepods fed algal diets and yeast deficient in HUFAs. *Tisbe holothuriae* and another *Tisbe* sp. from Nova Scotia have been shown to synthesize significant amounts of 20:5*n*3 (6%) and 22:6*n*3 (12%) when fed *Dunaliella tertiolecta*, which is HUFA limited, but contains large amounts of the precursor 18:3*n*-3 (Norsker and Støttrup 1994; Nanton and Castell 1998). This suggests that *Tisbe sp.* contains the Δ-5, Δ-6 desaturase and elongase enzymes necessary for the conversion of shorter chain *n*-3 polyunsaturated fatty acid to the essential fatty acids, DHA and EPA (Norsker and Støttrup 1994; Nanton and Castell 1998). No significant difference in egg production was found when *Tisbe sp.* was switched to a diet of HUFA-rich *Rhodomonas baltica* (Nanton and Castell 1998). Watanabe et al. (1978) found that *Tigriopus sp.*, another harpacticoid species, contained high levels of *n*-3 HUFA (12% DHA and 7% EPA) in its lipids even when fed with baker's yeast, a diet deficient in HUFAs. Based on the results of this experiment, *N. lacustris* exhibits the same bioconversion capability as other species of marine harpacticoid copepod.

Fish larvae require live feeds that contain high levels of essential fatty acid such as EPA and DHA (Watanabe 1982). The yolks of wild marine larvae contain a DHA-to-EPA ratio of about 2:1, which may suggest a requirement for high levels of DHA to EPA in first-feeding larvae (Parrish et al. 1994). Increasing the ratio of DHA to EPA improves the survival of marine fish larvae (Bell et al. 1985). Watanabe (1993) found that increasing the ratio of DHA-to-EPA increased the survival of marine fish larvae for yellowtail (*Seriola quinqueradiata*), striped jack (*Pseudocaranx dentex*), striped knifejaw (*Oplegnathus fasciatus*), red sea bream (*Pagrus major*), and Japanese flounder (*Paralichthys olivaceous*). Increasing the ratio of DHA to EPA from 1:10 to 1:2 in enriched *Artemia* fed to turbot (*Scophthalmus maximus* L.) larvae significantly increased their survival (Bell et al. 1985).

Nitokra lacustris exceeded the target DHA-to-EPA ratio of 2:1 on both diets. Copepods fed live algae had a DHA-to-EPA ratio of 3.9:1, while copepods fed formulated feed had a ratio of 2.7:1. The higher ratio in copepods fed live algae was most likely due to the reduced amount of EPA. If a ratio of DHA to EPA greater than 2:1 and an absolute amount of EPA greater than 3% of the total fatty acids were the determining factors for a successful live fish food, copepods fed formulated feed would be a slight improvement over copepods fed live algae. More importantly, there was no loss of fatty acid quality when *N. lacustris* were switched from live algae to formulated feed.

Artemia and rotifers, which are the more common live food organisms for marine fish culture, have little or no ability to bioconvert shorter chain *n*-3 polyunsaturated fatty acids (PUFA) into longer-chain HUFAs 20:5*n*3 (EPA) and 22:6*n*3 (DHA) (Lubzens et al. 1985; Howell and Tzoumas 1991; Smith et al. 2002). *Artemia* may even retroconvert DHA obtained through enrichment into EPA (Navarro et al. 1999). Copepods have higher amounts of HUFA and DHA than enriched *Artemia* (Norsker and Støttrup 1994; Støttrup et al. 1998; Shields et al. 1999).

Calanoid copepods, another live food alternative for marine fish larvae, may have limited bioconversion capabilities when fed algae deficient in HUFAs (Sargent and Henderson 1986; Fraser et al. 1989). Calanoid copepods may contain the Δ-5 desaturase and elongase enzymes necessary for the conversion of 18:3*n*3 to 20:5*n*3, but not the Δ-6 desaturase enzyme necessary for the final conversion of 20:5*n*3 to 22:6*n*3 (Sargent and Henderson 1986; Fraser et al. 1989). The population growth rate of calanoid copepods may also be affected by a diet deficient in HUFAs through impacts on reproduction (Støttrup and Jensen 1990; Jonasdottir et al. 1995). Calanoid copepods may have to incorporate the *n*-3 HUFA directly from their phytopolankton diet (Fraser et al. 1989; Støttrup and Jensen 1990). Benthic harpacticoid copepods do not require microalgae with HUFA to maintain a normal reproduction rate or a high DHA-to-EPA ratio (Norsker and Støttrup 1994; Nanton and Castell 1998). Therefore, the use of harpacticoid copepods for culture may be more efficient than calanoid copepods, *Artemia*, and rotifers, all of which require DHA enrichment to maintain a favorable DHA-to-EPA ratio for marine fish larvae.

CONCLUSION

The methods presented in this chapter may provide the framework for further investigations into other harpacticoid copepod species. To adapt any copepod species to aquaculture, it will be necessary to understand the instantaneous daily rates of increase of the populations and the conditions under which the species will thrive. This understanding will allow adequate planning of time and resources to reach the desired yield of copepods in the hatchery. It has been shown that even a short-term replacement or supplementation with copepods can have a beneficial effect on fish development. For example, typical defects in Atlantic halibut (*Hippoglossus hippoglossus*) can be avoided by feeding copepods during a critical window of time from 2 to 3 weeks after first feeding (Næss and Lie 1998).

The use of formulated feed to grow *N. lacustris* is an excellent model of production for harpacticoid copepods in hatcheries. The flatfish aquaculture industry has made the most use of harpacticoid copepods to date in the culture of turbot and halibut and other similar species (Heath and Moore 1997; Nanton and Castell 1998; Støttrup 2000; Cutts 2003). Other fish species, however, may benefit from the use of harpacticoid copepods that have these favorable ratios of DHA to EPA. To improve the ability to use any live feed, it must be predictable and provide a consistent nutritional quality of food.

Many of the problems with rotifers and *Artemia* relate to the difficulty in transferring these essential fatty acids to the fish through the enrichment of the live feeds (Sorgeloos et al. 2001; Dhert et al. 2001; Smith et al. 2002). These enrichments are costly, have a short shelf life, and often use marine fish oil as a main ingredient. In overall aquaculture production, not just larval rearing, aquafeeds may use more than 40% of the current supply of fish oil (New and Wijkstrom 2002). As fisheries stocks decline and the price for marine fish oil increases, alternative HUFA resources will be needed. The use of harpacticoid copepods, which produce a consistent lipid profile whether fed vegetable oil or live algae, may help obviate the need for marine fish oil in the aquaculture industry by providing a predictable and high quality supply of highly unsaturated fatty acids.

ACKNOWLEDGMENTS

Ruth Watkins assisted in the setup of lipid analyses, and Jeff Hostetler commented on a draft of this manuscript. John Fleeger of Louisiana State University has also been quite helpful. We would also like to thank Cheng-Sheng Lee and his assistant Pat O'Bryen at the Aquaculture Interchange Program, as well as The Oceanic Institute, for their support in putting on the Culture of Copepods and Applications to Marine Finfish Larval Rearing Workshop, 5–8 May 2003, which led to the publication of this book.

REFERENCES

Aarnio, K. 2000. Experimental evidence of predation by juvenile flounder *Platichthys flesus*, on a shallow water meiobenthic community. *Journal of Experimental Marine Biology and Ecology* 246:125–138.

Bell, M.V., Henderson, R.J., and Sargent, J.R. 1985. Changes in the fatty acid composition of phospholipids from turbot (*Scophthalmus maximus*) in relation to dietary polyunsaturated fatty acid deficiencies. *Comparative Biochemistry and Physiology B* 81:193–198.

Boyd, L.C., Drye, N.C., and Hansen, A.P. 1999. Isolation and characterization of whey phospholipids. *Journal of Dairy Science* 82:2550–2557.

Carli, A., Mariottini, G.L., and Pane, L. 1995. Influence of nutrition on fecundity and survival in *Tigriopus fulvus* Fischer (Copepoda, Harpacticoida). *Aquaculture* 134:113–119.

Christie, W. 1989. *Gas Chromatography and Lipids: A Practical Guide*. Ayr, Scotland: The Oily Press Ltd., 307 pp.

Cutts, C.J. 2003. Culture of harpacticoid copepods: Potential as live feed for rearing marine fish. *Advances in Marine Biology* 44:295–316.

Decho, A.W. and Fleeger, J.W. 1988. Ontogenetic feeding shifts in the meiobenthic harpacticoid copepod *Nitocra lacustris*. *Marine Biology* 97:191–197.

Dhert, P., Rombaut, G., Suantika, G., and Sorgeloos, P. 2001. Advancement of rotifer culture and manipulation techniques in Europe. *Aquaculture* 200:129–146.

Fábregas, J., Otero, A., Dominguez, A., and Patino, M. 2001. Growth rate of the microalga *Tetraselmis suecica* changes the biochemical composition of *Artemia* species. *Marine Biotechnology* 3:256–263.

Fraser, A.J., Sargent, J.R., and Gamble, J.C. 1989. Lipid class and fatty acid composition of *Calanus finmarchicus* (Gunnerus), *Pseudocalanus sp.* and *Temora longicornis* Müller from a nutrient enriched seawater enclosure. *Journal of Experimental Marine Biology and Ecology* 130:81–92.

Heath, P.L. and Moore, C.G. 1997. Rearing Dover sole larvae on *Tisbe* and *Artemia* diets. *Aquaculture International* 5:29–39.

Howell, B.R. and Tzoumas, T.S. 1991. The nutritional value of *Artemia* nauplii for larval sole, *Solea solea* (L.) with respect to their (n-3) HUFA content. *European Aquaculture Society Special Publication* 15:63–65.

Jonasdottir, S.H., Fields, D., and Pantoja, S. 1995. Copepod egg production in Long Island Sound, USA, as a function of the chemical composition of seston. *Marine Ecology Progress Series* 119:87–98.

Lubzens, E., Marko, A., and Tietz, A. 1985. De novo synthesis of fatty acids in the rotifer, *Brachionus plicatilis*. *Aquaculture* 47:27–37.

Miliou, H. and Moraitou-Apostolopoulou, M. 1991. Effects of 7 diets on the population-dynamics of laboratory cultured *Tisbe holothuriae* Humes (Copepoda, Harpacticoida). *Helgolander Meeresuntersuchungen* 45:345–356.

Næss, T. and Lie, Ø. 1998. A sensitive period during first feeding for the determination of pigmentation pattern in Atlantic halibut, *Hippoglossus hippoglossus* L., juveniles: the role of diet. *Aquaculture Research* 29:925–934.

Næss, T., Germain-Henry, M., and Naas, K.E. 1995. First feeding of Atlantic halibut (*Hippoglossus hippoglossus*) using different combinations of *Artemia* and wild zooplankton. *Aquaculture* 130:235–250.

Nanton, D.A. and Castell, J.D. 1998. The effects of dietary fatty acids on the fatty acid composition of the harpacticoid copepod, *Tisbe sp.*, for use as a live food for marine fish larvae. *Aquaculture* 163:251–261.

Navarro, J.C., Henderson, R.J., McEvoy, L.A., Bell, M.V., and Amat, F. 1999. Lipid conversions during enrichment of *Artemia*. *Aquaculture* 174:155–166.

New, M.B. and Wijkstrom, U.N. 2002. Use of fishmeal and fish oil in aquafeeds: further thoughts on the fish-meal trap. FAO Fisheries Circular No. 975. Food and Agriculture Organization of the United Nations (FAO), Rome, Italy. 68 pp.

Norsker, N.H. and Støttrup, J.G. 1994. The importance of dietary HUFAs for fecundity and HUFA content in the harpacticoid, *Tisbe holothuriae* Humes. *Aquaculture* 125:155–166.

O'Keefe, S.F. 2002. Nomenclature and classification of lipids. In *Food Lipids*, edited by C.C. Akoh and D.B. Min, pp. 1–40. New York, New York, USA: Marcel Dekker, Inc.

Okuyama, H., Kobayashi, T., and Watanabe, S. 1996. Dietary fatty acids—The *n*-6/*n*-3 balance and chronic elderly diseases. Excess linoleic acid and relative n-3 deficiency syndrome seen in Japan. *Progress in Lipid Research* 35:409–457.

Parrish, C.C., Castell, J.D., Brown, J.A., Boston, L., Strickland, J.S., and Somerton, D.C. 1994. Fatty acid composition of Atlantic halibut eggs in relation to fertilization. *Bulletin of the Aquaculture Association of Canada* 94:36–38.

Payne, M.F. and Rippingale, R.J. 2001. Intensive cultivation of the calanoid copepod *Gladioferens imparipes*. *Aquaculture* 201:329–342.

Prisco, A.R., de la Rosa, S.B.G., and de Astarloa, J.M.D. 2001. Feeding ecology of flatfish juveniles (*Pleuronectiformes*) in Mar Chiquita coastal lagoon (Buenos Aires, Argentina). *Estuaries* 24:917–925.

Rhodes, A. 2003. Methods for mass culture of *Nitokra lacustris*, a marine harpacticoid copepod. In *The Big Fish Bang: Proceedings of the 26th Annual Larval Fish Conference*, edited by H.I. Browman and A.B. Skiftesvik, pp. 449–465. Bergen, Norway, 22–26 July 2002. Bergen, Norway: Institute of Marine Research.

Sargent, J.R. and Henderson, R.J. 1986. Lipids. In *The Biological Chemistry of Marine Copepods*, edited by E.D.S Corner and S.C.M. O'Hara, pp. 59–108. Oxford, England: Clarendon Press.

Schipp, G.R., Bosmans, J.M.P., and Marshall, A.J. 1999. A method for hatchery culture of tropical calanoid copepods, *Acartia spp. Aquaculture* 174:81–88.

Shields, R.J., Bell, J.G., Luizi, F.S., Gara, B., Bromage, N.R., and Sargent, J.R. 1999. Natural copepods are superior to enriched *Artemia* nauplii as feed for halibut larvae (*Hippoglossus hippoglossus*) in terms of survival, pigmentation and retinal morphology: relation to dietary essential fatty acids. *Journal of Nutrition* 129:1186–1194.

Smith, G.G., Ritar, A.J., Phleger, C.F., Nelson, M.M., Mooney, B., Nichols, P.D., and Hart, P.R. 2002. Changes in gut content and composition of juvenile *Artemia* after oil enrichment and during starvation. *Aquaculture* 208:137–158.

Sorgeloos, P., Dhert, P., and Candreva, P. 2001. Use of the brine shrimp, *Artemia spp.*, in marine fish larviculture. *Aquaculture* 200:147–159.

Støttrup, J.G. 2000. The elusive copepods: their production and suitability in marine aquaculture. *Aquaculture Research* 31:703–711.

Støttrup, J.G. and Jensen, J. 1990 Influence of algal diet on feeding and egg production of the calanoid copepod *Acartia tonsa* Dana. *Journal of Experimental Marine Biology and Ecology* 141:87–105.

Støttrup, J.G., Shields, R., Gillespie, M., Gara, M.B., Sargent, J.R., Bell, J.G., Henderson, R.J., Tocher, D.R., Sutherland, R., Næss, T., Mangor Jensen, A., Naas, K., van der Meeren, T., Harboe, T., Sanchez, F.J., Sorgeloos, P., Dhert, P., and Fitzgerald, R. 1998. The production and use of copepods in larval rearing of halibut, turbot and cod. *Bulletin of the Aquaculture Association of Canada* 4:41–46.

Volkman, J.K., Jeffrey, S.W., Nichols, P.D., Rogers, G.I., and Garland, C.D. 1989. Fatty acid and lipid composition of 10 species of microalgae used in mariculture. *Journal of Experimental Marine Biology and Ecology* 128:219–240.

Watanabe, T. 1982. Lipid nutrition in fish. *Comparative Biochemistry and Physiology B* 73:3–15.

Watanabe, T. 1993. Importance of docosahexaenoic acid in marine larval fish. *Journal of the World Aquaculture Society* 24: 152–161.

Watanabe, T., Arakawa, T., Kitajima, C., Fukusho, K., and Fujita, S. 1978. Nutritional quality of living feed from the viewpoint of essential fatty acids for fish. *Bulletin of the Japanese Society of Scientific Fisheries* 44:1223–1227.

Watanabe, T., Kitajima, C., and Fujita, S. 1983. Nutritional value of live organisms used in Japan for mass propagation of fish: a review. *Aquaculture* 34:115–143.

Zhukova, N.V. and Aizdaicher, N.A. 1995. Fatty-acid composition of 15 species of marine microalgae. *Phytochemistry* 39:351–356.

7
A Brief Review of Mass Culture of Copepods Used for Fish Food in Japanese Mariculture and a Proposed Plan to Use High Biomass Natural Populations of Brackish-Water Copepods

Shin-ichi Uye

ABSTRACT

This chapter consists of two parts: a brief review of studies on mass culture of copepods as food for larval fish in Japan and a proposed plan to use extremely high biomass and production of natural brackish-water copepods for marine fish farming. In the first part previous attempts to collect natural copepod populations and mass culture of *Tigriopus japonicus* and *Acartia tsuensis* in Japanese hatcheries are reviewed, and associated problems are discussed. Although *T. japonicus* appears to be a most promising species, the mass culture of this harpacticoid copepod is no longer applied at present, mainly due to development of artificially formulated feed for fish larvae. The idea of using wild copepod populations discussed in the second part of the chapter stems from findings of very high biomass (range: 11.1–369 mg carbon [C]/m^3; mean: 59.2 mg C/m^3) and production rate (range: from 2.0 to 44.7 mg C/m^3/d; mean: 9.6 mg C/m^3/d) of copepods in a brackish-water lagoon (i.e., the Honjo District; area: 16.2 km^2, average depth: 5.1 m, water volume: 8.25 $\times$ 10^7 m^3) in the Lake Shinji and Lake Nakaumi brackish-water system. The predominant copepod species in this lagoon are *Acartia hudsonica*, *A. sinjiensis*, *Eurytemora pacifica*, *Oithona davisae*, *Pseudodiaptomus inopinus*, and *Sinocalanus tenellus*, and each population showed a pronounced seasonal variation. If an intensive larviculture system is employed, an exploitation of 10% of the daily copepod production (or about 2% of the biomass) from Honjo District is capable of feeding at least 3.2 million red sea bream (*Pagrus major*) or Japanese flounder (*Paralichthys olivaceus*) and 8.0 million ayu (*Plecoglossus altivelis*) larvae annually. Thus, deliberate use of productivity of natural copepod population as food for fish larvae is both ecologically and economically feasible.

INTRODUCTION

In 1930, the first scientific paper on successful rearing of marine fish larvae, that is, red sea bream (*Pagrus major*) and black sea bream (*Acanthopagrus shulegeli*), in Japan was published (Kajiyama and Nishioka 1930). After World War II, due to increased demand for expanding marine fish farming or aquaculture, the production of a large quantity of fish larvae or fish seed was required. Early studies in the 1950s and 1960s on the rearing of marine fish larvae were reviewed by Fujita (1973), who pointed out that because of a lack of knowledge about suitable larval food, a variety of food types, for example, protozoans, rotifers, bivalve larvae, brine shrimp nauplii, cladocerans (marine and freshwater), copepods, barnacle nauplii, mysids, and minced fish meat, were tested at random during those decades. Through a process of trial and error, researchers empirically learned that the necessary conditions for the mass production of suitable foods for fish larvae are (a) edibility, (b) proper nutrition, and (c) ease in obtaining the necessary amount.

Since the mid-1960s, mass cultured rotifers, *Brachionus plicatilis* and *B. rotundiformis*, have been used to rear larvae of finfish and shellfish in Japan. Both species of rotifers are closely associated, and *B. rotundiformis* has also been called small (S) type *B. plicatilis* (Fukusho and Okauchi 1984; Fu and Hirayama 1991). These rotifers were formerly regarded as pests, because their blooms and concomitant consumption of dissolved oxygen in brackish-water eel culture ponds caused the eels to suffocate (Ito 1959). Ito (1960) succeeded in acclimatizing them from brackish-water to normal seawater salinity in 1960. Since then, artificial propagations of this strain have been carried out in many Japanese hatcheries. The mass culture of *Brachionus* spp. is now in operation worldwide, and the majority of mass production of marine fish larvae depends on artificially cultured *Brachionus* spp. Through some improvements (food, nutrition, harvesting techniques, and stocking), the density of cultures attained are usually higher than several hundred rotifers per milliliter, and the techniques for the mass culture of *Brachionus* spp. have been established almost everywhere (Japan Sea-Farming Association 2002). The next target is to culture copepods on a large scale.

Unlike *Brachionus* spp., mass culture of copepods is still in a very early stage of development in Japan. There have been no prominent advancements in either research or practical use of copepods for fish mariculture since the 1970s, with the exception of studies on *Acartia tsuensis* by Ohno (1991). The major reasons for this slow development largely stem from the biological characteristics of copepods; many have a slow growth rate, unstable population conditions, and are harvested in small quantities, as well as require labor-intensive care. In addition, artificially formulated feed has been developed for larval and juvenile fish. It is now commonly used and is often dispensed by automatic feeding machines to save labor costs. In most Japanese hatcheries, the major dietary sequence for common marine fish larvae (e.g., red sea bream and Japanese founder, *Paralichthys olivaceus*) whose total annual production is in the order of hundreds of millions of individuals is cultured *Brachionus* spp., followed by *Artemia* nauplii and crumbled formulated feed. In some hatcheries, field-collected copepods are used together with *Artemia* nauplii.

Copepods are a valuable source of proteins, amino acids, lipids, fatty acids, vitamins, and minerals (Watanabe et al. 1983), and they can be an inexpensive replacement for expensive *Artemia* and formulated feed. If, like *Brachionus* spp., copepods could be obtained with ease, demand for copepods would be high in hatcheries. Young nauplii of small copepods, i.e., smaller than *Brachionus* spp., are needed for the production of fish larvae that have small gapes, such as the groupers and snappers (Fukuhara 1989; Doi et al. 1994). In this chapter, the author reviews the previous studies on mass culture of copepods as food for fish larvae in Japan and proposes an idea to use extremely high biomass and production of natural brackish-water copepod populations for marine fish seed production.

REVIEW OF UTILIZATION OF COPEPODS IN JAPANESE MARINE FARMING

Utilization of Field-Collected Copepods

Some hatcheries have been using field-collected copepods as food for fish larvae since the mid-1960s. In the beginning, wild copepods were collected in conventional plankton nets in straits or narrow passages where the tidal current is strong enough (about 2 m/sec) to keep the nets open. Fushimi et al. (1968) harvested several hundred wet weight grams of copepods daily at Ondo Strait, Hiroshima Prefecture, and used them as food for red sea bream larvae.

In coastal waters where the tidal current is slow, light traps, which lure positively phototactic copepods, have been employed. Copepods that aggregate in the vicinity of underwater electric lamps at night can be pumped or air-lifted into floating cages with netting of proper mesh size or to nearby land-based tanks. Yamashita and Arakawa (1974) invented a light trap consisting of an underwater lamp (500 W), a submersible pump (delivery rate: 80 L/min; depth: 1 m), and a floating cage (0.9 m × 0.9 m × 0.9 m, with 150-μm mesh net), which they operated daily from 1700 to 0900 from late October 1973 to early February 1974, in Nomo Bay, Nagasaki Prefecture. The wet weight of zooplankton caught daily varied from 0.045 kg to 2.045 kg with a mean of 0.385 kg, totaling 40 kg over the period (Fig. 7.1). Larger harvests were obtained during neap tides, when the aggregation of zooplankton was more intense, possibly due to the slower tidal current. Copepods (primarily of the genera *Acartia* and *Oithona*) were the major constituents (mean: 95%), with larvae of malacostracans, balanids, and polychaetes also in the catch.

As shown in Fig. 7.1, there are extremely large fluctuations in the size of the catch, demonstrating that the food supply from field-collected zooplankton is unreliable and unstable. Therefore, field-collected copepods are used as supplements to *Artemia* nauplii during the dietary transition period from *Brachionus* spp. to artificially formulated feed. When harvested copepods are not required, they can be kept frozen until needed.

Mass Culture of Copepods

Intensive studies were conducted mainly in the 1970s by Japanese planktologists and aquaculture scientists to establish procedures for the technical development of the mass culture of copepods (Anraku and Hirano 1973). In 1972, the Fisheries Agency started a project entitled, "Searching for suitable species and mass culture of zooplankton as food for the early stage of fish seed in marine fish farming." In this project, the following 13 copepod species were recommended for mass cultivation: *Acartia clausi* (*A. hudsonica* or *A. omorii*, according to current classification), *A. longiremis*, *Eurytemora pacifica*, *Euterpina acutifrons*, *Microsetella norvegica*, *Oithona brevicornis* (*O. davisae*, according to current classification), *O. nana*, *O. similis*, *Pseudodiaptomus inopinus*, *P. marinus*, and *Tigriopus japonicus* (Omori 1973; Nihon Suisanshigen Hogokyoukai 1979). Among these, *T. japonicus* was the only species that was cultured on a large scale and ac-

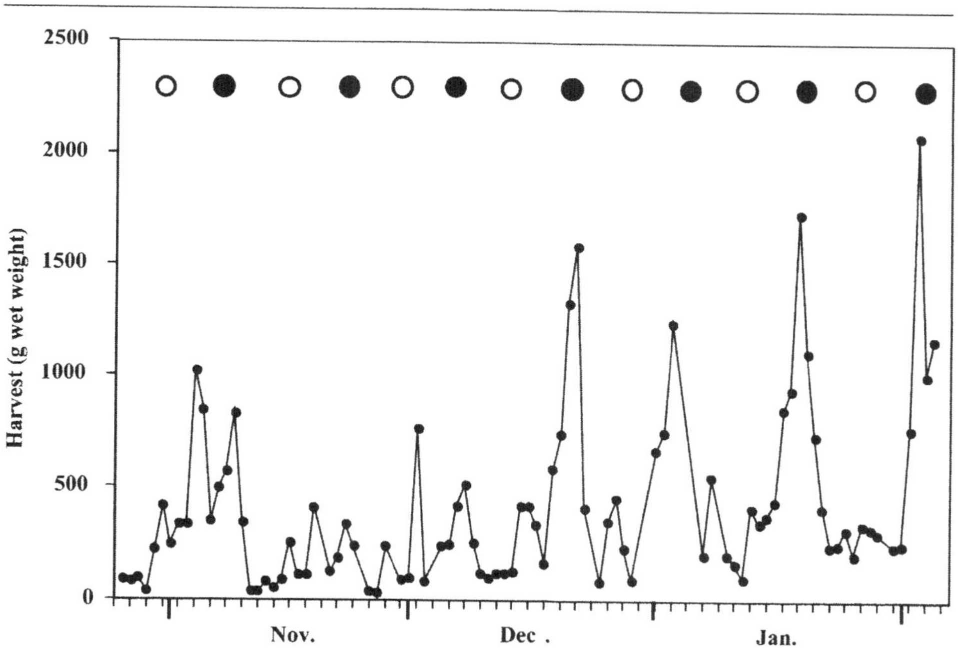

Figure 7.1. Daily variations in the harvest of natural zooplankton populations (gram wet weight) from Nomo Bay, Nagasaki Prefecture, 1973–1974. Samples were collected from late October to early February using a light trap. Filled and open circles denote the spring and neap tide, respectively. (Redrawn from Yamashita and Arakawa 1974).

tually used as food in marine fish farming (Kitajima 1973; Fukusho et al. 1977, 1978; Nihon Suisanshigen Hogokyoukai 1979). For the other species, rearing was attempted only in the laboratory (Iwasaki and Kamiya 1977; Iwasaki et al. 1977). There have been difficulties in maintaining their mass cultivation because of low and unstable population growth, the demand for large amounts of cultured algae, and frequent water changes.

Two most commonly used copepods for mass culture for fish mariculture are *Tigriopus japonicus* and *Acartia tsuensis*.

Tigriopus japonicus

T. japonicus is a harpacticoid copepod inhabiting tide pools on rocky shores along the Japanese coast (Ito 1970), with extremely wide temperature and salinity tolerances, often attaining very high population densities (many thousands of individuals [ind]/L) (Tokioka and Suzuki 1939; Matsutani 1961). It takes 8–10 days from hatching to molting to adult at 23–25°C (Fig. 7.2; Nihon Suisanshigen Hogokyoukai 1979), and mature females spawn their eggs in a sac (average clutch size: about 50 eggs) at 2-day intervals at 24°C (Koga 1970). This species is also characterized by broad feeding habits, ingesting a wide spectrum of organic particles (bacteria, protozoans, phytoplankton, macro algae, formulated feed for fish, detritus; Kitajima 1973).

Mass cultivation of *T. japonicus* was attempted in indoor and outdoor tanks of different sizes using various feeds such as the eustigmatophyte *Nannochloropsis oculata*, yeasts, the green seaweed *Ulva*, and formulated fish/shrimp feed. One of the largest-scale trials was conducted in 200-m^3 outdoor concrete tanks (15 m × 10 m × 1.4 m) at Nagasaki Prefectural Fisheries Experimental Station, from April to August 1976, by Fukusho et al. (1977). *Brachionus* spp. and *N. oculata* were already present in the tanks before inoculation of *T. japonicus*, and 5–10 kg wet weight of baker's yeast was also given daily during the cultivation period. In the tank, *N. oculata* and yeast were the major food for *T. japonicus* nauplii and *Brachionus* spp., the latter of which was food for *T. japonicus* copepodids and adults. The density of the *T. japonicus* population was monitored daily (range: 165–2,715 ind/L; mean: 1,414 ind/L). Part of the population was collected almost daily in June and July by pumping water into a cage with 300-μm mesh netting, which re-

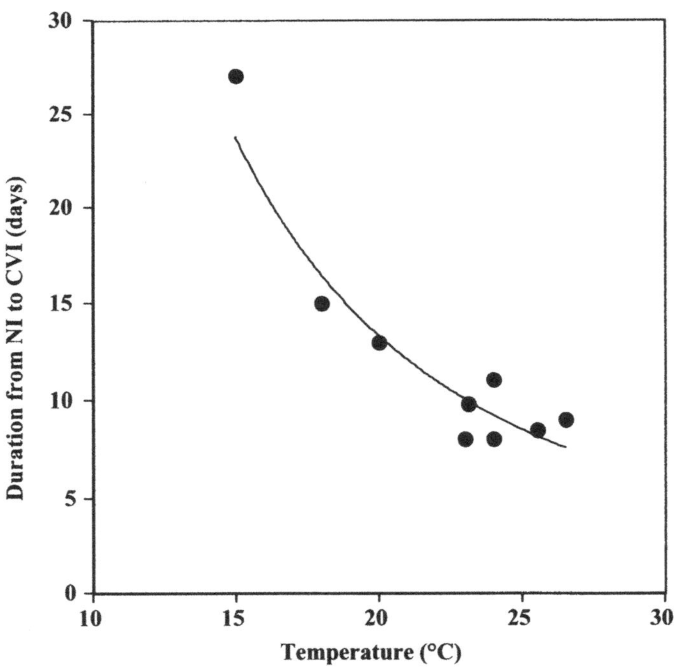

Figure 7.2. Relationship between temperature and the duration of life-cycle stages, from NI to CVI, of *Tigriopus japonicus*. (Redrawn from Nihon Suisanshigen Hogokyoukai 1979).

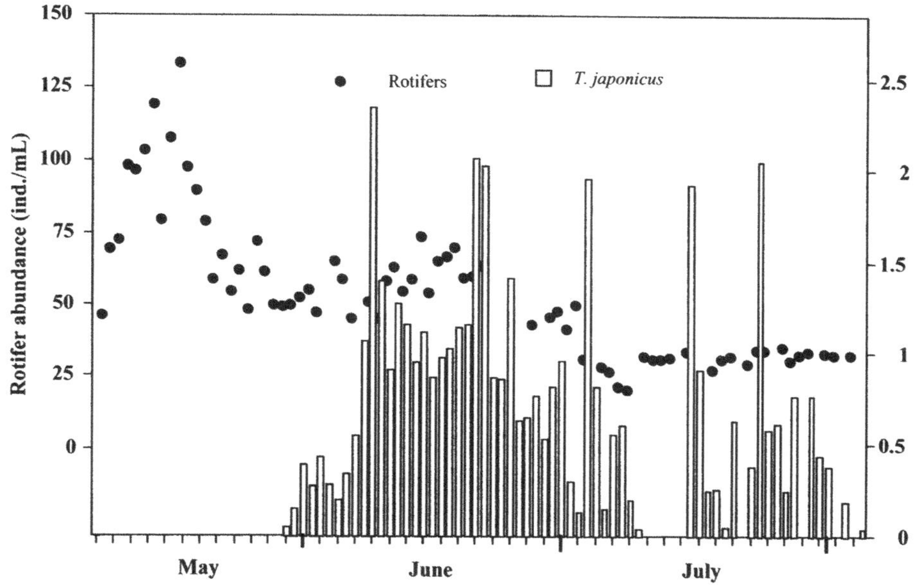

Figure 7.3. Daily variations in the harvest of *Tigriopus japonicus* (columns) and the density of the rotifer *Brachionus plicatilis* in a 200-m³ concrete tank at Nagasaki Prefectural Fisheries Experimental Station from May to August 1976. (Redrawn from Fukusho et al. 1977).

tained only late-stage copepodids and adults (Fig. 7.3). On average, about 0.5 kg wet weight of *T. japonicus* was harvested daily, amounting to 43.2 kg for the entire culture period. Another collection was made using the same tanks from October 1977 to January 1978, with a daily mean harvest of 1.6 kg wet weight and a total harvest of 65.5 kg (Fukusho et al. 1978).

The above-mentioned results demonstrate that *T. japonicus* is certainly a promising copepod species for mass production, although its benthic nature and hard exoskeleton are unfavorable characteristics as a diet for early-stage fish larvae (Nihon Suisan Shigen Hogokyoukai 1979). To this author's knowledge, *T. japonicus* is no longer being mass cultured in Japanese hatcheries. Ease of the use of synthesized feed, in addition to saving tank space and reduced labor costs are the major reasons why mass cultivation of this species has ceased.

Acartia tsuensis

A. tsuensis is a planktonic calanoid copepod distributed in estuarine and coastal waters of southern Japan and southeastern Asia (Ohno 1991; Toledo et al. 1999). In earthen ponds at Momoshima Station of the Japan Sea-Farming Association, Hiroshima Prefecture, this species is present in the plankton only in summer, when the water temperature is > 20°C (Ohno 1991). It takes about 8 days from spawning to molt to adult stage at 27.5°C (Fig. 7.4), and adult females spawn 8–14 eggs daily at 22.5–30.0°C (Takahashi and Ohno 1996).

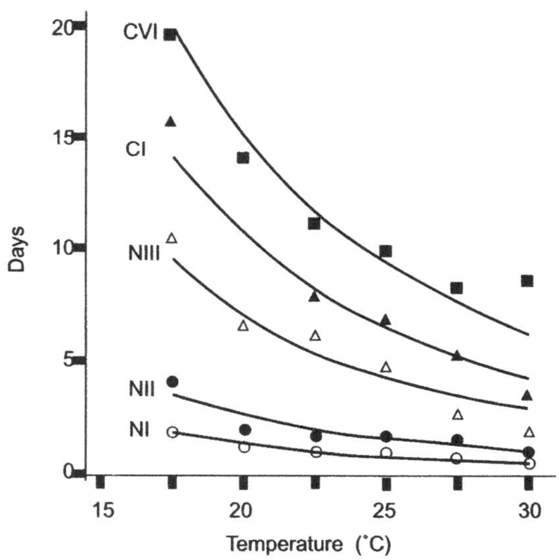

Figure 7.4. Relationship between temperature and median duration (in days) from egg to each developmental stage of *Acartia tsuensis*. (Redrawn from Takahashi and Ohno 1996).

At Momoshima Station, *A. tsuensis* was mass cultured in a 24-m^3 outdoor tank in which seawater from the adjacent sea had been introduced and fertilized with poultry manure and chemical agricultural fertilizer (Ohno 1991). The copepods fed on the available *N. oculata* as well as the introduced phytoplankton from the seawater. The abundance of each developmental stage of *A. tsuensis* was monitored for 50 days (Fig. 7.5). Although four distinctive cohorts were recognized at intervals of about 2 weeks, the development of the last cohort failed. The highest densities of nauplii, copepodids, and adults were 1,300, 590, and 280 ind/L, respectively, and the mean density (569 ind/L) was much lower than that observed in the mass culture of *T. japonicus* (1,414 ind/L; Fukusho et al. 1977). It was notable that eggs and nauplii were few when adults were abundant. This indicates that recruitment of the new population was inhibited by the cannibalism of adults. Hence, the most significant disadvantage of omnivorous species in the copepod genera such as *Acartia*, *Centropages*, and *Sinocalanus* for use for mass culture is cannibalism (Hada and Uye 1991; Liang et al. 1994; Uye and Liang 1998). For *A. tsuensis*, a high, daily, exploitation rate (up to 30% of the standing stock) was possible (Fig. 7.6), since the exploitation reduced mortality of the offspring due to cannibalism.

The propagation of *A. tsuensis* was also tested using outdoor earthen ponds at Momoshima Station, where the pond water was artificially fertilized to induce phytoplankton growth, for extensive mass seed productions of red sea bream (Ohno 1991). In the pond, not only *A. tsuensis* but also other copepod species (*A. omorii*, *A. sinjiensis*, *Paracalanus* sp., *Pseudodiaptomus marinus*, and unidentified harpacticoids) propagated and were consumed by fish larvae. The larvae could also feed on benthic animals such as amphipods and polychaetes. Due to the more natural feeding conditions, red sea bream produced in the ponds looked healthier with more vivid body coloration

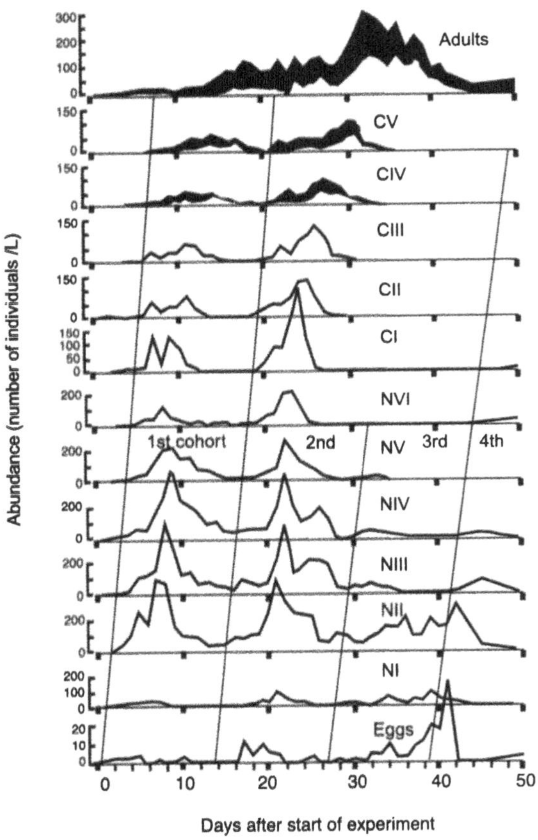

Figure 7.5. Temporal changes in stage-specific abundance (ind/L) of *Acartia tsuensis* in a 24-m^3 tank at Momoshima Station of the Japan Sea-Farming Association, Hiroshima Prefecture. Dark portions in CIV–CVI denote abundance of males. (Redrawn from Ohno 1991).

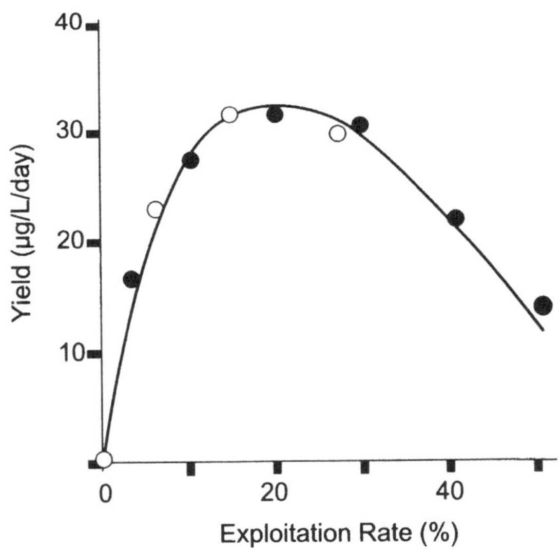

Figure 7.6. Relationship between daily exploitation rate (percentage of population biomass) and mean daily yield (microgram per liter per day) of *Acartia tsuensis* cultured in 24-m^3 tanks (open circles) and in 30-L containers (filled circles). (Redrawn from Ohno 1991).

and were stronger than those reared in the intensive method using *Brachionus* spp. and formulated feed (Maruyama et al. 1986). Due primarily to high operating costs (high labor costs but low numbers of fish seed produced), extensive culture of fish larvae using earthen ponds is no longer done in Japan. Mass culture of *A. tsuensis*, either in concrete tanks or large earthen ponds, is also not being pursued.

POTENTIAL USE OF NATURAL BRACKISH-WATER COPEPODS AS FOOD FOR FISH LARVAE

Brackish waters are generally recognized as having the highest biological production among aquatic ecosystems due to the high nutrient supply from adjacent land and rivers (Odum 1971). A 2-year investigation of mesozooplankton in a brackish-water lagoon (called the Honjo Disrict) in the Lake Shinji and Lake Nakaumi brackish-water system in Japan revealed that copepod biomass and production rates were among the highest thus far reported in the literature (Uye et al. 2004). In 1997, when this investigation was started, reclamation of the Honjo District for agricultural use was a controversial political issue. The author was opposed to the reclamation plan, but before taking any action against the plan, needed to show that the economic value of this water body was not inferior to that of reclaimed land. The author therefore proposed using the Honjo District as a field for mass seed production of economically important fishes, which would add new economic value to the Honjo District. Details of the seasonal variations in terms of taxonomic composition, abundance, biomass, and production rates of total mesozooplankton have been reported elsewhere (Uye et al. 2004). In the remainder of this chapter, details of seasonal variations for copepods, the major constituents of the zooplankton, are presented. Based on the production of copepods, an estimate is made of the quantity of fish seed that can be produced.

Study Site, Sampling, and Methods Used to Estimate Production Rate

Two estuarine lagoons, Lake Shinji and Lake Nakaumi, are the largest remaining brackish-water ecosystems in Japan (Fig. 7.7). In 1981, a

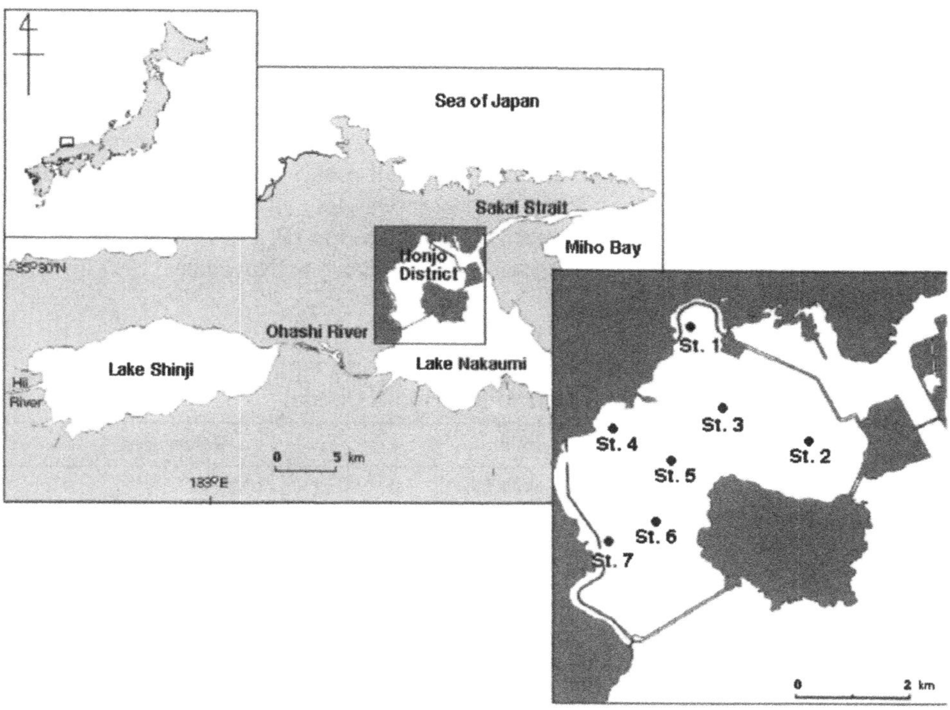

Figure 7.7. Map of the Honjo District with locations of sampling stations. St, station.

northwestern part of Lake Nakaumi (called the Honjo District; area: 16.2 km^2, average depth: 5.1 m; water volume: 8.25×10^7 m^3) was enclosed with the construction of a bank as part of a reclamation project for agricultural use. Water exchange between the Honjo District and Lake Nakaumi per se is limited only through a narrow waterway along the western shore. Although fisheries, mainly by set nets, are in operation, landings have been greatly reduced since the construction of the embankment.

Mesozooplankton were collected with vertical hauls of a plankton net (mouth diameter: 0.225 m; length: 1 m; mesh size: 100 µm) fitted with a flow meter (Rigosha & Co., Ltd., Tokyo, Japan) from the bottom to the surface at seven stations in the Honjo District (Station 1–7; depth: 3–9 m) monthly from June 1997 to May 1999 (Fig. 7.7). Plankton samples were immediately fixed with 5% buffered formalin, and later they were identified to species or genus level, except for harpacticoids, and at least 200 specimens were counted under a stereoscopic binocular microscope. Specific body dimensions (Uye 1982) were measured using a video micrometer (VM-10, Olympus), and automatically converted, using a personal computer, to carbon (C) weights using predetermined length-weight regressions (see Uye and Shimazu 1997 for details).

The production rate (P, mg C/m^3/d) of each taxonomic group was estimated based on its biomass (B, mg C/m^3) and an empirically determined potential (or maximum under nonfood limitation) specific growth rate (g, per day): $P = B \times g$. The specific growth rates in relation to temperature have been determined previously for various copepod species, mainly from the Inland Sea of Japan (Table 7.1).

Environmental Variables (Temperature, Salinity, Dissolved Oxygen Concentration, and Chlorophyll a Concentration)

Since environmental variables were similar among the seven stations, Station 3 was selected as a representative of the Honjo District. Seasonal variations of temperature, salinity, dissolved oxygen concentration, and chlorophyll *a* concentration at the surface (0.5 m) and the bottom (6.0 m) are shown in Fig. 7.8. Temperature fluctuated from 6 to 29°C, and salinity from 9.6 to 19.2 practical salinity units (psu) at the surface (mean: 14.7 psu) and from 10.5 to 22.7 psu at the bottom (mean: 16.7 psu). The dissolved oxygen (DO) concentrations were usually oversaturated at the surface, while low oxygen levels were prominent on the bottom during summer and fall. Chlorophyll *a* concentration fluctuated irregularly, with means of 4.7 and 3.0 µg/L at the surface and on the bottom, respectively.

Copepod Abundance, Biomass, and Production Rate

The regional variation in copepod abundance within the Honjo District was not large; the coefficient of variation around the means for the seven stations was usually < 50%. The abundance was annually bimodal, with a large peak in the fall and a small peak in early summer, although this pattern was less apparent in the second half of the study period (Fig. 7.9). It varied, from 114.3×10^4 ind/m^3 in November 1997 to 2.37×10^4

Table 7.1. Relationships between specific growth rate and temperature used for the calculation of production rate of copepods in the Honjo District

Species	Regression equation	References
Acartia hudsonica[a]	$g = 0.059 \exp(0.084T)$	Uye (1980)
Acartia sinjiensis	$g = 0.022 \exp(0.032T)$	Uye (unpublished data)
Oithona davisae	$g = 0.025 \exp(0.11T)$	Uye and Sano (unpublished data)
Paracalanus sp.	$g = 0.094 \exp(0.067T)$	Uye (1991)
Pseudodiaptomus inopinus[b]	$g = 0.033 \exp(0.11T)$	Uye et al. (1983)
Sinocalanus tenellus	$g = 0.043 \exp(0.12T)$	Kimoto et al. (1986)
Others	$g = 0.078 \exp(0.062T)$	Uye (unpublished data)

Note: Growth rate is calculated per day; temperature is in degrees Centigrade.
[a]Assumed to be the same as *A. omorii*.
[b]Assumed to be the same as *P. marinus*.

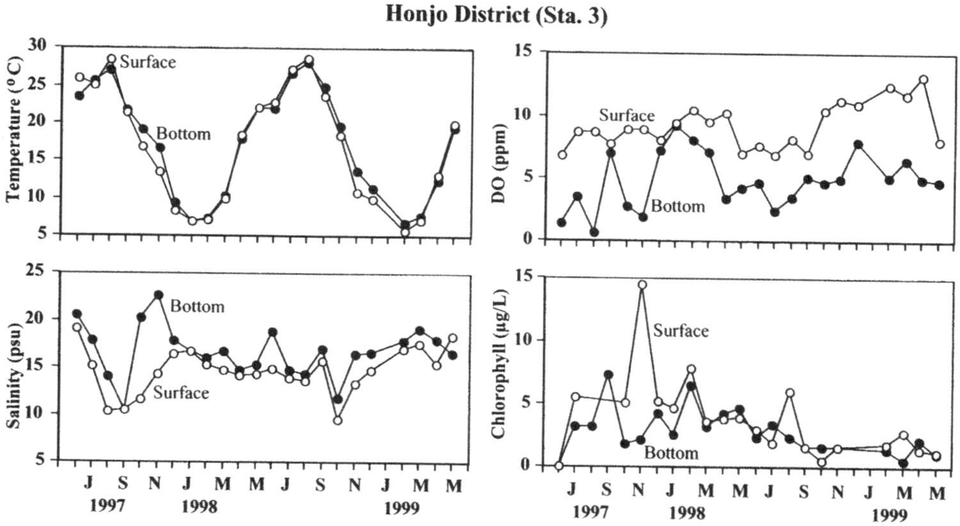

Figure 7.8. Monthly variations in temperature (°C), salinity (psu), dissolved oxygen concentration (ppm), and chlorophyll *a* concentration (µg/L) at Station 3 in the Honjo District from June 1997 to May 1999. DO, dissolved oxygen.

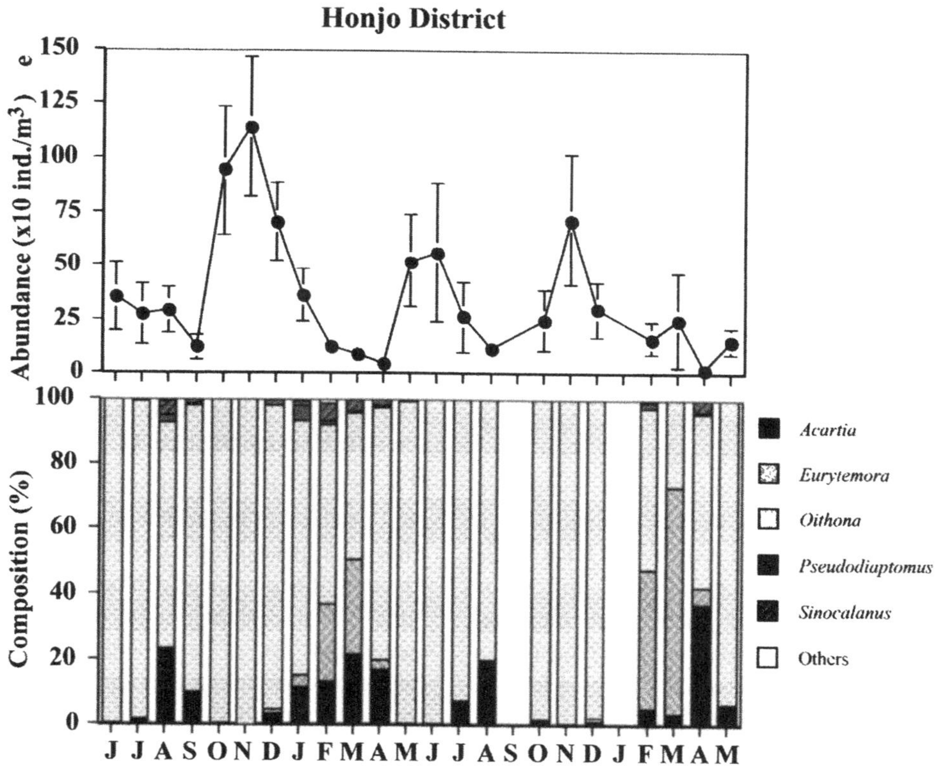

Figure 7.9. Monthly variations in mean abundance in number of individuals per cubic meter (ind/m^3) and taxonomic composition of copepods in the Honjo District, June 1997–May 1999.

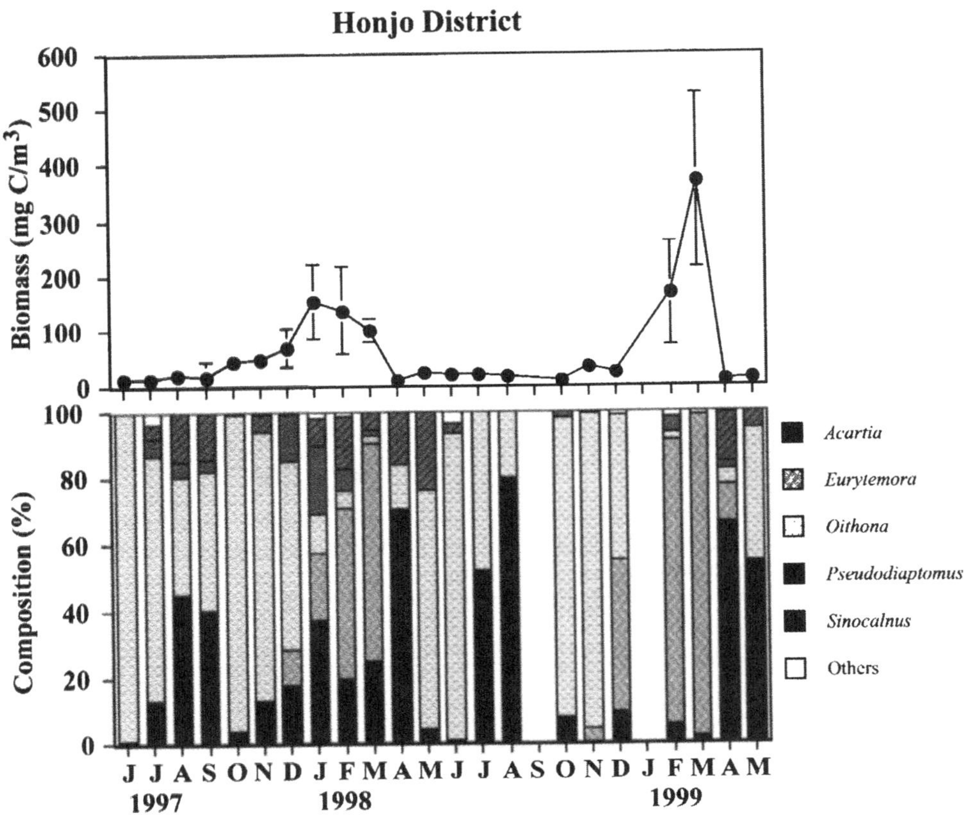

Figure 7.10. Monthly variations in mean biomass in milligrams of C per cubic meter and taxonomic composition of copepods in the Honjo District, June 1997–May 1999.

ind/m^3 in April 1999, with an overall mean of 35.0×10^4 ind/m^3. Copepods represented 94.4% of the mesozooplankton. A small cyclopoid copepod *O. davisae* dominated, except during winter, when *E. pacifica* and *A. hudsonica* were more abundant. In summer, another *Acartia* species (*A. sinjiensis*) occurred. *Sinocalanus tenellus* and *P. inopinus* occurred rather irregularly.

Unlike the seasonal variation in abundance, copepod C biomass showed an annual peak in winter (Fig. 7.10). It changed from 11.1 to 369.1 mg C/m^3, with a mean of 59.2 mg C/m^3. Copepods accounted for 83.4% of the mesozooplankton community biomass. The contribution of *O. davisae* to total C biomass became less important because of its small body size. In winter, *A. hudsonica* and *E. pacifica* accounted for > 50% of the mesozooplankton biomass. A large biomass peak (around 95%) in March 1999 was attributed primarily to *E. pacifica*.

Due to the positive effect of temperature on the specific growth rate, the production rate of copepods was computed to be higher despite the lower biomass in the summer (Fig. 7.11). It ranged from 2.0 to 44.7 mg C/m^3/d, with a mean of 9.63 mg C/m^3/d.

How Many Fish Seeds Can be Produced in the Honjo District?

The extremely high mesozooplankton biomass and production in the Honjo District led the author to propose using it as food for mass seed production of economically important finfishes (Uye et al. 2004). In this study, the quantity of fish seeds that could be produced is recalculated, based on copepod production. The extensive seed production method, namely, stocking fish larvae at low densities in large, outdoor earthen ponds (Ohno 1991), is inappropriate for marine fishes in this case, due to the low salinity found in the Honjo District. Thus, the author proposes an intensive system, which requires the construction of

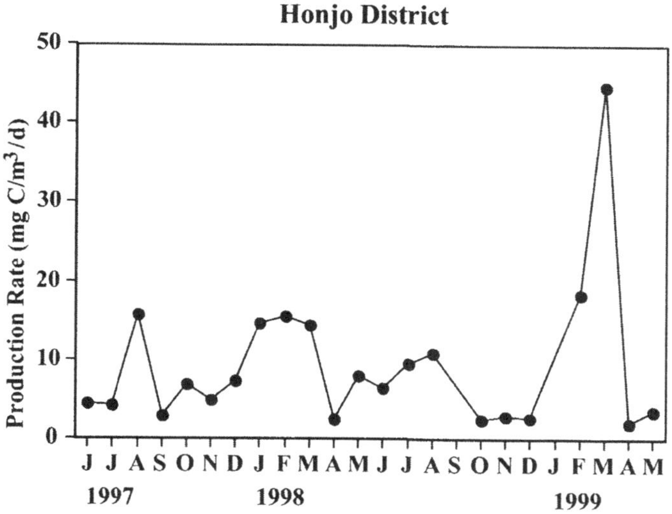

Figure 7.11. Monthly variations in mean production rate in milligrams of C per cubic meter per day of copepods in the Honjo District, June 1997–May 1999.

a hatchery, on the bank between Honjo District and the outer area connecting the Sakai Strait (salinity: > 25 psu), with indoor concrete tanks, as adopted by many sea-farming stations in Japan. This plan would also require facilities to collect large amounts of copepods, using light traps or large streamer-like plankton nets, and deliver the copepods by pumping them to each rearing tank on land.

Red sea bream, the most common fish species for intensive seed production in Japan, ingests copepods during larval stages from 10 to 40 mm body length (Tanaka et al. 1987). According to Kitajima (Suisan Shigen Hogokyoukai 1979), a larval red sea bream of 30 mm body length (about 390 mg wet weight), the target size in hatchery rearing, ingests approximately 250 mg wet weight of copepods daily. Using a wet weight to C conversion of 0.1 for copepods (Hirota 1981), the daily food requirement is 25 mg C per fish. The author assumes that the overall mean biomass and production rate of copepods (59.2 mg C/m^3 and 9.63 mg C/m^3/d, respectively) is attainable in the Honjo District (water volume: 8.25×10^7 m^3) throughout the seed production period (from May to July for red sea bream). If 10% of the copepod production (about 2% of biomass) were exploitable daily, the harvest (79.5 kg wet weight) would be equivalent to the diet for 3.2 million 30-mm-long seeds. A similar procedure may also be applicable to Japanese flounder, whose seed-production season largely overlaps that of red sea bream (Takashima 1997).

Ayu (*Plecoglossus altivelis*) is also a species recommended for seed production. This diadromous fish spawns eggs in late fall in rivers, and the hatched larvae are driven downstream to coastal marine waters to feed mainly on copepods until they migrate upstream in spring and early summer. The seed production season is from December to February. A larval ayu of 50 mm body length (about 800 mg wet weight), the target seed size, consumes 100 mg of wet weight diet per day (Takashima 1997). Using the method described above, it is estimated that at least 8.0 million seeds of ayu can be produced.

The numbers given above for seed production of red sea bream and ayu are apparently underestimates, since the estimations are based on the production of a single fish cohort, which grows rather synchronously and requires the most abundant food just before the target size. Production using several different cohorts at the same time, however, can enhance the total number of seeds produced. In addition to live copepods, copepods that have been collected during nonproduction seasons and frozen can also be used as supplementary food. Previous studies (Miyadi 1962;

Table 7.2. Comparison of annual mean biomass and production rate of mesozooplankton among various estuarine and coastal marine waters

Location	Mean biomass (mg C/m³)	Production rate (mg C/m³/d)	Reference
Fukuyama Harbor, Japan	39.1	6.85	Uye and Liang 1998[a]
Inland Sea of Japan, Japan	20.0	2.83	Uye and Shimazu 1997
Honjo District, Japan	47.4	—	Uye et al. 2000
	59.2	9.63	Current study[a]
Lake Nakaumi, Japan	37.3	—	Uye et al. 2000
Lake Shinji, Japan	80.2	—	Uye et al. 2000
Long Island Estuary, USA	33.2	—	Turner et al. 1983
Narragansett Bay, USA	28.0	—	Hulsizer 1976
	—	27.3–32.1	Durbin and Durbin 1981[b]
Jakle's Lagoon, USA	—	9.0	Landry 1978[c]
Newport Estuary, USA	21.5	—	Fulton 1984[a]
Nueces Estuary, USA	16.8	—	Buskey 1993
Westershelde Estuary, The Netherlands	35.5	4.4	Escaravage and Soetaert 1995

[a]For copepods only.
[b]For *Acartia hudsonica* and *A. tonsa* from March to December.
[c]For *Acartia clausii*.

Uye et al. 2000) demonstrated that the brackish-water system of Lakes Shinji and Nakaumi supports a high standing stock of mesozooplankton (Table 7.2). The major reasons for higher zooplankton biomass in this system might be a larger food supply and fewer occurrences of planktivorous fish (*Harengus zunasi*, *Englauris japonica*, and *Hemiramphus sajori*). Larvae and juveniles of these fish species migrate from Miho Bay, in the Sea of Japan, in spring/early summer and dwell in this food-rich brackish system until they migrate back in late fall (Miyadi 1962; Koshikawa 1999; Ishitobi et al. 2000), indicating that the predation pressure by fish is higher in warmer seasons than during the cold seasons.

Judging from the estimates above, the proposed plan to use the Honjo District as a field for mass seed production of economically important fishes seems to be highly feasible. This plan adds some new economical value to the Honjo District, where current operations are limited to set-net fisheries being operated inefficiently and offers an option for efficient utilization of this enclosed water body.

CONCLUSION

Currently, in Japanese fish hatcheries, copepods are rarely used as food, despite previous attempts and a growing body of knowledge about mass-culturing copepods (such as *T. japonicus* and *A. tsuensis*). Most hatcheries have set a goal of producing more larvae of economically important fish species (red sea bream, Japanese flounder, and ayu) using intensive larviculture systems that require fewer workers. Due to their heavy reliance on machinery for feeding larvae and cleaning tanks, these systems employ a much-reduced labor force. As a consequence, there are few workers available at hatcheries to culture copepods using large outdoor tanks. Despite the high cost of *Artemia* resting eggs, *Artemia* nauplii are used more in the hatcheries. Locally available live copepods are used occasionally, in addition to frozen copepods. As was described above, the mass culture of copepods, particularly calanoid copepods, requires intensive labor to control water quality and the food level in the culture tank and to minimize cannibalism among these types of copepods. It would appear to be difficult to incorporate mass culture of copepods into the current routine of hatchery operations in Japan.

The author proposes construction of a hatchery adjacent to the Honjo District, which supports a very large copepod biomass, and the use of freshly caught copepods as food for fish larvae. Natural copepods ingest a variety of food types and are thought to be nutritionally suitable for growth of fish larvae. Based on estimates provided above, more than 10 million fish larvae can be produced annually by exploiting approximately 2% of the copepod biomass in the Honjo District. Hence, this proposal seems to be both ecologically and economically feasible. In addition, high copepod biomass has been found not only in the Honjo District, but also throughout the entire Lake Shinji–Lake Nakaumi brackish-water system. The author further proposes establishing a venture business to harvest copepods commercially from these brackish-water lakes and to transport them to hatcheries, similar to the practices of some privately owned companies in Taiwan (C.-F. Liu personal communication, May 2003). To transport live copepods to remote hatcheries, a method of maintaining the viability of the copepods for extended periods (e.g., at least 24 hours) is needed. On the other hand, no particular innovation is needed for producing and transporting frozen copepods.

The simplified dietary sequence (e.g., *Brachionus-Artemia* nauplii-synthetic feed) applied currently in Japanese hatcheries may be responsible for the production of fish larvae with slightly different morphology, coloration, and behavior from the wild ones. These larvae are thought to be more susceptible to harsh feeding conditions and predators when they are released in the field. To improve upon these features, it is necessary to use more natural food. For the first-feeding fish larvae, mass-cultured *Brachionus* spp. may be irreplaceable. *Artemia* nauplii, however, may be replaced by copepods if the commercially harvested copepods are less expensive than *Artemia*. Brackish-water copepods are often highly productive (fast development, high fecundity, tolerant of adverse environmental conditions) and occur in enormously high biomass, as found in the Honjo District. More attention to the locally available brackish-water copepods and extensive examination of their biological characteristics are needed.

ACKNOWLEDGMENTS

The author would like to thank Dr. M. Aizaki of Shimane University and S. Nakai and H. Yoshida of Hiroshima University for assistance in the zooplankton study in the Honjo District. This study was partially supported by research grants from the Japan Society for the Promotion of Science (JSPS, No. 12 NP0201).

REFERENCES

Anraku, M. and Hirano, R. 1973. Plankton symposium: cultivation and mass culture of zooplankton. Objective of planning. *Bulletin of the Plankton Society of Japan* 20:1–2. (In Japanese with English abstract)

Buskey, E.J. 1993. Annual pattern of micro- and mesozooplankton abundance and biomass in a subtropical estuary. *Journal of Plankton Research* 15:907–924.

Doi, M., Khono, H., Ohno, A., and Taki, Y. 1994. Development of mixed-feeding state larvae of red snapper, *Lutjanus argentimaculatus*. *Suisanzoshoku* 42:471–476.

Durbin, A.G. and Durbin, E.G. 1981. Standing stock and estimated production rates of phytoplankton and zooplankton in Narragansett Bay, Rhode Island. *Estuaries* 4:24–41.

Escaravage, V. and Soetaert, K. 1995. Secondary production of the brackish copepod communities and their contribution to the carbon flues in the Westershelde estuary (The Netherlands). *Hydrobiologia* 311:103–114.

Fu, Y. and Hirayama, K. 1991. Genetic divergence between S and L type strains of the rotifer *Brachionus plicatilis*. *Journal of Experimental Marine Biology and Ecology* 151:43–56.

Fujita, S. 1973. Importance of zooplankton mass culture in producing marine fish seed for fish farming. *Bulletin of the Plankton Society of Japan* 20:49–53. (In Japanese with English abstract)

Fukuhara, O. 1989. A review of the culture of grouper in Japan. *Bulletin of Nansei Regional Fisheries Research Laboratory* 20:47–57.

Fukusho, K., Hara, O., Iwamoto, H., and Kitajima, C. 1977. Mass production of the copepod, *Tigriopus japonicus*, in combination with rotifer, *Brachionus plicalitis*, feeding baking yeast and using large-scale outdoor tanks. *Report of Nagasaki Prefectural Fisheries Experimental Station* 3:33–40. (In Japanese with English abstract)

Fukusho, K., Iwamoto, H., Seikai, T., and Kitajima, C. 1978. Effects of initial supply of *Chlorella* sp. on the copepod *Tigriopus japonicus* population in combination with rotifer *Brachionus plicatilis*, feeding bakers' yeast. *Report of Nagasaki Prefectural Fisheries Experimental Station* 4:47–56. (In Japanese with English abstract)

Fukusho, K. and Okauchi, M. 1984. Seasonal isolation between two strains of rotifer of *Brachionus plicatilis* in an eel-culture pond. *Bulletin of the Japanese Society of Scientific Fisheries* 50:909.

Fulton, R.S., III. 1984. Distribution and community structure of estuarine copepods. *Estuaries* 7:38–50.

Fushimi, T., Kitajima, C., and Ouchi, A. 1968. Studies on red sea bream seed production. I. Rearing from egg to larval stage. *Bulletin of Hiroshima Prefectural Fisheries Experimental Station* 1:37–48. (In Japanese)

Hada, A. and Uye, S. 1991. Cannibalistic feeding behavior of the brackish-water copepod *Sinocalanus tenellus*. *Journal of Plankton Research* 13:155–166.

Hirota, R. 1981. Dry weight and chemical composition of the important zooplankton in the Setonaikai (Inland Sea of Japan). *Bulletin of the Plankton Society of Japan* 28:19–24. (In Japanese with English abstract)

Hulsizer, E.E. 1976. Zooplankton of lower Narragansett Bay, 1972–1973. *Chesapeake Science* 17:260–270.

Ishitobi, Y., Hiratsuka, J., Kuwabara, H., and Yamamuro, M. 2000. Comparison of fish fauna in three areas of adjacent eutrophic estuarine lagoons with different salinities. *Journal of Marine Systems* 26:171–181.

Ito, T. 1959. Sudden change of water quality in brackish-water eel ponds by the rotifer *Brachionus plicatilis*. *Suisanzoshoku* 6:86–97. (In Japanese)

Ito, T. 1960. On the culture of mixohaline rotifer *Brachionus plicatilis* O.F. Muller in the sea water. *Report of Faculty of Fisheries, Prefectural University of Mie* 3:708–740. (In Japanese with English abstract)

Ito, T. 1970. The biology of a harpacticoid copepod, *Tigriopus japonicus* Mori. *Journal of Faculty of Science, Hokkaido University, Series VI* 17:474–500.

Iwasaki, H. and Kamiya, S. 1977. Cultivation of marine copepod, *Pseudodiaptomus marinus* Sato. *Bulletin of the Plankton Society of Japan* 24:44–54. (In Japanese with English abstract)

Iwasaki, H., Katoh, H., and Fujiyama, T. 1977. Cultivation of marine copepod, *Acartia clausi* Giesbrecht. I. Factors affecting the generation time and egg production. *Bulletin of the Plankton Society of Japan* 24: 55–61. (In Japanese with English abstract)

Japan Sea-Farming Association. 2002. A Guidebook for Mass Culture of *Brachionus*. Tokyo, Japan: Japan Sea-Farming Association. 137 pp.

Kajiyama, E. and Nishioka, U. 1930. On the successful rearing of sea bream larvae. *Suikenshi* 25:35–40. (In Japanese)

Kimoto, K., Uye, S., and Onbe, T. 1986. Growth characteristics of a brackish-water copepod *Sinocalanus tenellus* in relation to temperature and salinity. *Bulletin of the Plankton Society of Japan* 33:43–57.

Kitajima, C. 1973. Experimental trials on mass culture of copepods. *Bulletin of the Plankton Society of Japan* 20:54–60. (In Japanese with English abstract)

Koga, F. 1970. On the life history of *Tigriopus japonicus* Mori (Copepoda). *Journal of the Oceanographic Society of Japan* 26:12–21. (In Japanese with English abstract)

Koshikawa, T. 1999. A study of fish of Honjo Area in Lake Nakaumi: Data analysis of wholesale of fish. *Laguna* 6:157–164. (In Japanese with English abstract)

Landry, M. R. 1978. Population dynamics and production of a planktonic copepod, *Acartia clausii*, in a small temperate lagoon on San Juan Island, Washington. *Internationale Revue der Gesumten Hydrobiologie* 63:77–119.

Liang, T., Uye, S., and Onbe, T. 1994. Production and loss of eggs in the calanoid copepod *Centropages abdominalis* Sato in Fukuyama Harbor, the Inland Sea of Japan. *Bulletin of the Plankton Society of Japan* 41:131–142.

Maruyama, K., Tsumura, S., and Morioka, T. 1986. Test of healthiness of larval red sea bream (*Pagrus major*). I. Comparison between larvae from extensive culture and intensive culture. *Saibaigiken* 15:157–167. (In Japanese)

Matsutani, K. 1961. Studies on the heat resistance of *Tigriopus japonicus*. *Publications of Seto Marine Biological Laboratory* 9:379–411.

Miyadi, D. 1962. Report on the ecological study for fishes in relation to the Nakaumi reclamation and freshening project. Shimane Prefecture, Japan. 226 pp. (In Japanese)

Nihon Suisanshigen Hogokyoukai. 1979. *Mass Culture of Zooplankton as Food for Fish Larvae*. Tokyo, Japan: Nihon Suisanshigen Hogokyoukai. 142 pp. (In Japanese)

Odum, E.P. 1971. *Fundamentals of Ecology*. Philadelphia, Pennsylvania, USA: W.B. Saunders. 546 pp.

Ohno, A. 1991. Fundamental study on the extensive seed production of the red sea bream, *Pagrus major*. Special Research Report No. 2, Japan Sea-farming Association, Tokyo, Japan, 110 pp. (In Japanese with English abstract)

Omori, M. 1973. Cultivation of marine copepods. *Bulletin of the Plankton Society of Japan* 20:3–11. (In Japanese with English abstract)

Takashima, T. 1997. *Marine Life Farming*. Tokyo, Japan: Seizando. 258 pp. (In Japanese)

Takahashi, T. and Ohno, A. 1996. The temperature effect on the development of calanoid copepod, *Acartia tsuensis*, with some comments on morphogenesis. *Journal of Oceanography* 52:125–137.

Tanaka, M., Ueda, H., and Sudo, H. 1987. Significance of near-bottom copepod aggregation as food resources for the juvenile red sea bream in Shijiki Bay. *Fisheries Science* 55:1545–1552.

Tokioka, T. and Suzuki, T. 1939. A glimpse upon the biology of *Tigriopus japonicus* Mori. *Physiology and Ecology* 6:49–54. (In Japanese with English abstract)

Toledo, J.D., Golez, M.S., Doi, M., and Ohno, A. 1999. Use of copepod nauplii during early feeding stage of grouper *Epinephelus coioides*. *Fisheries Science* 65:390–397.

Turner, J.T., Bruno, S.F., Larson, R.J., Staker, R.D. and Sharma, G.M. 1983. Seasonality of plankton assemblages in a temperate estuary. *Marine Ecology (Berlin)* 4:81–99.

Uye, S. 1980. Development of neritic copepods *Acartia clausi* and *A. steueri*. I. Some environmental factors affecting egg development and the nature of resting eggs. *Bulletin of the Plankton Society of Japan* 27:1–9.

Uye, S. 1982. Length-weight relationships of important zooplankton from the Inland Sea of Japan. *Journal of the Oceanographic Society of Japan* 38:149–158.

Uye, S. 1991. Temperature-dependent development and growth of the planktonic copepod *Paracalanus* sp. in the laboratory. *Bulletin of the Plankton Society of Japan Special Volume*:627–636.

Uye, S., Iwai, Y., and Kasahara, S. 1983. Growth and production of the inshore marine copepod *Pseudodiaptomus marinus* in the central part of the Inland Sea of Japan. *Marine Biology* 73:91–98.

Uye, S. and Liang, D. 1998. Copepods attain high abundance, biomass and production in the absence of large predators but suffer cannibalistic loss. *Journal of Marine Systems* 15:495–501.

Uye, S., Nakai, S., and Aizaki, M. 2004. Potential use of extremely high biomass and production of copepods in an enclosed brackish water body in Lake Nakaumi, Japan, for mass seed production of fishes. *Zoological Studies*. 43:165–172.

Uye, S. and Shimazu, T. 1997. Geographical and seasonal variations in abundance, biomass and estimated production rates of meso- and macrozooplankton in the Inland Sea of Japan. *Journal of Oceanography* 53:529–538.

Uye, S., Shimazu, T., Yamamuro, M., Ishitobi, Y., and Kamiya, H. 2000. Geographical and seasonal variations in mesozooplankton abundance and biomass in relation to environmental parameters in Lake Shinji-Ohashi River-Lake Nakaumi brackish-water system, Japan. *Journal of Marine Systems* 26:193–207.

Watanabe, T., Kitajima, C., and Fujita, S. 1983. Nutritional value of live organisms used in Japan for mass propagation of fish: a review. *Aquaculture* 34:115–143.

Yamashita, K. and Arakawa, T. 1974. Zooplankton harvest by a light trap and their frozen storage. Report of Aquaculture-II (Nagasaki Prefectural Fisheries Experimental Station), pp. 76–78. (In Japanese, cited in Nihon Suisanshigen Hogokyoukai, 1979)

8

Behavioral Characteristics of Copepods That Affect Their Suitability as Food for Larval Fishes

Edward J. Buskey

ABSTRACT

Most fish larvae have evolved to feed selectively on complex natural assemblages of microzooplankton prey, and motion patterns are an important component in the recognition of particles as potential food items. This is why live (swimming) prey are often critical to the survival of larval fish. In response to fish predation many zooplankton species have also evolved adaptations to make them less visible to potential predators, including increased transparency and intermittent locomotion. In addition, calanoid copepods have evolved some of the most effective escape behaviors found among aquatic organisms. Some copepods can respond rapidly (< 3 milliseconds) to the slight hydrodynamic disturbance caused by an approaching predator, rapidly accelerating to speeds of up to 800 body lengths per second.

Part of the challenge in choosing copepod species for aquaculture applications is to select species that are easier for larval fish to recognize and that have less effective adaptations against small visual predators. Since copepods are preyed on by a wide array of predators ranging from protozoa to whales, they have evolved a wide variety of behaviors that may make some species less vulnerable to one type of predator but more vulnerable to others. In addition, environmental conditions, including light and turbulence, may modify the effectiveness of some of these adaptations. For example, many copepod species are difficult to see in light with a natural angular distribution (low image contrast), but become more easily visible under collimated illumination. In addition, intermediate levels of turbulence not only increase encounter rates between predator and prey, but also inhibit escape responses of copepods through production of background hydrodynamic noise, which masks the approach of the predator, and through habituation of the copepods to small hydrodynamic disturbances.

INTRODUCTION

The science of rearing larval marine fish in captivity would be greatly simplified if marine fish larvae would readily consume artificial diets; the only challenge would be determining the nutritional requirements and correct size of particles for optimal growth of the species in question. Most fish larvae, however, have evolved to feed selectively on complex natural assemblages of zooplankton prey, and size, visibility, and motion patterns are important components in the recognition of particles as potential food items (Buskey et al. 1993). As a consequence, live (swimming) prey are often critical to the survival of larval fish (Watanabe et al. 1983). Many zooplankton species, however, have also evolved adaptations to avoid predation. Therefore, to successfully culture larval and juvenile fish, it is important to consider the factors affecting this predator–prey interaction and to choose conditions and prey species that favor the larval fish predator and maximize growth and survival of cultured species.

Artemia nauplii, which can be conveniently hatched from easily stored cysts, have been extensively used in larval fish culture (Leger et al.

1986). Due to the large size of *Artemia* nauplii, some fish species are first fed the easily cultured rotifer *Brachionus plicatilis* followed by *Artemia* nauplii. Supplementing *Artemia* diets with wild-collected zooplankton has led to better survival and growth of larval fish (e.g., Witt et al. 1984). These results have led to doubts regarding the nutritional suitability of *Artemia* for larval fish, especially in terms of its fatty acid composition (Watanabe et al. 1983), and raised interest in copepods as a potential alternative food for rearing larval fish. In nature, copepods are a major dietary component for marine fish larvae (May 1970; McMichel and Peten 1989). They have also been used in aquaculture as food for marine fish larvae (Houde 1973) and have proven a superior food source for rearing fish and crustaceans in intensive culture (Watanabe et al. 1983; Støttrup and Norsker 1997; McEvoy et al. 1998). The great number of species and sizes potentially available for aquaculture (Delbare et al. 1996) is further evidence of their suitability as a food source.

Cultivation of planktonic copepod species, mainly calanoids, has generally been limited in practical application by the low densities achieved in culture (Støttrup and Norsker 1997). Although species that protect their eggs by carrying them in clusters may experience lower mortality (Kiørboe and Sabatini 1994), in general, copepod species that free-spawn their eggs have higher fecundity than species in which females carry their eggs in clusters (Mauchline 1998). Overcrowding can decrease fecundity in some species (Miralto et al. 1996) or result in cannibalism of nauplii (Lonsdale et al. 1979; Ohno et al. 1990). While harpacticoid copepods have been cultured in high densities, their benthic nature may make them less readily available to fish larvae. The nauplii of some harpacticoids, however, exhibit positive phototaxis, compared to the negative phototaxis of adults, making them easy to harvest from the surface layers of a culture (Strøttup and Norsker 1997). Recent studies with the swarm-forming cyclopoid copepod *Dioithona oculata* suggest that this species may be a good candidate for high-density culture (Hernandez Molejon and Alvarez-Lajonchere 2003). This species lives in swarms that can reach densities of over 50,000 copepods/L (Buskey et al. 1996).

While the major factors affecting the decision of which copepod species to culture for aquaculture applications have rightfully centered around their ease of culture and potential for mass culture, it is also worthwhile to consider the behavioral characteristics of various copepod species and how these characteristics may affect their availability as prey to larval fish. A wide range of predators feed on copepods, ranging from protozoa to whales (Greene 1985), and the adaptations that help protect copepods from one type of predator might be ineffective against other groups of predators. Copepod predators can be classified in several ways. There are active predators, such as fish that seek out their prey, and passive predators, such as corals or jellyfish that rely on copepods swimming into tentacles or being carried there by currents. Active predators can also be classified by the dominant sensory systems used to locate prey, making for a natural division between visual and nonvisual predators. Most planktivorous fish are visual predators (Hunter 1981; Blaxter 1991) that feed on individual prey by rapidly sucking in the water surrounding them (Lauder 1983), although a few species of planktivorous fish may feed primarily by filtering plankton from the water (Durbin and Durbin 1975); some species are capable of both feeding modes (e.g., O'Connell 1972) or switch between them during ontogeny (e.g., Drenner et al. 1982a). Visual predators typically feed selectively on the largest and most conspicuous zooplankton species, based on body size, pigmentation, and prey motion (Brooks and Dodson 1965; Zaret and Kerfoot 1975; Zaret 1980). For filter-feeding fish, it might be expected that these planktivores would randomly sample their food environment within a particular size range determined by the morphology of their feeding structures. Prey avoidance, however, can significantly alter prey selection in filter-feeding fish, causing them to feed more successfully on smaller, less evasive prey (Drenner and McComas 1980; Drenner et al. 1982b).

Fish are often not the dominant predators on zooplankton (Cushing 1983; Bollens 1988; Dagg and Govoni 1996). Therefore, invertebrate predators may be a major selective force in the evolution of antipredator adaptations, and their impact on copepod populations must be considered at least briefly. Under some circumstances invertebrates may be the dominant predators on zooplankton in marine environments (e.g., Ohman 1986; Behrends and Schneider 1995) and may therefore play an important role in the evolution of antipredator adaptations. Invertebrate predators on zooplankton can be distinguished as either cruising or ambush predators (Greene 1985). Cruising predators tend to prey upon slow-moving prey, while ambush predators, which rely on the swimming behavior of the prey to initiate encounters (Gerritsen 1978), tend to favor prey with higher swimming speeds. Invertebrate predators can also be characterized according to the means by which they capture their prey (i.e., raptorial versus entangling predators). Although some raptorial invertebrate planktivores rely upon vision to locate prey (e.g., crustaceans with compound eyes and good visual acuity), many rely upon mechanoreception (e.g., chaetognaths [Newbury 1972] and predatory copepods [Strickler 1975]). Thus the encounter radius for the predator can be defined by its optical or mechanosensory range, resulting in selection for more visually or hydrodynamically (respectively) conspicuous prey. The encounter radius for entangling predators, on the other hand, is defined by the volume occupied by tentacles (Madin 1988). In those systems where they are the dominant predators on zooplankton (e.g., Kremer 1979; Deason 1982; Purcell 1992), entangling predators may favor prey with higher swimming speeds and therefore higher encounter rates with predators.

Zooplankton that have evolved in extreme or temporary habitats are often exposed to fewer predators, and, as a consequence, may have evolved fewer adaptations to avoid predation. In addition, they may also be easier to culture because of their adaptations to extreme or changing environments. For example, the brine shrimp *Artemia salina* is widely used as live food in large part because it is easily raised from dormant cysts, has broad tolerance to varying temperature and salinity conditions, and has the added advantage of being a highly visible prey with virtually no escape behavior (Buskey et al. 1993). Similarly, tide pool copepods, such as the harpacticoid *Tigriopus californicus,* are often easier to culture than many coastal marine species (Fukusho 1980), in part because of their tolerance to highly fluctuating tide pool environments. Cyclopoid copepods of the genus *Apocyclops*, which are characteristic of coastal saline pools and ponds (Reid et al. 2002), are also extremely easy to culture (Buskey unpublished data) and may be of value in aquaculture (James and Al-Khars 1986).

To understand the characteristics that make a copepod species suitable for larval fish culture, it is advantageous to consider the biology and ecology of the predator–prey interactions between larval and juvenile fish and copepods. A convenient framework for considering this interaction is to consider the four primary components of a predator–prey interaction (encounter, attack, capture, and ingestion) as suggested by Holling (1959) (see, for example, Gerritsen and Strickler 1977; Greene 1983; Ohman 1988). By studying each of these components and the biological and environmental factors that affect them, a better understanding of the overall predator–prey interaction may be gained.

FACTORS AFFECTING ENCOUNTER PROBABILITY

For a predator–prey interaction to occur, the predator must encounter the prey either through direct contact with feeding structures or by remotely detecting its prey using one or more sensory systems. Gerritsen and Strickler (1977) modeled the factors affecting predator–prey encounter rates in aquatic systems, including the speed of the predator, the speed of the prey, and the distance at which a predator is able to detect the presence of prey (encounter radius). In its simplest form this model assumes a random distribution of prey, random swimming direction, and no water motion. Of the variables used in this model (population density, speeds of predator and prey, encounter radius), encounter radius has the greatest effect on encounter probabilities. In a later study examining the effects of swimming direction, it was determined that cruising predators can maximize their encounter rates by swimming orthogonally to the predominant prey direction, while prey can minimize their encounter rates by swimming parallel to predators (Gerritsen 1980). This encounter model was later modified to include the effects of turbulent mixing on encounter rates (Rothschild and Osborn 1988).

In calculation of encounter rates, predator and prey movement speeds are theoretically of equal importance. Larval fish are primarily cruising raptorial feeders whose routine swimming speeds are often in the range of one to three body lengths per second (Miller et al. 1988). Their copepod prey have similar swimming speeds (in terms of body length), ranging from less than one to about five body lengths per second (summarized in Mauchline 1998). Since copepods are generally much smaller than their fish predators, their absolute swimming speeds are generally much lower than those of their predators, and prey swimming speeds have relatively little effect on encounter rates. This is especially true for early naupliar stages of copepods, which (due to their small size) would be the logical choice as food for larval fish in aquaculture applications. For copepods there is a general pattern of increased swimming speed with both body size and developmental stage (Fig. 8.1). Early naupliar stages of copepods are often nonfeeding (Gauld 1959; Marshall 1973); in some genera (mainly deep sea or predatory species), none of the naupliar stages feed (Sekiguchi 1974). These nonfeeding developmental stages tend to conserve energy reserves and make themselves less conspicuous to predators by swimming less vigorously than feeding nauplii; very low swimming speeds have generally been found among early naupliar stages of copepods (Gerritsen 1978).

Another factor that may affect encounter frequencies is an uneven, heterogeneous distribution of prey. Early studies of larval fish feeding emphasized the concept that certain minimum thresholds of food concentration were necessary for the survival of larval fish and that these often occurred in spatially limited, high-density patches, such as the chlorophyll maximum layer (Lasker 1975). Taken to the extreme, however, exceptionally high prey densities, such as copepod swarms, may have a negative impact on visual predators such as larval fish. Dense swarms of copepods have been reported from a wide range of marine habitats, including coastal bays, coral reefs, and among mangrove prop roots (Emery 1968; Ueda et al. 1983; Ambler et al. 1991). Even *Artemia*, an important food item for rearing larval fish, is known to form swarms within culture tanks (Gulbrandsen 2001), which in turn affects ingestion rates of larval fish (Gulbrandsen 1996). The adaptive value of swarming behavior to copepods is thought to include enhanced mating opportunities, reduced dispersion by currents, and protection from predators (Hamner and Carlton 1979).

Although aggregated prey may initially be more conspicuous to visual predators (Vine 1973; Treisman 1975), especially within the confines of a tank for rearing larval fish, feeding on individ-

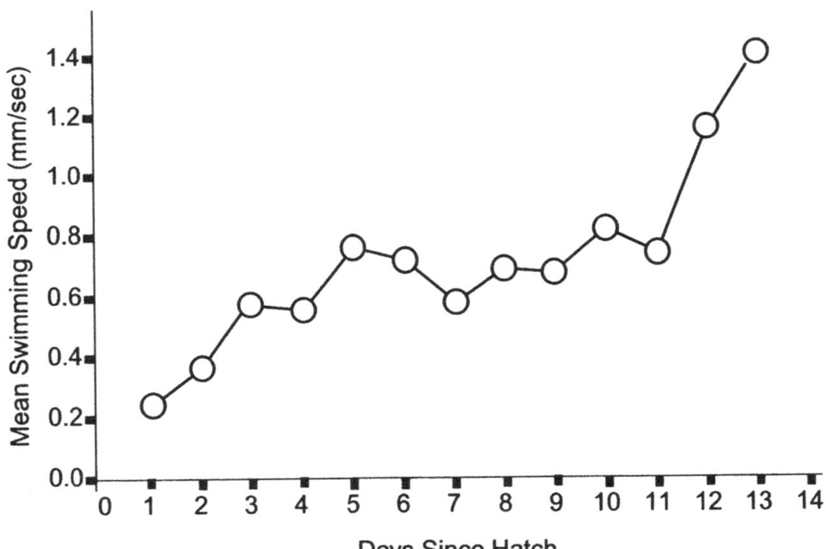

Figure 8.1. Changes in mean swimming speed over the developmental stages of the calanoid copepod *Acartia tonsa* from naupliar stage NI on day 1 through adult copepods on day 13. From Buskey (1994) with permission of Kluwer Academic Publishers.

ual prey within a swarm may be more difficult than feeding on isolated individuals because of the confusion effect caused by a great number of uniform moving targets (Neill and Cullen 1974). This confusion effect is enhanced by increasing the number of individuals in an aggregation (Neill and Cullen 1974), through high density within a swarm (Millinski 1977a, 1977b; Morgan and Ritz 1984), and by uniformity of individuals within a swarm, since visually conspicuous swarm members are most at risk (Ohguchi 1978). While swarming behavior may be an effective defense against visual planktivores, it may make them more vulnerable to nonvisual entangling predators such as cubozoan medusae that can use photobehavior to locate swarms (Buskey 2003).

Turbulence may also have an important effect on encounter rates between planktonic predators and their prey. Early attention was given to the role of vertical mixing in breaking down patchy areas of high food concentration, which in turn could lead to decreased feeding and survival of fish larvae (Lasker 1975). Rothschild and Osborn (1988) modeled the effects of turbulence on contact rates between planktonic organisms as an extension of the encounter rate model of Gerritsen and Strickler (1977). Mackenzie et al. (1994) modeled the effects of wind-induced turbulence on encounter rates between fish larvae and their prey and on capture success of prey by larval fish. They predicted a dome-shaped relationship for the effects of turbulence on feeding rate, with moderate levels of turbulence leading to the greatest enhancement of feeding rate. Field studies have confirmed the role of moderate turbulence in enhancing the feeding rates of larval fish (e.g., Sundby et al. 1994).

FACTORS AFFECTING ATTACK PROBABILITY

What characteristics of zooplankton prey cause them to be recognized as potential food for larval fish? Most predator–prey interactions in the plankton that involve remote detection of prey are mediated through vision or mechanoreception. There is relatively little evidence for chemoreception being involved in detection of zooplankton prey, although there is evidence for its use in mate location among copepods (Katona 1973). Larval and juvenile fish are primarily diurnal, visual predators (Hunter 1981), and feed selectively when presented with prey choices (e.g., Checkley 1982; Govoni et al. 1986), indicating that certain prey characteristics are more likely to induce attacks by larval fish.

Prey size has been widely recognized as an important factor in the feeding of planktivorous fish. Since Brooks and Dodson (1965) demonstrated body size–related predation by planktivorous fish, there has been a wealth of information published on this aspect of visual predation on zooplankton prey. In general, preferred prey organisms lie within a size range that varies with fish species and size, below which prey are too small either to be seen or to be energetically profitable to consume and above which there are physical limitations to prevent ingestion. For example, *Artemia* nauplii are too large for first-feeding larvae of some species to ingest, but small enough to escape detection by larger fish. Copepod nauplii occur in a wide range of sizes, with copepod species that carry eggs producing relatively larger eggs and nauplii than those species that free-spawn their eggs (Kiørboe and Sabatini 1994).

In addition to size, there is also evidence that both illumination and prey contrast can affect the feeding and growth of larval and juvenile fish (Vinyard and O'Brien 1976; Hinshaw 1985). Some zooplankton species are nearly transparent, and only the portions of their bodies with increased pigmentation may be visible (Zaret 1972; Zaret and Kerfoot 1975). The first, nonfeeding naupliar stages of copepods may have significantly lower image contrast than older feeding stages (Fig. 8.2). For zooplankton that live near the surface in clear waters, there may be a trade-off between increased risk of damage from ultraviolet light due to low pigmentation and higher risk of predation if protective pigments are produced (Hansson 2000). The distance at which prey are visible underwater depends on prey size, image contrast, ambient illumination, and the turbidity of the water (Eggers 1977; Aksnes and Giske 1993; Aksnes and Utne 1997). Because nearly transparent zooplankton have low image contrast when viewed against a bright background, planktivorous fish can enhance the contrast of their prey by searching for them at angles greater than 48.6 degrees from the vertical, which places their image just outside of Snell's window and makes them appear bright against a dark

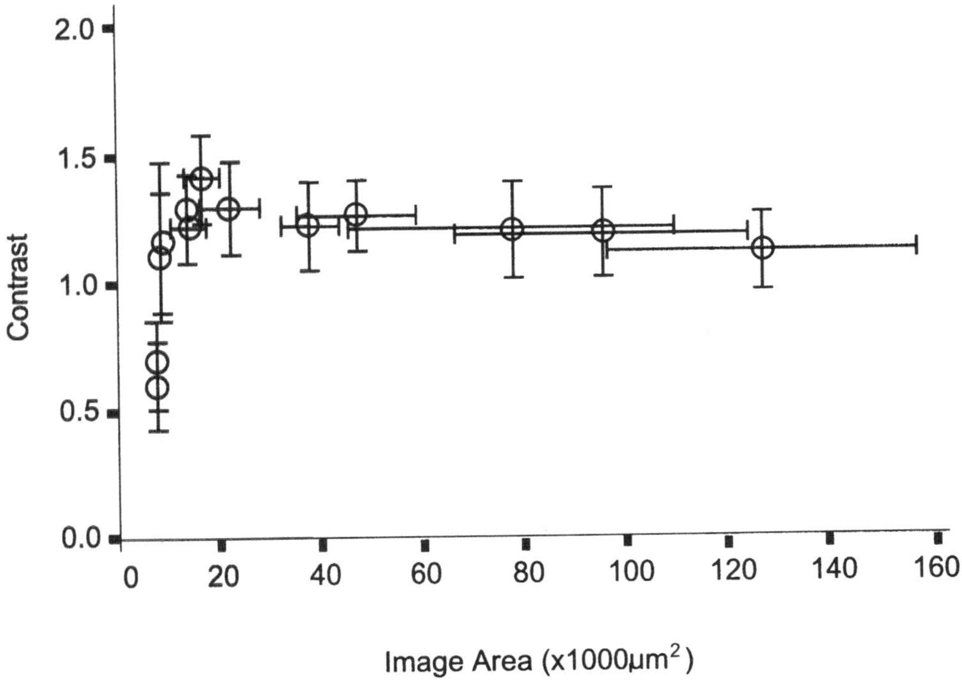

Figure 8.2. Changes in image area (visible size) and image contrast (image brightness minus background brightness, divided by background brightness) over developmental stages for the calanoid copepod *Acartia tonsa*, over 13 days of development from naupliar stage NI through adult. From Buskey (1994) with permission of Kluwer Academic Publishers.

background, similar to observing them under dark-field illumination with a microscope (Janssen 1981).

In coastal bays and estuaries, where the density of nonfood particles is very high, zooplankton prey may be difficult to recognize, based on size and image contrast alone. There is increasing evidence that prey recognition by planktivores may be strongly influenced by prey movement (Zaret 1980; Wright and O'Brien 1982). The steady, jerky swimming of *Artemia* nauplii appear to make it easily recognized as food by a wide range of larval fish. Most copepod nauplii exhibit intermittent locomotion (Buskey et al. 1993), with periods of jerky swimming interspersed with periods of quiescence (Fig. 8.3). This behavior pattern may not only make it more difficult for visual predators to keep track of their prey, but may also allow the nauplii to detect the approach of predators using mechanoreception during periods of no motion (Kramer and McLaughlin 2001). The activity levels of copepod nauplii differ greatly among species, and this may influence their conspicuousness as food for larval fish. For example, nauplii of *Parvocalanus crassirostris* hop with a frequency of about once per second, while nauplii of *Oithona plumifera* often go for periods of up to 10 seconds without moving (Buskey et al. 1993). In general, it would be expected that copepod nauplii with more continuous swimming behavior would be more easily recognized as prey by larval fish.

While there is considerable evidence that mechanoreception plays an important role in predation on zooplankton by invertebrate predators (e.g., Yen and Strickler 1996), the role of mechanoreception in predation by larval fish is less clear. There is evidence that some species of juvenile and adult fish feed raptorially on zooplankton in the dark, based on mechanoreception of hydrodynamic signals by the lateral line (Montgomery and Saunders 1985; Montgomery 1989; Janssen et al. 1995), as well as evidence that some species filter feed in the dark (Batty et al. 1986). There is also evidence that larval fish use their developing lateral line to detect zooplankton in the dark (Batty

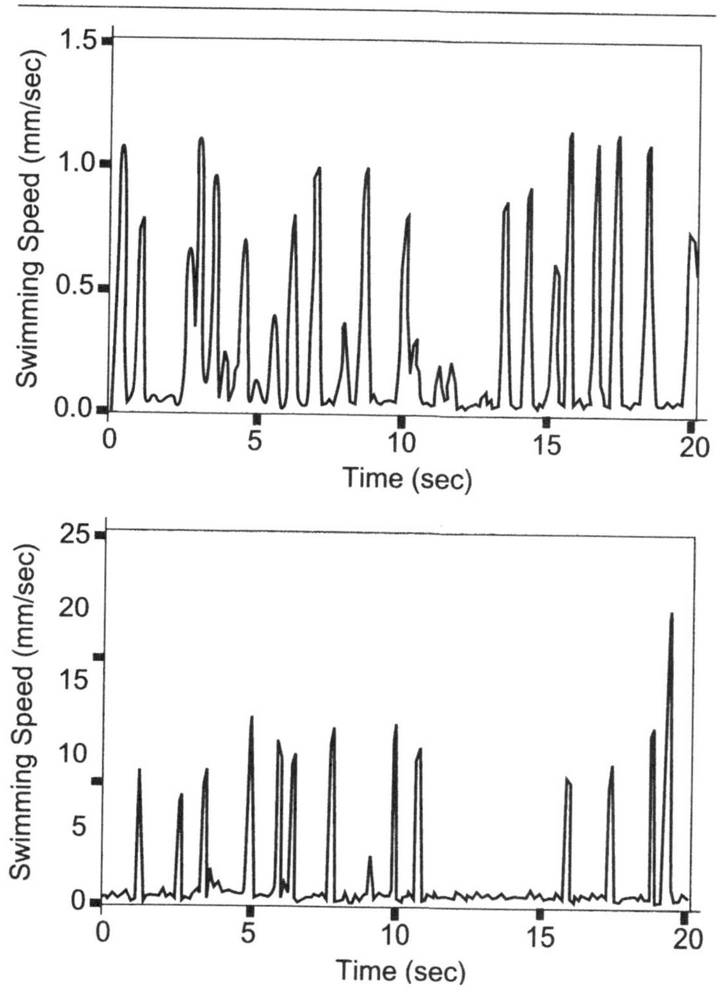

Figure 8.3. Representative records of swimming speed over time for the calanoid copepod *Acartia tonsa* NI nauplius (top panel) and adult female (bottom panel). Both stages exhibit a highly conspicuous, intermittent swimming pattern. From Buskey (1994) with permission of Kluwer Academic Publishers.

and Hoyt 1995; Salgado and Hoyt 1996), but the importance of this feeding mode and the extent to which it is used in various species of larval fish needs further study. Conditions that favor visual predation of fish larvae, along with prey possessing high visual conspicuousness, should lead to highest feeding rates in aquaculture applications.

FACTORS AFFECTING CAPTURE PROBABILITY

Once a copepod has been detected and attacked by a larval fish, the probability of capture is affected primarily by the escape behavior of the copepod. The vigorous escape behaviors of copepods represent important adaptations for avoiding predation and may play an important role in the selective feeding of planktivores (Drenner et al. 1978). These escape responses can be elicited by hydrodynamic (Hartline et al. 1999; Kiørboe et al. 1999; Lenz and Hartline 1999) and photic stimuli (Buskey et al. 1986; Buskey and Hartline 2003). It is generally accepted that water deformation is the primary effective stimulus to elicit an escape response by copepods (Haury et al. 1980; Yen et al. 1992; Kiørboe and Visser 1999). Rapid changes in light intensity (such as caused by a shadow during the day) or rapid increases in light intensity (such as the flash of a bioluminescent organism at night) trigger photophobic escape responses (Buskey et al. 1987). Chemosensory detection of predators by copepods may cause a reduction in swimming activity, perhaps making them less easily detected by planktivorous fish (Van Duren and Videler 1996), but chemoreception does not directly stimulate escape behavior.

Several factors may determine the effectiveness of copepod escape responses, including sensitivity to hydrodynamic stimuli, response latencies, and kinetic characteristics of the escape jump itself (Buskey et al. 2002). Several studies have attempted to determine hydrodynamic sensitivity thresholds (Viitasalo et al. 1998; Kiørboe and Visser 1999), but information on threshold sensitivity to hydrodynamic disturbances is only available for a small number of copepod species at present. Preliminary studies indicate that there are pronounced differences in hydrodynamic sensitivity among copepods (Fields and Yen 1997; Waggett and Buskey, unpublished data). There appears to be little change in sensitivity to hydrodynamic stimuli with increasing age and developmental state for nauplii and early copepodids of *Acartia tonsa*; the distance at which they respond to a standardized siphon flow is similar for all naupliar stages, although an increase in sensitivity is observed during the transition from the last naupliar stage (NVI) to the first copepodid stage (Fig. 8.4). Sensitivity thresholds are affected by the amount of ambient hydrodynamic noise in the environment, and escape responses may be reduced due to habituation in highly turbulent conditions (Hwang et al. 1994). Recent studies have demonstrated that planktivorous fish with low success capturing evasive prey (copepods) and high success at capturing nonevasive prey (*Artemia*) under still water conditions increase their success at capturing evasive prey and decrease their success at capturing nonevasive prey under turbulent conditions (Clarke et al., unpublished). This suggests that turbulence is either providing too much background noise for the copepod to distinguish the hydrodynamic signal associated with the approach of the predator or that the copepod has habituated to all small hydrodynamic fluctuations and does not respond to the predator's approach. Therefore, addition of moderate levels of turbulence to larval rearing tanks might increase the encounter rates

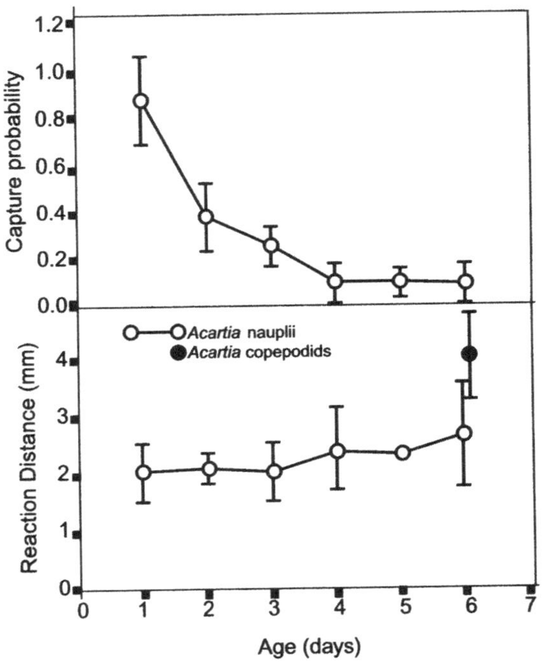

Figure 8.4. Changes with development in capture probability of *Acartia tonsa* naupliar stages by a siphon tube from NI nauplii on day 1 to NVI nauplii on day 6 (top panel), and mean reaction distance for nauplii exhibiting escape responses (bottom panel). Capture probability decreases with naupliar development, with little change in reaction distance, indicating that the strength of escape responses is increasing rapidly with development. From Buskey (1994) with permission of Kluwer Academic Publishers.

between predator and prey (MacKenzie et al. 1994; Sundby 1997) as well as reduce the ability of prey to detect their predators.

The time delay between the stimulus and behavioral response is a critical component of the escape response of an organism. Extremely short latencies (< 1 millisecond) to the first preparatory movement of an escape response to hydrodynamic stimuli have been demonstrated in copepods possessing myelinated nervous systems (Davis et al. 1999; Lenz and Hartline 1999; Lenz et al. 2000). These are among the fastest response times recorded for animals (Lenz et al. 2000) and are faster than even the Mauthner-mediated fast start reaction in adult zebra fish (5 milliseconds) (Eaton et al. 1977). Response latencies to photic stimuli are considerably longer; for adult *A. tonsa*, the minimum response latency to a hydrodynamic stimulus is 2 milliseconds (Buskey et al.

2002), while the minimum response latency to a photic stimulus is 30 milliseconds (Buskey and Hartline 2003). Visual response latencies for small fish are also approximately 30 milliseconds, providing a response time advantage to copepods that detect the approach of their predators with hydrodynamic signals over predators that detect their prey visually. Although fewer studies have focused on response latencies of early developmental stages of copepods, recent evidence indicates that response latencies to hydrodynamic stimuli are significantly longer for copepod nauplii than for adult stages. For example, the response latency to a hydrodynamic disturbance for a first naupliar stage *A. tonsa* averages about 27 milliseconds, compared to about 3.5 milliseconds for adults (Buskey et al. 2002; unpublished data).

In addition to quick response times, copepods also have extremely vigorous escape responses that are among the fastest of any aquatic organisms (Fig. 8.5). Maximum escape speeds of up to 840 mm/sec (approximately 800 body lengths/sec) have been recorded for the copepod *Acartia lilljeborgii*, with maximum accelerations exceeding 300 m/sec (Buskey et al. 2002). This compares with maximum-burst swimming speeds that can be as high as 400 mm/sec (approximately 80 body lengths/sec) for small larval fish (Williams et al. 1996). Copepod escape jumps are also characterized by a rapid initial change in direction of about 50 degrees (but ranging from 0 to 180 degrees;

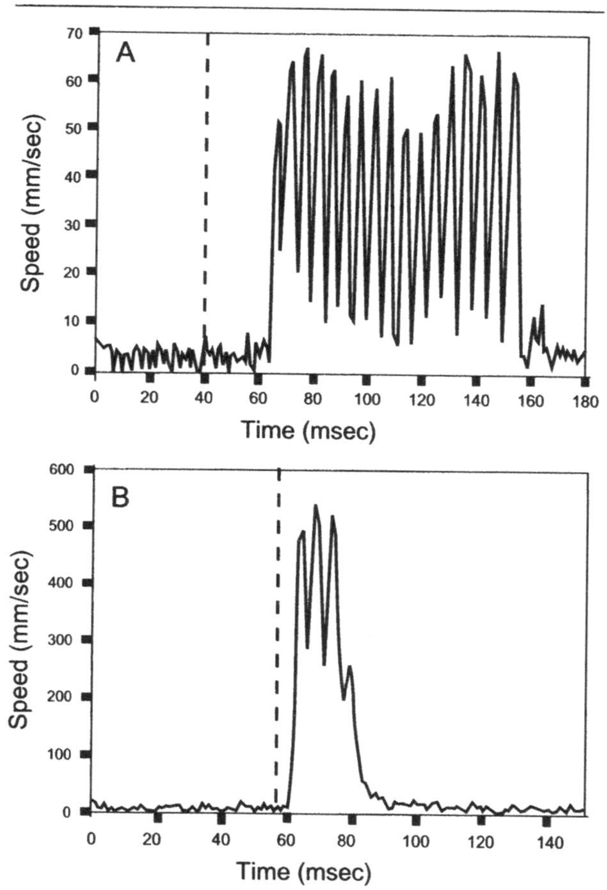

Figure 8.5. Representative records of swimming speed over time for escape responses of *A. tonsa* NI nauplius (top) and adult (bottom), based on high-speed video recordings at 1,000 frames per second. Vertical dashed line indicates onset of hydrodynamic stimulus. Bottom graph from Buskey et al. (2002) with permission of Inter-Research Science Publisher.

Buskey et al. 2002). This may make it very difficult for visual predators to anticipate the direction of the escape trajectory. Less is known about the detailed kinetics of escape responses of copepod nauplii, but they have effective escape responses to hydrodynamic stimuli (Yen and Fields 1992) and the effectiveness of their escape behavior increases with developmental stage, due mainly to an increase in the vigor of the escape response (Buskey 1994). While the maximum escape speeds of copepod nauplii are considerably slower than those found for adults (about 60 mm/sec for NI *A. tonsa* nauplii; Fig. 8.5), they are similar to adult speeds when scaled to their size (600 body lengths/sec for a 0.1-mm *A. tonsa* NI and 500 body lengths/sec for a 1-mm adult *A. tonsa*; Buskey et al. 2002). While the motion kinetics of escape behavior are well characterized for only a limited number of copepod species, as more species are studied, it is expected that a wide range of escape characteristics will be found among copepod species. For example, the swarm-forming copepod *D. oculata*, while a very active swimmer, seems to lack the capability for the vigorous escape response found in most planktonic copepod species (Buskey, unpublished data). When choosing copepod species for aquaculture, those with less effective escape behaviors might be favored.

FACTORS AFFECTING INGESTION

Once prey have been attacked and captured, there is still a small probability that the predator may reject the prey and release it. The reasons for prey rejection could include size (too big to ingest), morphological adaptations such as spines to deter predation, or chemical adaptations to make the prey toxic or unpalatable.

There are numerous examples of morphological adaptations to avoid predation among freshwater zooplankton, including spines on rotifers and helmets on cladocerans that are induced by the presence of predators (e.g., Stemberger and Gilbert 1987; Tollrian 1995). Although spination is a common characteristic of many marine planktonic organisms, spines have only rarely been shown to effectively deter predation by fish; for example, the prominent spines of crab zoea (Morgan 1989) and the setae of polychaete trochophores (Pennington and Chia 1984). Although planktonic copepods generally do not have sufficiently formidable spines to serve in an antipredator function, some harpacticoid copepods have stout, elongated caudal setae that cause planktivorous fish to reject them after capture (Buskey, unpublished data). There are no known reports of chemical defenses in copepods that would make them poisonous or distasteful to their predators. Chemical defenses have been found in marine invertebrate larvae (Lindquist and Hay 1996) and freshwater zooplankton (Kerfoot 1982).

CONCLUSION

While the ease of establishing highly productive cultures will be the main factor affecting the choice of copepod species to act as food for rearing of marine fish larvae, an awareness of the physical and behavioral characteristics of copepods that affect their ability to avoid predation, as well as environmental factors that reduce the effectiveness of antipredator characteristics, may lead to enhanced feeding rates of fish larvae and improved survival and growth. Copepod species from harsh or temporary environments may have wide environmental tolerances and lack sophisticated antipredator adaptations. Environmental conditions in larval rearing tanks may also be optimized in terms of lighting that provides high-contrast images of zooplankton and water circulation to provide enough turbulence to both increase encounter rates and deter escape responses of prey. These modifications may defeat some predator avoidance adaptations in copepods and enhance feeding rates of larvae.

ACKNOWLEDGMENTS

This research was funded by the National Science Foundation through grants OCE 9012161, OCE 9521375, OCE 9910608, and OCE 0324413. This is University of Texas Marine Science Institute Contribution Number 1301.

REFERENCES

Ambler, J.W., Ferrari, F.D., and Fornshell, J.A. 1991. Population structure and swarm formation of the copepod *Dioithona oculata* near mangrove cays. *Journal of Plankton Research* 13:1257–1272.

Aksnes, D.L. and Giske, J. 1993. A theoretical model of aquatic visual feeding. *Ecological Modeling* 67:233–250.

Asknes, D.L. and Utne, A.C.W. 1997. A revised model of visual range in fish. *Sarsia* 82:137–147.

Batty, R.S., Blaxter, J.H.S., and Libby, D.A. 1986. Herring (*Clupea harengus*) filter-feeding in the dark. *Marine Biology* 91:371–375.

Batty, R.S. and Hoyt R.D. 1995. The role of sense organs in the feeding behavior of juvenile sole and plaice. *Journal of Fish Biology* 47:931–939.

Behrends, G. and Schneider, G. 1995. Impact of *Aurelia aurita* medusae (Cnidaria, Scyphozoa) on the standing stock and community composition of mesozooplankton in the Kiel Bight (western Baltic Sea). *Marine Ecology Progress Series* 127:39–45.

Blaxter, J.H.S. 1991. Sensory systems and behavior of larval fish. In *Marine Biology: Its Accomplishment and Future Prospect,* edited by J. Mauchline and T. Nemoto, pp. 15–38. New York, New York, USA: Elsevier.

Bollens, S.M. 1988. A model of the predatory impact of larval marine fish on the population dynamics of their zooplankton prey. *Journal of Plankton Research* 10:887–906.

Brooks, J.L. and Dodson, S.I. 1965. Predation, body size and composition of plankton. *Science* 150:28–35.

Buskey, E.J. 1994. Factors affecting feeding selectivity of visual predators on the copepod *Acartia tonsa*: locomotion, visibility and escape responses. *Hydrobiologia* 292/293:447–453.

Buskey, E.J. 2003. Behavioral adaptations of the cubozoan medusa *Tripedalia cystophora* for feeding on copepod (*Dioithona oculata*) swarms. *Marine Biology* 142:225–232.

Buskey, E.J., Coulter, C.J., and Strom, S.L. 1993. Locomotory patterns of microzooplankton: potential effects on food selectivity of larval fish. *Bulletin of Marine Science* 53:29–43.

Buskey, E.J. and Hartline, D.K. 2003. High-speed video analysis of the escape responses of the copepod *Acartia tonsa* to shadows. *Biological Bulletin* 204:28–37.

Buskey, E.J., Lenz, P.H., and Hartline, D.K. 2002. Escape behavior of planktonic copepods in response to hydrodynamic disturbances: high speed video analysis. *Marine Ecology Progress Series* 235:135–146.

Buskey, E.J., Mann, C.G., and Swift, E. 1986. The shadow response of the estuarine copepod *Acartia tonsa. Journal of Experimental Marine Biology and Ecology* 103:65–75.

Buskey, E.J., Mann, C.G., and Swift, E. 1987. Photophobic responses in calanoid copepods: possible adaptive value. *Journal of Plankton Research* 9:857–870.

Buskey, E.J., Peterson, J.O., and Ambler, J.W. 1996. The swarming behavior of the copepod *Dioithona oculata*: *in situ* and laboratory studies. *Limnology and Oceanography* 41:513–521.

Checkley, D.M. 1982. Selective feeding by Atlantic herring (*Clupea harengus*) larvae on zooplankton in natural assemblages. *Marine Ecology Progress Series* 9:245–253.

Cushing, D.H. 1983. Are fish larvae too dilute to affect the density of their prey? *Journal of Plankton Research* 5:847–854.

Dagg, M.J. and Govoni, J.J. 1996. Is icthyoplankton predation an important source of copepod mortality in subtropical coastal waters? *Marine and Freshwater Research* 47:137–144.

Davis, A.D., Weatherby, T.M., Hartline, D.K., and Lenz, P.H. 1999. Myelin-like sheaths in copepod axons. *Nature* 398:571.

Deason, E.E. 1982. *Mnemiopsis leidyi* (Ctenophora) in Narragansett Bay, 1975–1980: Abundance, size composition and estimation of grazing. *Estuarine, Coastal and Shelf Science* 15:121–134.

Delbare, D., Dhert, P., and Lavens, P. 1996. Zooplankton. In *Manual on the Production and Use of Live Food for Aquaculture*, edited by P. Lavens and P. Sorgeloos, pp. 252–282. FAO Fisheries Technical Papers, No. 361. Rome, Italy: Food and Agriculture Organization of the United Nations.

Drenner, R.W., de Noyelles, F., Jr., and Kettle, D. 1982a. Selective impact of filter feeding gizzard shad on zooplankton community structure. *Limnology and Oceanography* 27:965–968.

Drenner, R.W. and McComas, S.R. 1980. The role of zooplankton escape ability and fish size selectivity in the selective feeding and impact of planktivorous fish. In *Evolution and Ecology of Zooplankton Communities*, edited by W.C. Kerfoot, pp. 587–593. Hanover, New Hampshire, USA: University Press of New England.

Drenner R.W., Strickler, J.R., and O'Brien, W.J. 1978. Capture probability: the role of zooplankton escape in the selective feeding of planktivorous fish. *Journal of the Fisheries Research Board of Canada* 35:1370–1373.

Drenner, R., Vinyard, G.L., Gophen, M., and McComas, S.R. 1982b. Feeding behavior of the cichlid *Sarotherodon galilaeum*: selective predation on Lake Kinneret zooplankton. *Hydrobiologia* 87:17–20.

Durbin, A.G. and Durbin, E.G. 1975. Grazing rates of the Atlantic menhaden *Brevoortia tyrannus* as a function of particle size and concentration. *Marine Biology* 33:265–277.

Eaton, R.C., Bombardieri, R.A., and Meyer, D.L. 1977. The Mauthner-initiated startle response in teleost fish. *Journal of Experimental Biology* 66:65–81.

Eggers, D.M. 1977. The nature of prey selection by planktivorous fish. *Ecology* 58:46–59.

Emery, A. 1968. Preliminary observations of coral reef plankton. *Limnology and Oceanography* 13:293–303.

Fields, D.M. and Yen, J. 1997. The escape behavior of marine copepods in response to a quantifiable fluid mechanical disturbance. *Journal of Plankton Research* 19:1289–1304.

Fukusho, K. 1980. Mass production of a copepod, *Tigriopus japonicus*, in combination culture with a rotifer *Brachionus plicatilus*, fed yeast as a source. *Bulletin of the Japanese Society of Scientific Fisheries* 46:625–629.

Gauld, D.T. 1959. Swimming and feeding in crustacean larvae: the nauplius larva. *Proceedings of the Zoological Society of London* 132:31–50.

Gerritsen, J. 1978. Instar-specific swimming patterns and predation of planktonic copepods. *Verhandlungen der Internationalen Vereinigung fur Theoretische und Angewandte Limnologie* 20:2531–2536.

Gerritsen, J. 1980. Adaptive responses to encounter problems. In *Evolution and Ecology of Zooplankton Communities*, edited by W.C. Kerfoot, pp. 52–62. Hanover, New Hampshire, USA: University Press of New England.

Gerritsen, J. and Strickler, J.R. 1977. Encounter probabilities and community structure in zooplankton: a mathematical model. *Journal of the Fisheries Research Board of Canada* 34:73–82.

Govoni, J.J., Ortner, P.B., Al Yamani, R., and Hill, L.C. 1986. Selective feeding of spot, *Leiostomus xanthurus*, and Atlantic croaker, *Micropogonius undulatus*, larvae in the northern Gulf of Mexico. *Marine Ecology Progress Series* 28:175–183.

Greene, C.H. 1983. Selective predation in freshwater zooplankton communities. *International Review of Hydrobiology* 68:297–315.

Greene, C.H. 1985. Planktivore functional groups and patterns of prey selection in pelagic communities. *Journal of Plankton Research* 7:35–40.

Gulbrandsen, J. 1996. Effects of spatial distribution of light on prey ingestion of Atlantic halibut larvae (*Hippoglossus hippoglossus* L.). *Journal of Fish Biology* 48:478–483.

Gulbrandsen, J. 2001. *Artemia* swarming — mechanisms and suggested reasons. *Journal of Plankton Research* 23:659–669.

Hamner, W.M. and Carlton, J. 1979. Copepod swarms: attributes and role in coral reef ecosystems. *Limnology and Oceanography* 24:1–14.

Hansson, L.-A. 2000. Induced pigmentation in zooplankton: A trade-off between threats from predation and ultraviolet radiation. *Proceedings of the Royal Society of London: Series B, Biological Sciences* 267:2327–2331.

Hartline, D.K., Buskey, E.J., and Lenz, P.H. 1999. Rapid jumps and bioluminescence elicited by controlled hydrodynamic stimuli in a mesopelagic copepod, *Pleuromamma xiphias*. *Biological Bulletin* 197:132–143.

Haury, L.R., Kenyon, D.E., and Brooks, J.R. 1980. Experimental evaluation of the avoidance reaction of *Calanus finmarchicus*. *Journal of Plankton Research* 2:187–202.

Hernandez Molejon, O.G. and Alvarez-Lajonchere, L. 2003. Culture experiments with *Oithona oculata* Farran, 1913 (Copepoda: Cyclopoida), and its advantages as food for marine fish larvae. *Aquaculture* 219:471–483.

Hinshaw, J.M. 1985. Effects of illumination and prey contrast on survival and growth of larval yellow perch *Perca flavescens*. *Transactions of the American Fisheries Society* 114:540–545.

Holling, C.S. 1959. The components of predation as revealed by a study of small mammal predation of the European pine sawfly. *Canadian Entomologist* 91:293–320.

Houde, E.D. 1973. Some recent advances and unsolved problems in the culture of marine fish larvae. *Proceedings of the World Mariculture Society* 3:83–112.

Hunter, J.R. 1981. Feeding ecology and predation of marine fish larvae. In *Marine Fish Larvae: Morphology, Ecology and Relation to Fisheries*, edited by R. Lasker, pp. 33–77. Seattle, Washington, USA: University of Washington Press.

Hwang, J.-S., Costello, J.H., and Strickler, J.R. 1994. Copepod grazing in turbulent flow: elevated foraging behavior and habituation of escape responses. *Journal of Plankton Research* 16:421–431.

James, C.M. and Al-Khars, A.M. 1986. Studies on the production of planktonic copepods for aquaculture. *Syllogeus* 58:333–340.

Janssen, J. 1981. Searching for zooplankton just outside Snell's window. *Limnology and Oceanography* 26:1168–1171.

Janssen, J., Jones, W.R., Whang, A., and Oshel, P.E. 1995. Use of the lateral line in particulate feeding in the dark by juvenile alewife (*Alosa pseudoharengus*). *Canadian Journal of Fisheries and Aquatic Sciences* 52:358–363.

Katona, S.K. 1973. Evidence for sex pheromones in planktonic copepods. *Limnology and Oceanography* 18:574–583.

Kerfoot, W.C. 1982. A question of taste: crypsis and warning coloration in freshwater zooplankton communities. *Ecology* 63:538–554.

Kiørboe, T. and Sabatini, M. 1994. Reproductive and life cycle strategies in egg-carrying cyclopoid and free-spawning copepods. *Journal of Plankton Research* 16:1353–1366.

Kiørboe, T., Saiz, E., and Visser, A. 1999. Hydrodynamic signal perception in the copepod *Acartia tonsa. Marine Ecology Progress Series* 179:97–111.

Kiørboe, T. and Visser, A. 1999. Predator and prey perception in copepods due to hydromechanical signals. *Marine Ecology Progress Series* 179:81–95.

Kramer, D.L. and McLaughlin, R.L. 2001. The behavioral ecology of intermittent locomotion. *American Zoologist* 41:137–153.

Kremer, P. 1979. Predation by the ctenophore *Mnemiopsis ledyi* in Narragansett Bay, Rhode Island. *Estuaries* 2:97–105.

Lasker, R. 1975. Field criteria for survival of anchovy larvae: the relation between inshore chlorophyll maximum layers and successful first feeding. *Fishery Bulletin of the U.S. Department of Commerce* 73:453–462.

Lauder, G.V. 1983. Prey capture hydrodynamics in fishes: experimental tests of two models. *Journal of Experimental Biology* 104:1–13.

Leger, P., Bengtson, D.A., Simpson, K.L., and Sorgeloos, P. 1986. The use and nutritional value of *Artemia* as a food source. *Oceanography and Marine Biology: an Annual Review* 24:521–623.

Lenz, P.H. and Hartline, D.K. 1999. Reaction times and force production during escape behavior of the calanoid copepod *Undinula vulgaris. Marine Biology* 133:249–258.

Lenz, P.H., Hartline, D.K., and Davis, A. 2000. The need for speed. I. Fast reactions and myelinated axons in copepods. *Journal of Comparative Physiology A* 186:337–345.

Lindquist, N. and Hay, M.E. 1996. Palatability and chemical defense of marine invertebrate larvae. *Ecological Monographs* 66:431–450.

Lonsdale. D.J., Heinle, D.R., and Siegfried, C. 1979. Carnivorous feeding behavior of the adult calanoid copepod *Acartia tonsa* Dana. *Journal of Experimental Marine Biology and Ecology* 36:235–248.

MacKenzie, B.R., Miller, T.J., Cyr, S., and Leggett, W.C. 1994. Evidence for a dome-shaped relationship between turbulence and larval fish ingestion rates. *Limnology and Oceanography* 39:1790–1799.

Madin, L.P. 1988. Feeding behavior of tentaculate predators: *in situ* observations and a conceptual model. *Bulletin of Marine Science* 43:413–429.

Marshall, S.M. 1973. Respiration and feeding in copepods. *Advances in Marine Biology* 11:57–120.

Mauchline, J. 1998. The biology of calanoid copepods. *Advances in Marine Biology* 33:1–701.

May, R.C. 1970. Feeding larval marine fishes in the laboratory: a review. *Reports of the California Cooperative Oceanic Fisheries Investigation* 14:76–83.

McEvoy, L.A., Naess, T., Bell, J.G., and Lie, O. 1998. Lipid and fatty acid composition of normal and malpigmented Atlantic halibut (*Hippoglossus hippoglossus*) fed enriched *Artemia*: a comparison with fry fed wild copepods. *Aquaculture* 163:237–250.

McMichel, R.H., Jr., and Peten, K.M. 1989. Early life history of spotted sea trout, *Cynoscion nebulosus* (Pisces: Sciaenidae) in Tampa Bay, Florida. *Estuaries* 12:98–110.

Miller, T.J., Crowder, L.B., Rice, J.A., and Marschall, E.A. 1988. Larval size and recruitment mechanisms in fishes: toward a conceptual framework. *Canadian Journal of Fisheries and Aquatic Sciences* 45:1657–1670.

Millinski, M. 1977a. Experiments on the selection by predators against spatial oddity of their prey. *Zeitschrift fur Tierpsychologie* 43:311–325.

Millinski, M. 1977b. Do all members of a swarm suffer the same predation? *Zeitschrift fur Tierpsychologie* 45:373–388.

Miralto, A., Ianora, A., Poulet, S.A., Romano, G., and Laabir, M. 1996. Is fecundity modified by crowding in the copepod *Centropages typicus*? *Journal of Plankton Research* 18:1033–1040.

Montgomery, J.C. 1989. Lateral line detection of planktonic prey. In *The Mechanosensory Lateral Line: Neurobiology and Evolution*, edited by S. Coombs, P. Gorner, and H. Munz, pp. 561–574. New York, New York, USA: Springer-Verlag.

Montgomery, J.C. and Saunders, A.J. 1985. Functional-morphology of the piper *Hyporhamphus ihi* with reference to the role of the lateral line in feeding. *Proceedings of the Royal Society of London: Series B, Biological Sciences* 224:197–208.

Morgan, S.G. 1989. Adaptive significance of spination in estuarine crab zoeae. *Ecology* 70:464–482.

Morgan, W.L. and D.A. Ritz. 1984. Effect of prey density and hunger state on capture of krill *Nyctiphanes australis* Sars, by Australian salmon, *Arripis trutta* (Block and Schneider). *Journal of Fish Biology* 24:51–58.

Neill, S.R.S.J. and Cullen, J.M. 1974. Experiments on whether schooling by their prey affects the hunting behavior of cephalopod and fish predators. *Journal of Zoology* 172:549–569.

Newbury, T.K. 1972. Vibration perception by chaetognaths. *Nature* 236:459–460.

O'Connell, C.O. 1972. The interrelation of biting and filtering in the feeding activity of the northern anchovy (*Engraulis mordax*). *Journal of the Fisheries Research Board of Canada* 29:285–293.

Ohguchi, O. 1978. Experiments on the selection against colour oddity of water fleas by three-spined sticklebacks. *Zeitschrift fur Tierpsychologie* 47:254–267.

Ohman, M.D. 1986. Predator-limited population growth of the copepod *Pseudocalanus* sp. *Journal of Plankton Research* 8:673–713.

Ohman, M.D. 1988. Behavioral responses of zooplankton to predation. *Bulletin of Marine Science* 43:530–550.

Ohno, A., Takahashi, T., and Yashuhiko, T. 1990. Dynamics of exploited populations of the calanoid copepod *Acartia tsuensis*. *Aquaculture* 84:27–39.

Pennington, J.T. and Chia, F. 1984. Morphological and behavioral defenses of trochophore larvae of *Sabellaria cementarium* (polychaeta) against four planktonic predators. *Biological Bulletin* 167:168–175.

Purcell, J.E. 1992. Effects of predation by the scyphomedusan *Chrysaora quinquecirrha* on zooplankton populations in Chesapeake Bay. *Marine Ecology Progress Series* 87:65–76.

Reid, J.W., Hamilton, R., IV, and Duffield, R.M. 2002. First confirmed New World record of *Apocyclops dengizicus* (Lepeshkin), with a key to the species of *Apocyclops* in North America and the Caribbean region (Copepoda: Cyclopidae). *Jeffersoniana* 10:1–25.

Rothschild, B.J. and Osborn, T.R. 1988. Small-scale turbulence and plankton contact rates. *Journal of Plankton Research* 10:465–474.

Salgado, S.D. and Hoyt, R.D. 1996. Early behavior formation in fathead minnow larvae, *Pimephales promelas*: implications for sensory function. *Marine and Freshwater Behaviour and Physiology* 28:91–106.

Sekiguchi, H. 1974. Relation between ontogenetic vertical migration and mandibular gnathobase in pelagic copepods. *Bulletin of the Faculty of Fisheries Mie University* 1:1–10.

Stemberger, R.S. and Gilbert, J.J. 1987. Multiple species induction of morphological defenses in the rotifer *Keratella testudo*. *Ecology* 68:370–378.

Støttrup, J.G. and Norsker, N.H. 1997. Production and use of copepods in marine fish larviculture. *Aquaculture* 155:231–247.

Strickler, J.R. 1975. Intra- and interspecific information flow among planktonic copepods: receptors. *Verhandlungen der Internationalen Vereinigung fur Theoretische und Angewandte Limnologie* 19:2951–2958.

Sundby, S. 1997. Turbulence and ichthyoplankton: influence on vertical distribution and encounter rates. *Scientia Marina* 61:159–176.

Sundby, S., Ellertsen, B., and Fossum, P. 1994. Encounter rates between first-feeding cod larvae and their prey during moderate to strong turbulent mixing. *ICES Marine Science Symposium* 198:393–405.

Tollrian, R. 1995. Predator-induced morphological defenses: costs, life history shifts and maternal effects in *Daphnia pulex*. *Ecology* 76:1691–1705.

Treisman, M. 1975. Predation and the evolution of gregariousness. I. Models for concealment and evasion. *Animal Behaviour* 23:779–800.

Ueda, H., Kuwahara, A., Tanaka, M., and Azeta, M. 1983. Underwater observations on copepod swarms in temperate and subtropical waters. *Marine Ecology Progress Series* 11:165–171.

Van Duren, L.A. and Videler, J.J. 1996. The trade-off between feeding, mate seeking and predator avoidance in copepods: behavioral responses to chemical cues. *Journal of Plankton Research* 18:805–818.

Viitasalo, M., Kiørboe, T., Flinkman, J., Pederson, L., and Visser, A.W. 1998. Predation vulnerability of planktonic copepods: consequences of predator foraging strategies and prey sensory abilities. *Marine Ecology Progress Series* 175:129–142.

Vine, I. 1973. Detection of prey flocks by predators. *Journal of Theoretical Biology* 40:207–210.

Vinyard, G.L. and O'Brien, W.J. 1976. Effects of light and turbidity on the reactive distance of bluegill (*Lepomis macrochirus*). *Journal of the Fisheries Research Board of Canada* 33:2845–2849.

Watanabe, T., Katajima, C., and Fujita, S. 1983. Nutritional value of live organisms used in Japan for mass propagation of fish: a review. *Aquaculture* 34:115–143.

Williams, P.J., Brown, J.A., Gotceitas, V., and Pepin, P. 1996. Developmental changes in escape response performance of five species of marine larval fish. *Canadian Journal of Fisheries and Aquatic Sciences* 53:1246–1253.

Witt, U., Quantz, G., Kuhlmann, O., and Kattner, G. 1984. Survival and growth of turbot larvae, *Scophthalmus maximus* L. reared on different food organisms with special regard to long-chain polyunsaturated fatty acids. *Aquacultural Engineering* 3:177–190.

Wright, D.I. and O'Brien, W.J. 1982. Differential location of *Chaoborous* larvae and *Daphnia* by fish: the importance of motion and visible size. *American Midland Naturalist* 108:68–73.

Yen, J. and Fields, D.M. 1992. Escape response of *Acartia hudsonica* (Copepoda) nauplii from the flow field of *Temora longicornis* (Copepoda). *Ergebnisse der Limnologie/Advances in Limnology. Stuttgart* 36:123–134.

Yen, J., Lenz, P.H., Gassie, D.V., and Hartline, D.K. 1992. Mechanoreception in marine copepods: electrophysiological studies of the first antennae. *Journal of Plankton Research* 14:495–512.

Yen, J. and Strickler, J.R. 1996. Advertisement and concealment in the plankton: what makes a copepod hydrodynamically conspicuous? *Invertebrate Biology* 115:191–205.

Zaret, T.M. 1972. Predators, invisible prey, and the nature of polymorphism in the Cladocera (Class Crustacea). *Limnology and Oceanography* 17:171–184.

Zaret, T.M. 1980. The effect of prey motion on planktivore choice. In *Evolution and Ecology of Zooplankton Communities*, edited by W.C. Kerfoot, pp. 599–603. Hanover, New Hampshire, USA: University Press of New England.

Zaret, T.M. and Kerfoot, W.C. 1975. Fish predation on *Bosmina longirostris*: body size selection versus visibility selection. *Ecology* 56:232–237.

9

Suitability of the Copepod *Gladioferens imparipes* for Intensive Cultivation for Aquaculture

Robert J. Rippingale and Michael F. Payne

ABSTRACT

For successful economic cultivation of finfish, it is important that live food be available for early fish larvae. The benefits of providing copepods as live food are well known, but reliable supply of copepods continues to be a challenge for the industry. This chapter discusses the relative benefits of working with different types of copepods, with special attention to the development of culture techniques and fish-feeding trials using the calanoid copepod *Gladioferens imparipes*. It is argued that the natural history and evolution of different copepods makes some species more amenable to intensive cultivation than others, and even for those that can be cultivated, some have attributes that make them more useful than others in different branches of the aquaculture industry. This chapter focuses on aspects of copepod feeding and nutrition, life cycle, and behavior in relation to techniques for intensive cultivation.

INTRODUCTION

The reproductive strategy of many marine fish involves large numbers of eggs, with a low probability of survival to maturity. Soon after hatching, reserves in the yolk become depleted, and the survival of the fish depends on appropriate food being available in the immediate environment. The importance of copepod nauplii as the natural first food for larval fish in the sea is widely reported (e.g., discussion by McKinnon et al. 2003). These nauplii, by grazing on living algal cells, accumulate lipids from algal cell membranes and other complex biochemicals that are beneficial for the developing fish, and many fish larvae preferentially select copepod nauplii as food.

In the culture of marine finfish, extensive use is made of brine shrimp and rotifers as food. These animals, although not the natural diet of marine fish, can be biochemically enriched and successfully used with many species of fish. Although copepods from intensive culture have been shown to support healthy development of many fishes, the economics of supply favors the use of *Artemia* and rotifers rather than copepods in most enterprises. Development of cost-effective copepod cultivation could increase the live-food options available to aquaculture and allow rearing of fish with specific requirements for copepods. Of the many fish species, only a few are suitable as aquaculture targets, and in the same way, of thousands of copepod species, only a few are likely to be suitable to cultivate as live food in aquaculture. It is important for aquaculture that the feasibility of cultivation be investigated for a wide range of potentially useful live-food organisms and important for ecosystem management that locally occurring species be used in aquaculture. This would minimize the risks associated with organisms being released into new habitats.

Corresponding author: R.J. Rippingale, 15 Stuart Crescent, Lesmurdie, Western Australia 6076, Australia; E-mail: Rippingalerob@hotmail.com.

Here we discuss the suitability of *Gladioferens imparipes* (Thomson), a copepod endemic to the southwest of Western Australia (Thomson 1946). The genus *Gladioferens* (Order Calanoida, Family Centropagidae) is represented by five species occurring in inland coastal waters of Australia and New Zealand: *G. symmetricus*, *G. inermis*, *G. pectinatus*, *G. spinosus*, and *G. imparipes* (Bayly 1963). *Gladioferens imparipes* can be maintained in intensive cultivation (Payne and Rippingale 2001b), and studies to date indicate that it can be beneficial in the diet of larval fish (Payne et al. 1998; Payne and Rippingale 2000b; Payne et al. 2001). Here the authors summarize their understanding of the biology of the animal and attempt to explain how attributes that can be related to survival in the estuarine habitat make *G. imparipes* a convenient animal for use in aquaculture. Most of the important information is published, but some is the result of student investigations that are not published. Some information is from informal observations during years of keeping the copepods in culture. Comparisons are made between *G. imparipes* and other copepods that are intensively cultivated.

THE NATURAL HABITAT OF *G. IMPARIPES*

The estuary of the Swan River in Western Australia is the type locality for *G. imparipes* (Thomson 1946). The physical, chemical, and biological conditions in this estuary, and other estuaries in southwest Western Australia, where *G. imparipes* live, are characterized by marked differences associated with the seasons (Stephens and Imberger 1996; Thompson and Hosja 1996; Hamilton et al. 2001; Twomey and John 2001). Warm, dry summers and cool, wet winters result in several weeks of low salinity fluvial flow in winter. Sustained flow rates up to 1 m/sec in the upper reaches of the Swan estuary are typical (Rippingale 1994). After cessation of winter rainfall, marine water enters the estuary causing both longitudinal and vertical salt wedges. By the end of summer, marine conditions prevail in the lower estuary, and water with salinity up to 30 practical salinity units (psu) penetrates at least 65 km from the mouth. Further upstream, a few kilometers of the estuary remain at low salinity throughout summer. Still further upstream, sand bars with a few small pools at the base of the Darling Scarp represent the transition between the riverbed descending from the plateau (dry in summer) and the estuary on the coastal plain. Seasonal changes in the hydrology of the Swan estuary are clearly illustrated by Stephens and Imberger (1996).

During summer, tide changes result in horizontal ebb and flow of water along the entire estuary, with cycles up to 12 hours (Rippingale 1994). Barometrically induced pseudo-tides can exaggerate or minimize the astronomical tide (Hamilton et al. 2001). Water temperature in the estuary ranges from 10–12°C (winter) to 20–28°C (summer).

Biological productivity of the plankton community in the estuary is profoundly influenced by the seasonal hydrology. During winter, silt-laden water supports almost no primary production in the water column (Twomey and John 2001), and estuarine zooplankton is flushed downstream. When fluvial flow ceases in spring, conditions of temperature, photoperiod, and light intensity favor the growth of phytoplankton. This growth is supported by nutrients raised from rich interstitial water of the sediments by the intruding salt wedge. Each year, a series of ecological changes occur as algal productivity is followed by grazing animals and by fish. Physical, chemical, and biological conditions within the estuary continue to change until the next winter rain repeats the flushing process. Some changes in physical conditions in the estuary, particularly changes in salinity, local changes in dissolved oxygen, and to a lesser extent, changes in water temperature, impose restrictions on the distribution of organisms. Fluctuations in nutrient availability result in changing patterns of algal production, with occasional algal blooms resulting in deterioration of water quality and consequent changes in the community structure at the micro- and macrofaunal level.

The natural history of *G. imparipes* in the Swan River estuary has been described elsewhere (Rippingale and Hodgkin 1974b). In summary, the animal may occur in a wide range of physical conditions. A population occurs in low salinity cool water (< 5 psu, 12°C) in the lower estuary in winter. Where these animals occur in deep water (> 10 m), individuals will cross a salinity discontinuity from < 5 to > 30 psu during diurnal vertical migration. Others, in shallow water, remain in low salinity. As saline conditions return to the entire water column in the middle estuary, and spring blooms of microalgae occur, *G. imparipes* becomes a pioneer herbivore exploiting the algal production. As a salt wedge moves up the estuary in spring, populations of other zooplankton develop, including the copepod *Sulcanus conflictus,* in salinity up to 25 psu, and the coastal marine copepod *Acartia* sp. (*Acartiura*), in salinity > 20 psu. As omnivores, both *S. conflictus* and *Acartia* sp. predate the nauplii of *G. imparipes*. In addition to the pressure of predation from other copepods, predation probably occurs from the 17 species of fish that spawn in the estuary (Kanandjembo et al. 2001) during spring and summer. By the middle of summer *G. imparipes* persists as a refugee population in the highest reaches of the estuary, where neither species of copepod predators establish a population.

In the natural environment *G. imparipes* survives through a wide range of physical conditions, exploits different food resources, and persists despite predation on all developmental stages. Observations of natural populations indicate that breeding may occur over a salinity range of 2–35 psu and a temperature range of 12–28°C. Individuals can tolerate rapid changes in salinity, as shown by vertical migration across a distinct salinity discontinuity. Adult *G. imparipes* can hold to various substrates and not be relocated by moderate currents (Rippingale 1994), but can also move freely in open water. In different phases of the seasonal changes in estuarine biota, *G. imparipes* maintain populations when different assemblages of phytoplankton are available as food, and individual adults survive during times of low natural productivity.

ASPECTS OF *G. IMPARIPES* BIOLOGY

Gladioferens imparipes feeds exclusively on microparticles. In a suspension of unicellular phytoplankton they feed almost continuously and regularly release distinct fecal pellets.

Gladioferens imparipes has a high potential population growth rate (Rippingale and Hodgkin 1974a). Sex ratios of 1:1 occur at maturity, when males are conspicuous by the asymmetrical antennae. Reproduction follows the "embryo brood" strategy rather than the "embryo broadcast" strategy. Mature females may carry a brood of 25–45 embryos enclosed in a distinct membrane (Fig. 9.1). Repeated mating is necessary, with each spermatophore providing for two broods of embryos. Temperature, salinity, and diet influence the rate of growth to maturity and the rate of nauplius production. At 25°C, salinity 27 psu, and with *Isochrysis galbana* (T-ISO) as food, average maturation time for females was 10.4 days (Payne and Rippingale 2000a) and 25 nauplii per female per day were produced (Payne and Rippingale 2000a). Hatching of nauplii causes rupture of the membrane containing the embryo brood, and newly released nauplii swim almost immediately. On hatching, nauplii contain energy reserves as lipid droplets (Fig. 9.1). Nauplii commence feeding within hours and remain in water close to the surface. In the natural habitat, diurnal vertical migration commences when animals reach midcopepodid stages (Rippingale and Hodgkin 1974a).

Figure 9.1 illustrates how, in conditions of abundant food, adult females may carry a maturing brood of embryos, carry a younger cohort of embryos in the reproductive tract, and may have substantial reserves of lipid in the body cavity that represent > 40% of the dry mass (Payne et al. 2001). In addition, with abundant food, individuals of all developmental stages have the digestive tract packed with newly collected food and often with an unvoided fecal pellet. These copepods must be valuable food items for predatory fish.

In the natural habitat a high rate of reproduction and population growth allows *G. imparipes,* as a pioneer herbivore, to exploit spring blooms of microalgae ahead of other herbivores. As obligate small particle feeders, adult *G. imparipes* have no predatory impact on their own nauplii. Because females carry embryos until they hatch, nauplii start independent life in an immediate environment that sustained the parent. Photopositive behavior of *G. imparipes* nauplii maximizes the chance of their remaining in well-oxygenated water where phytoplankton may be abundant. Late copepodid and adult *G. imparipes* rarely occur in open water during daylight. Where water is shallow, they may occur among shoreline vegetation, but where water is deeper, diurnal migration is pronounced, and they remain close, or attached, to the bottom during daylight.

The tips of sensillae on the dorsal prosome of late copepodid and adult *G. imparipes* are able to

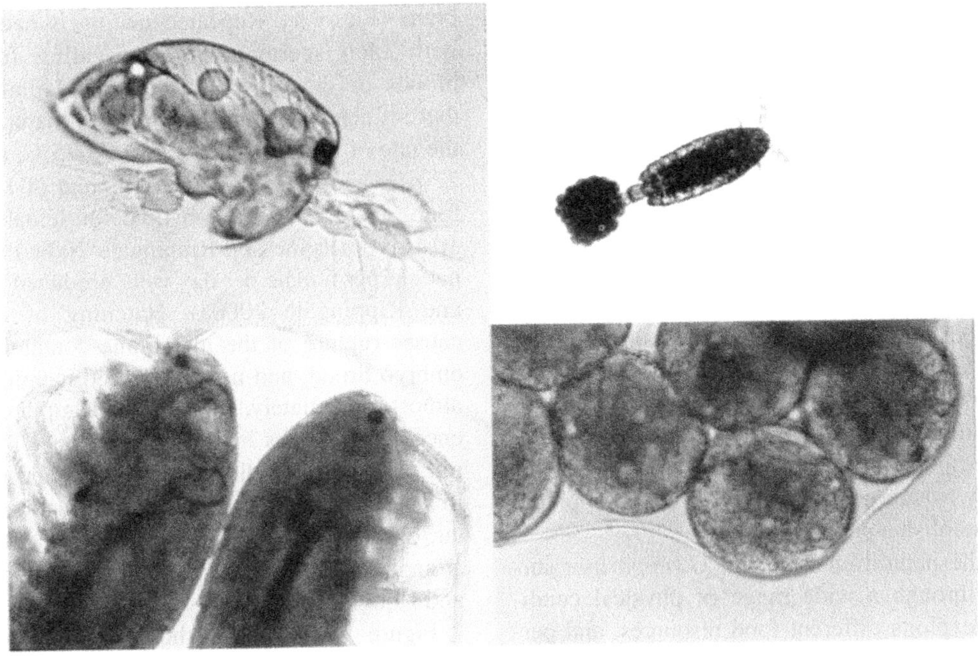

Figure 9.1. Photomicrographs of *Gladioferens imparipes.* Clockwise from top left: early nauplius (length 125 μm) with internal lipid drops; adult female (prosome length approximately 900 μm) with embryos in the reproductive tract and in an external sac; embryos contained in a sac; adults internally lipid stained with Sudan IV.

make contact with firm substrates and hold the animals stationary (Rippingale 1994). This behavior is likely to influence the distribution of the copepods in the estuary. Animals in the night plankton may be moved up to 5 or 6 km by tidally induced currents, but by day adults may hold position against the tide, and hence, be relocated with respect to a particular mass of water. This mechanism may explain upstream movement of a population ahead of the intruding salt wedge in spring and explain the retention of a small population upstream, beyond the reach of fully planktonic copepod predators in summer.

GLADIOFERENS IMPARIPES IN CULTIVATION

Gladioferens imparipes is amenable to artificial cultivation. The authors have kept them, at various times since 1972, for use in teaching and research. A manual of instructions is based on this experience (Rippingale and Payne 2001). Many attributes of *G. imparipes* that contribute to survival in the natural habitat allow the animals to be kept at high population density in relatively low-cost artificial cultivation systems without very close control of water conditions. Some of these are discussed below and summarized in Table 9.1.

Salinity Tolerance

Cultures of *G. imparipes* can easily be kept at salinities between 9 and 35 psu (Payne and Rippingale 2001a) or a wider range. Such a wide tolerance allows cultures to be kept without close control of salinity, but more importantly, it allows rapid salinity change to be used to control some invasive organisms. While routinely maintaining *G. imparipes* at salinity 27–35 psu, the authors have been able to remove harpacticoid copepods and other invertebrates by rapidly lowering the salinity to approximately 15 psu (for 10 minutes and then reducing it to 5 psu) or lower for as long as necessary to kill marine invaders. The cultivated copepods are then returned to higher salinity in two stages. A similar procedure can be used to isolate *G. imparipes* from freshly collected plankton containing a mix of species.

Table 9.1. Some biological features of *G. imparipes* with implications for using the animals in aquaculture

Biological features	Implications for aquaculture
Active in a wide range of salinity, < 5 – > 35 psu	Cultures can be adapted for use in different salinities.
Tolerate sudden changes in salinity	Salinity change can be used to suppress invasive organisms.
Survives in temperature range from 7 to 28°C	Development rate can be controlled by temperature. Animals can be temporarily stored at 7–8°C.
Feeds only on small particles	Commonly cultured phytoplankton is suitable as food. No cannibalism occurs.
Nauplius production rate up to 25 nauplii/female/day	Potential for large numbers of nauplii from healthy cultures
Healthy nauplii swim toward directional light.	A light trap collects clean nauplii with little disturbance to cultures. Healthy nauplii are self-selected for collection. Close age/size classes of nauplii can be collected.
Tolerant of population density up to at least 1 adult/mL	Less tolerant of crowding than most harpacticoid copepods
Adults can hold to hard surfaces and continue feeding	May contribute to high growth rates as energy is conserved
First nauplius 125 μm long, 65 μm wide	Suitable size as food for larvae of many fishes
Nauplii are not transparent	Conspicuous targets for visual predators
All stages may have large lipid stores	Allows animals to be stored for periods without food Valuable food items for fish
Adults are not transparent, particularly if well fed	Conspicuous food items for visual predators
Mature females carry embryos in a brood sac.	Conspicuous and valuable food items for visual predators
Adult behavior includes:	
slow gliding swimming,	Vulnerable to visual predators
rapid jerky swimming,	Alerts attention of predators
attaching to hard surfaces	Conspicuous to predators if lighting is suitable

Physiological Robustness

The physiological robustness of *G. imparipes* allows animals to survive conditions of low water quality that may inadvertently occur in culture systems of low technology. They tolerate being briefly kept at high population density without damaging each other, they survive periods of hours at low levels of dissolved oxygen, and they survive (but not thrive) in water with a high organic load. The authors have maintained copepod cultures with at least 1,000 adults/L with failure occurring very rarely.

Feeding on Small Particles

Because they feed exclusively on small particles, adult *G. imparipes* can be kept in the same water as nauplii without losses to cannibalism. Different marine microalgae can be used as food, including those in genera that are routinely used in aquaculture. As examples, *Isochrysis*, *Chaetoceros*, *Heterocapsa*, *Rhodomonas*, *Pavlova*, and *Dunaliella* (*tertiolecta*) have all been used as food for the copepods, with growth to maturity and reproduction occurring with each alga, but at different rates (Payne and Rippingale 2000a). In maintaining copepod cultures that are not required at maximum production, copepods survive with occasional feeding. Lipid reserves, as shown in Fig. 9.1, accumulated during a period of abundant food, may sustain adult animals for at least 14 days at 25°C and longer at lower temperatures (e.g., Payne and Rippingale 2001a).

Holding Behavior

The behavior of holding to surfaces with dorsal prosomal sensillae by late copepodid and adult *G. imparipes* is unusual among the calanoid copepods. The authors speculate that the behavior contributes to the success of this animal in cultivation. While remaining still for extended periods but not sinking to the bottom, energy is not expended in overcoming surface drag, as is the case for animals that sink if they are passive and have to expend energy as they swim to maintain a position in the water column. Along the continuum of different swimming and feeding behaviors described for copepods by Greene (1988), *G. imparipes* fit at the extreme end of energy-conserving passivity.

In a cylindrical culture vessel, it is simple to generate a gentle rotational flow of water with airlift. When *G. imparipes* hold to the walls of such a vessel they are ventilated by the current. This may slightly decrease the energy required for generating feeding currents. Also, while attached to surfaces, *G. imparipes* minimize the likelihood of collisions between individuals that could otherwise cause damage or stress.

When adult *G. imparipes* have shown atypical behavior by *not* holding to the walls of containers but continually swimming, microscope observations have indicated ciliates living on the exoskeleton. These copepods have also had damaged sensillae. Conspicuous holding behavior can be used as one diagnosis of the health of a culture.

Reproduction

Maturing eggs or embryos in the reproductive tract and embryo broods carried externally by female *G. imparipes* are easy to observe in healthy animals that attach to the walls of a glass container. Observing a sample collected from a larger culture in a glass beaker allows the intensity of reproductive activity and general health of a culture to be rapidly assessed by observing the proportion of females carrying broods, the size of the brood, and the proportion of females with conspicuous reproductive tracts.

Responses to Light

Where the water is deep in the natural habitat, vertical migration by adult *G. imparipes* keeps them in very low light intensity or continuous darkness. Subsets of natural populations living in shallow water are unable to avoid exposure to daylight. These animals remain healthy. Although darkness is not necessary for their well being, keeping cultures in darkness minimizes infestations of invading algae, either in the water, on the culture vessel walls, or on the exoskeletons of the animals.

Although the normal behavior of nauplii is to move toward a source of light, they remain healthy in complete darkness, and the authors speculate that some energy may be conserved in the absence of a strong stimulus to swim and aggregate in one section of a culture vessel. In intensive cultivation, the innate photopositive behavior of nauplii can be used to separate them from other age classes and remove them from a culture (Payne and Rippingale 2001b). This is an effective method of collection, which, if carried out daily, ensures that close-age cohorts of nauplii are available for use. As an additional benefit, separation of nauplii with a light trap allows new cultures to be started with a close cohort of nauplii that have been selected for good health and vigor by their rapid response to the light.

Response to Low Temperature

High growth rates can be obtained from cultures of *G. imparipes* at 25°C. Relative inactivity but continued survival at 4–8°C (Payne and Rippingale 2001a) allows for convenient storage of animals when an active culture is not required. It also allows cohorts of nauplii to be collected on successive days and accumulated in a refrigerator, where development is retarded. Large numbers of nauplii may then be available later for a short critical period.

CULTURE PROCEDURES FOR *G. IMPARIPES*

In recent years the authors have developed culture systems for *G. imparipes*, starting with vessels of a few liters and gradually refining systems using 30 L, 60 L, 500 L, and eventually 1,000 L (Rippingale and MacShane 1991; Payne and Rippingale 2001b). The objective in the most recent work was to provide a reliable supply of healthy nauplii for use as food for larval fish. Routine procedures were automated for a double 500-L system using recycled water and daily collection of nauplii. Although nauplii were produced sufficiently for experimental work (Payne and Rippingale 2001b), the numbers collected (878 nauplii per liter per day) fell below the potential of 12,500 nauplii per liter per day calculated from the initial stocking density (1 nauplius per milliliter), the expectation of 100% survival, and nauplius production rate of 25 nauplii per female per day (Payne and Rippingale 2000a). Further refinement of procedures could eventually reduce this gap and make copepod cultivation a more realistic economical proposition. The largest cultures attempted, with 1,000 L of water, involved flow through rather than recycled water.

GLADIOFERENS IMPARIPES AS FOOD FOR LARVAL FISHES

Size is a critical factor in the suitability of organisms as food for larval fishes. Early nauplii of *G. imparipes*, of width 67 μm and length 126 μm (Payne and Rippingale 2001a), are small enough for the gape of many larval fishes and similar to the size of nauplii of other species that have been used to rear fish (Toledo et al. 1999). Older nauplii of *G. imparipes* are taken by larger larvae and by many small fish. Juvenile and adults of small fish species readily ingest the adult copepods as food.

A variety of marine fish have been successfully reared to juvenile stage with *G. imparipes* either as the only food or a component of a mixed diet with *Artemia* or rotifers. These include pink snapper *Pagrus auratus* and West Australian dhufish *Glaucosoma hebraicum* (Payne et al. 2001) and various syngnathid fishes including West Australian seahorse *Hippocampus subelongatus* (Payne and Rippingale 2000b), coral seahorses *H. barbouri* (Payne 2003), and spotted pipefish *Stigmatopora argus* (Payne et al. 1998). For fish capable of ingesting adult copepods, a very beneficial food item would be found in a reproductive female with lipid reserves, food in the digestive tract, embryos in the reproductive tract, and a large external brood of embryos.

Also critical to the value of live food items for larval fish is the biochemical profile of the food items. The authors have shown that the tissues of *G. imparipes* can have a profile of fatty acids, including the essential omega-3 fatty acids, that reflects the profile of the algal food on which the copepod culture was maintained (Payne et al. 1998). Thus, copepods fed with a combination of the microalgae T-Iso (*Isochrysis* sp.) and *Nannochloropsis oculata* have a high content of the essential fatty acids docosahexaenoic acid, DHA, and eicosapentaenoic acid, EPA (Payne et al. 2001).

COMPARISON WITH OTHER COPEPODS

Of the three main groups of free-living marine copepods, harpacticoids are generally more amenable to cultivation than cyclopoids or calanoids. There is an extensive literature relating to the cultivation of harpacticoids (Sun and Fleeger 1995; Støttrup and Norsker 1997; Nanton and Castell 1998), much of it with a focus on providing live food in aquaculture. Harpacticoid copepod cultures can be very productive, high population densities can be sustained, and culture systems can be automated (e.g., Sun and Fleeger 1995).

Most harpacticoid copepods are epibenthic animals. They live on sediments or on substrates provided by vegetation or sessile animals. Most tend to move away from directional light, which probably affords some protection from visual predators. The behavior of harpacticoids has implications for their use in aquaculture. Fish that feed from sediments may thrive with access to these copepods, but those that feed primarily in open water are less likely to be successful. Exceptions to this may be the pelagic harpacticoid copepod *Euterpina acutifrons*, which has been used successfully to rear pelagic larvae of the mahi mahi *Coryphaena hippurus* (Kraul et al. 1992), and *Nitokra lacustris*, in which the early copepodid stage swims freely and is of suitable size for larval fish predators (A. Rhodes personal communication, May 2003).

Harpacticoid copepods feed mainly by browsing across surfaces rather than by collecting particles from open water. They survive with access to organic debris and do not usually depend upon live algae as food, although live algae can be a source of organic material. In intensive culture, harpacticoid copepods may be highly productive with a diet of mixed algae, and the biochemical profile of their tissues may make them suitable as food for fish (Norsker and Støttrup 1994; Støttrup and Norsker 1997). At least some harpacticoids may be able to elongate carbon chains and synthesize the fatty acids that are essential in the diet of fishes (Norsker and Støttrup 1994; Nanton and Castell 1998). If this is the case, those fatty acids are probably required by the copepod. Ability to elongate carbon chains is consistent with harpacticoid copepods having diets of organic debris that probably lacks algal cell membranes with highly unsaturated fatty acids. In contrast, many calanoid copepods either graze directly on algae or herbivorous microzooplankton. These latter copepods, with long-chain fatty acids normally in the diet, lack the ability to produce them.

As with harpacticoids, understanding of the biology of calanoid copepods has been advanced through use of animals in cultivation (e.g., Davis 1983), and, as with harpacticoids, some species are more amenable to cultivation than others. Copepods from open-ocean or deep marine water occur in relatively constant environments with consistently high water quality. To be kept in culture, these animals would require carefully controlled conditions to keep them healthy. They are unlikely to be cultivated in low-technology systems at the high population density required for aquaculture. Of the calanoid copepods that can be readily cultivated, those from coastal water, inlets, lagoons, and estuaries are the most likely to meet the requirements of intensive cultivation for aquaculture. These copepods survive naturally in changeable habitats and therefore tolerate a range of environmental conditions. Examples of these copepods include various species of *Acartia* (Toledo et al. 1997; Schipp et al. 1999), *Eurytemora* (Katona 1970; Witt et al. 1984; Shields et al. 1999), *Pseudodiaptomus* (Doi et al. 1997; Toledo et al. 1999), and some tropical species from the Family Paracalanidae (McKinnon et al. 2003).

Acartia spp. are widespread and often the most abundant calanoid copepods in coastal or estuary water plankton (Kimmerer 1991; Irigoien and Castel 1995; Liang and Uye 1996), where they feed as omnivores on a spectrum of food types, including their own eggs and nauplii (e.g., Dam et al. 1994; Liang and Uye 1996; Uye and Liang 1998). *Acartia* broadcast subitaneous or diapausing embryos, either singly or in short strings, as a response to food being available (Kiørboe et al. 1985; White and Roman 1992). Adult *Acartia* spp. tend to produce modest lipid reserves. They direct food into reproduction rather than storage. In the coastal marine environment, *Acartia* spp. copepods are probably important items in the diet of larval fish.

Acartia spp. copepods have been maintained in culture systems designed around the biological requirements and behavior of the animals (Støttrup et al. 1986). An important requirement for keeping these animals at high population density is to separate newly released embryos from adults and thereby minimize losses to cannibalism.

The size of first-stage nauplii of *Acartia* spp., at length 120 μm (Hopcroft and Roff 1998), makes them accessible as food for many fish larvae, and larger fish might predate the adult copepods. *Acartia* spp. occur in predator-rich coastal waters where low visibility of the animals (Kimmerer 1991) and frequent rapid movement (Saiz and Alcaraz 1992; Tiselius 1992) probably provide some defenses against predators. As individual food items for fish, adult female *Acartia* spp. are probably less easily captured than copepods that are visually conspicuous and present slow-moving targets. As individual food items, they are also probably less valuable than copepods that store large lipid reserves and carry broods of embryos.

Although they are from distinct families (Huys and Boxshall 1991), close comparisons can be made between members of the genera *Gladioferens*, from Australia and New Zealand, *Eurytemora* from the Northern Hemisphere, and some species of *Pseudodiaptomus*. These animals are probably similar in their response to the conditions of cultivation. Obvious similarities can be seen in size, gross morphology, reproductive strategy, life cycle, and general ecology, with one significant difference: *G. imparipes* has dorsal sensillae on the prosome by which it attaches to surfaces.

Eurytemora affinis, including the synonymous *E. hirundo* and *E. hirundoides* (Busch and Brenning 1992), is widely distributed in estuaries, and the recent extension of its distribution in waters of North America, including freshwater (Saunders 1993), indicates a robust physiology. Although not normally occurring in marine conditions, *E. affinis* occur in a wide range of salinity, and individuals that are well fed are able to survive in seawater (Rippingale and Crossland 1993). Cultivation of *Eurytemora* spp., especially *E. affinis*, has been successful (Katona 1970), and these copepods can be used in aquaculture. The distribution of *E. affinis* within estuaries differs from that of *G. imparipes*. Without dorsal sensillae or other means of holding to surfaces, *E. affinis* are more passive under the influence of water movement. Castel and Veiga (1990) describe retention of *E. affinis hirundoides* in the Gironde estuary of France as being caused by the same mechanisms by which suspended sediment is retained.

Various species of *Pseudodiaptomus* occur in coastal and estuarine water. In southern Africa, *P. hesseii* and *P. charteri* thrive in estuaries in a wide range of salinity (Grindley 1963; Jerling and Wooldridge 1991), and *P. hessei* can be readily maintained in cultivation. As evidence of robustness, *P. inopinus*, described as the "Asian copepod," has recently spread as an invader in North America (Cordell et al. 1992), and another Pseudodiaptomid, *P. forbesis*, has invaded a niche in San Francisco Bay, USA (Kimmerer and Orsi 1996). As for *G. imparipes*, the attributes that allow successful exploitation of variable environments probably qualify both *E. affinis* and some *Pseudodiaptomus* spp. copepods as candidates for intensive cultivation.

Various authors have discussed the ecological significance of embryo brooding and embryo broadcasting by copepods. Ohman and Townsend (1998) show that for copepods of intermediate size, there are far fewer species in the marine environment that carry broods than species that broadcast. They suggest that the brooding strategy would make females of intermediate-sized copepods so vulnerable to visual predators that they are not common in the coastal marine environment. It is consistent with the argument of Ohman and Townsend (1998) that copepods of the genera *Eurytemora*, *Pseudodiaptomus,* and *Gladioferens*, which carry very conspicuous broods of embryos, occur mainly in relatively turbid waters of estuaries where some protection from visual predators is afforded. Vuorinen et al. (1983) and Vuorinen (1987) describe the vulnerability to predators of *Eurytemora* spp. carrying broods, but show that some reduction in predation is achieved by diurnal vertical migration, although this is at a cost of reduced production by time spent in cooler water. Although Redden and Daborn (1991) observed that some subitaneous embryos of *E. herdmani* survived passage through the gut of predatory fish, this does not significantly alter the usual understanding that predation by fish is fatal for copepods, that natural selection might be expected to favor mechanisms that provide protection from predators, and that predators can exclude prey from environments in which they might be particularly vulnerable. For copepods of intermediate size, it is likely that embryo brooding is only an effective strategy in relatively turbid water. For *G. imparipes*, although vulnerable to predators in open water, some protection is afforded by the behavior pattern through which the animals remain inconspicuous while attached to surfaces.

Although embryo brooding by female calanoid copepods may make them vulnerable to visual predators, it has some advantages. Unhatched embryos of broadcast spawners are at risk of mortality if they sink into anaerobic sediments (Marcus and Lutz 1994) or other unfavorable environments. Embryo-brooding copepods avoid these risks. Although rates of embryo production vary widely, copepods that carry broods are generally less fecund than broadcast spawners (Ianora 1998). They have slower rates of embryo production, but each embryo has a greater chance of survival (Kiørboe and Sabatini 1994). Production of up to 25 nauplii per female per day by *G. imparipes* (Payne and Rippingale 2000a) is much less than reported for many embryo broadcasters, but there is insufficient data available to compare the production rates of embryo biomass. For use as live food in aquaculture, nauplii of copepods must be small enough for the gape of the fish, but also of sufficient biomass to be of food value. Nauplii produced in cultures of *G. imparipes* show conspicuous lipid droplets (Fig. 9.1). Maternal investment ensures high survival of these embryos, at least in the absence of predators. Whatever the relative advantages of embryo brooding, it seems that the genera *Gladioferens*, *Eurytemora*, and *Pseudodiaptomus* have species that are successful with this strategy in their estuarine environments. It is a wry consequence of the value of embryo brooders as food for fish that copepods that would be valuable and vulnerable food items are excluded from the range of most marine fish and survive in estuaries. In aquaculture it should be possible to reverse the trends of natural history and provide fish with the most vulnerable copepods, the most well-fed copepods, and thereby achieve unnaturally high rates of fish survival.

Many copepods are able to produce either subitaneous or resting (or diapausing) embryos. The ability to respond to some environmental cue and suspend the development of embryos provides many advantages. Diapausing embryos may persist in sediments during conditions that are unfavorable for active animals or may provide for retention of populations in zones of advection (Peterson 1998). Diapause may occur in the embryos of copepods that broadcast and those that brood (Lindley 1990), and to focus on copepods suitable for intensive cultivation, diapause occurs in *Acartia* spp. (Liang and Uye 1996) and in *E. affinis* (Ban 1992a; Ban 1992b; Ban and Minoda 1992). In aquaculture, opportunity to collect diapausing embryos may be convenient for storing copepods during periods when cultures are not required, but the environmental cues for producing them must be understood. If a high rate of production of active nauplii is required from a culture, conditions that favor production of resting eggs must be avoided. Ban (1992a) reported that *E. affinis* produced a significant proportion of diapausing embryos in response to crowded conditions or short photoperiods. This has implications for the management of cultures of this copepod. No evidence for diapause or other resting stage has been found for *G. imparipes*. Females release free-swimming nauplii under a wide range of conditions.

CONCLUSION

Related animals have much of their genetic legacy in common, but by definition, each species is also unique. The unique features are the product of independent evolution, often in unique habitats. For some species of copepods, the attributes of physiology and life cycle that allow them to survive in naturally variable conditions also allow them to be maintained in intensive cultivation. For these animals, culture systems can be easily designed to provide for the biological requirements and tolerances of the animal. In general, copepods from coastal and estuary water seem to be amenable to intensive cultivation, but they may not be equally useful as live food in aquaculture. Those with highly developed strategies to escape predators may not be the most suitable copepods on which to focus for intensive cultivation. The experience of the authors with *G. imparipes* suggests that copepods with similar attributes should be thoroughly investigated for use in aquaculture in the locations where the animals naturally occur. Risks associated with the release of potentially invasive copepods in different parts of the world should be avoided.

REFERENCES

Ban, S. 1992a. Effects of photoperiod, temperature, and population density on induction of diapause egg production in *Eurytemora affinis* (Copepoda: Calanoida) in Lake Ohnuma, Hokkaido, Japan. *Journal of Crustacean Biology* 12:361–367.

Ban, S. 1992b. Seasonal distribution, abundance and viability of diapause eggs of *Eurytemora affinis* (Copepoda: Calanoida) in the sediment of lake Ohnuma, Hokkaido. *Bulletin of Plankton Society of Japan* 39:41–48.

Ban, S. and Minoda, T. 1992. Hatching of diapausing eggs of *Eurytemora affinis* (Copepoda: Calanoida) collected from lake-bottom sediments. *Journal of Crustacean Biology* 12:51–56.

Bayly, I.A.E. 1963. A revision of the coastal water genus *Gladioferens* (Copepoda: Calanoida). *Australian Journal of Marine and Freshwater Research* 14:194–217.

Busch, A. and Brenning, U. 1992. Studies on the status of Eurytemora affinis (Poppe, 1880) (Copepoda, Calanoida). *Crustaceana* 62:13–38.

Castel, J. and Veiga, J. 1990. Distribution and retention of the copepod *Eurytemora affinis hirundoides* in a turbid estuary. *Marine Biology* 107:119–128.

Cordell, J.R., Morgan, C.A., and Simenstad, C.A. 1992. Occurrence of the Asian calanoid copepod *Pseudodiaptomus inopinus* in the zooplankton of the Columbia River estuary. *Journal of Crustacean Biology* 12:260–269.

Dam, H.G., Peterson, H.T., and Bellantoni, D.C. 1994. Seasonal feeding and fecundity of the calanoid copepod *Acartia tonsa* in Long Island Sound: is omnivory important to egg production? *Hydrobiologia* 292/293:191–199.

Davis, C.S. 1983. Laboratory rearing of marine calanoid copepods. *Journal of Experimental Marine Biology and Ecology* 71:119–133.

Doi, M., Toledo, J.D., Golez, M.S.N., Santos, M.D.L., and Ohno, A. 1997. Preliminary investigation of feeding performance of larvae of early red-spotted grouper, *Epinephelus coioides*, reared with mixed zooplankton. *Hydrobiologia* 358:259–263.

Greene, C.H. 1988. Foraging tactics and prey-selection patterns of omnivorous and carnivorous calanoid copepods. *Hydrobiologia* 167:295–302.

Grindley, J.R. 1963. The Psuedodiaptomidae (Copepoda; Calanoida) of Southern African water, including a new species, *Pseudodiaptomus charteri*. *Annals of the South African Museum* 46:373–391.

Hamilton, D.P., Chan, T., Robb, M., Pattiaratchi, C., and Herzfeld, M. 2001. The hydrology of the upper Swan River Estuary with focus on an artificial destratification trial. *Hydrological Processes* 15:2465–2480.

Hopcroft, R.R. and Roff, J.C. 1998. Zooplankton growth rates: the influence of size in nauplii of tropical marine copepods. *Marine Biology* 132:87–96.

Huys, R. and Boxshall, G.A. 1991. *Copepod Evolution.* London, UK: The Ray Society. 468 pp.

Ianora, A. 1998. Copepod life history traits in subtemperate regions. *Journal of Marine Systems* 15:337–449.

Irigoien, X. and Castel, J. 1995. Feeding rates and productivity of the copepod *Acartia bifilosa* in a highly turbid estuary; the Gironde (SW France). *Hydrobiologia* 311:115–125.

Jerling, H.L. and Wooldridge, T.H. 1991. Population dynamics and estimates of production for the calanoid copepod *Pseudodiaptomus hessei* in a warm temperate estuary. *Estuary and Coastal Research Science* 33:121–135.

Kanandjembo, A.N., Potter, I.C. and Plattel, M.E. 2001. Abrupt shifts in the fish community of the hydrologically variable upper estuary of the Swan River. *Hydrological Processes* 15:2503–2517.

Katona, S.K. 1970. Growth characteristics of the copepods *Eurytemora affinis* and *E. herdmani* in laboratory cultures. *Helgolander wiss. Meeresunters* 20:373–384.

Kimmerer, W.J. 1991. Predatory influences on copepod distributions in coastal waters. *Bulletin of the Plankton Society of Japan* Special Volume of the Proceeding of the Fourth International Conference on Copepoda:161–174.

Kimmerer, W.J. and Orsi, J. 1996. Changes in the zooplankton of the San Francisco Bay Estuary since the introduction of the clam *Potamocorbula amurensis.* In *San Francisco Bay: The Ecosystem,* edited by J.T. Hollibaugh, pp. 403–424. San Francisco, California, USA: Pacific Division of the American Association for the Advancement of Science.

Kiørboe, T., Møhlenberg, F. and Hamburger, K. 1985. Bioenergetics of the planktonic copepod *Acartia tonsa*: relationship between feeding, egg production and respiration, and composition of specific dynamic action. *Marine Ecology Progress Series* 26:85–97.

Kiørboe, T. and Sabatini, M. 1994. Reproductive and life cycle strategies in egg-carrying cyclopoid and free-spawning calanoid copepods. *Journal of Plankton Research* 16:1353–1366.

Kraul, S., Nelson, A., Brittain, K., Ako, H., and Ogasawara, A. 1992. Evaluation of live feeds for larval and postlarval Mahimahi *Coryphaena hippurus. Journal of the World Aquaculture Society* 23:299–307.

Liang, D. and Uye, S. 1996. Population dynamics and production of the planktonic copepods in a eutrophic inlet of the Inland Sea of Japan. II. *Acartia omorii. Marine Biology* 125:109–117.

Lindley, J.A. 1990. Distribution of overwintering calanoid copepod eggs in sea-bed sediments around southern Britain. *Marine Biology* 104:209–217.

Marcus, N.H. and Lutz, R.V. 1994. Effects of anoxia on the viability of subitaneous eggs of planktonic copepods. *Marine Biology* 121:83–87.

McKinnon, A.D., Duggan, S., Nichols, P.D., Rimmer, M.A., Semmens, G., and Robino, B. 2003. The potential paracalanid copepods as live feeds in aquaculture. *Aquaculture* 223:89–106.

Nanton, D.A. and Castell, J.D. 1998. The effects of dietary fatty acids on the fatty acid composition of the harpacticoid copepod, *Tisbe* sp., for use as a live food for marine fish larvae. *Aquaculture* 163:251–261.

Norsker, N.H. and Støttrup, J.G. 1994. The importance of dietary HUFAs for fecundity and HUFA content in the harpacticoid, *Tisbe holothuriae* Humes. *Aquaculture* 125:155–166.

Ohman, M.D. and Townsend, A.W. 1998. Egg strings in *Euchirella pseudopulchra* (Aetideidae) and comments on constraints on egg brooding on planktonic marine copepods. *Journal of Marine Systems* 15:61–69.

Payne, M. 2003. Rearing the coral seahorse, *Hippocampus barbouri*, on live and inert prey. In *Marine Ornamental Species: Collection, Culture and Conservation*, edited by J.C. Cato and C.L. Brown, pp. 289–296. Ames, Iowa, USA: Iowa State Press.

Payne, M.F. and Rippingale, R.J. 2000a. Evaluation of diets for culture of the calanoid copepod *Gladioferens imparipes. Aquaculture* 187:85–96.

Payne, M.F. and Rippingale, R.J. 2000b. Rearing West Australian seahorse, *Hippocampus subelongatus*, juveniles on copepod nauplii and enriched *Artemia. Aquaculture* 188:353–361.

Payne, M.F. and Rippingale, R.J. 2001a. Effects of salinity, cold storage and enrichment on the calanoid copepod *Gladioferens imparipes. Aquaculture* 201:251–262.

Payne, M.F. and Rippingale, R.J. 2001b. Intensive cultivation of the calanoid copepod *Gladioferens imparipes. Aquaculture* 201:329–342.

Payne, M.F., Rippingale, R.J., and Cleary, J.J. 2001. Cultured copepods as food for West Australian dhufish (*Glaucosoma hebraicum*) and pink snapper (*Pagrus auratus*) larvae. *Aquaculture* 194:137–150.

Payne, M.F., Rippingale, R.J., and Longmore, R.B. 1998. Growth and survival of juvenile pipefish (*Stigmatopora argus*) fed live copepods with high and low HUFA content. *Aquaculture* 167:237–245.

Peterson, W. 1998. Life cycle strategies of copepods in coastal upwelling zones. *Journal of Marine Systems* 15:313–326.

Redden, A.M. and Daborn, G.R. 1991. Viability of subitaneous copepod eggs following fish predation on egg-carrying copepods. *Marine Ecology Progress Series* 77:307–310.

Rippingale, R.J. 1994. A calanoid copepod *Gladioferens imparipes*, holding to surfaces. *Hydrobiologia* 292/293:351–360.

Rippingale, R.J. and Crossland, P.A.M. 1993. Food availability and salinity tolerance in the copepod *Eurytemora affinis* Poppe. *Archives of Hydrobiology/Supplement 75* 3/4:357–362.

Rippingale, R.J. and Hodgkin, E.P. 1974a. Population growth of a copepod *Gladioferens imparipes* Thomson. *Australian Journal of Marine and Freshwater Research* 25:351–360.

Rippingale, R.J. and Hodgkin, E.P. 1974b. Predation effects on the distribution of a copepod. *Australian Journal of Marine and Freshwater Research* 25:81–91.

Rippingale, R.J. and MacShane, M.G. 1991. A calanoid copepod for intensive cultivation. *Memoirs of the Queensland Museum* 31:457.

Rippingale, R.J. and Payne, M.F. 2001. *Intensive Cultivation of a Calanoid Copepod Gladioferens imparipes*. 67 pp. On-line version, www.aims.gov.au/pages/research/hatchery-feeds/pdf/copepod-culture-manual.pdf

Saiz, E. and Alcaraz, M. 1992. Free-swimming behaviour of *Acartia clausi* (Copepoda:calanoida) under turbulent water movement. *Marine Ecology Progress Series* 80:229–236.

Saunders, J.F. 1993. Distribution of *Eurytemora affinis* (Copepoda:Calanoida) in the Southern Great Plains, with notes on zoogeography. *Journal of Crustacean Biology* 13:564–570.

Schipp, G.R., Bosmans, J.M.P., and Marshall, A.J. 1999. A method for hatchery culture of tropical calanoid copepods, *Acartia* spp. *Aquaculture* 174:81–88.

Shields, R.J., Bell, J.G., Luizi, F.S., Gara, B., Bromage, N.R., and Sargent, J.R. 1999. Natural copepods are superior to enriched *Artemia* nauplii as feed for halibut larvae (*Hippoglossus hippoglossus*) in terms of survival, pigmentation and retinal morphology: relation to dietary essential fatty acids. *Journal of Nutrition* 129:1186–1194.

Stephens, R. and Imberger, J. 1996. Dynamics of the Swan River estuary: the seasonal variability. *Marine and Freshwater Research* 47:517–529.

Støttrup, J.G. and Norsker, N.H. 1997. Production and use of copepods in marine fish larviculture. *Aquaculture* 155:231–248.

Støttrup, J.G., Richardson, K., Kirkegaard, E., and Pihl, N.J. 1986. The cultivation of *Acartia tonsa* Dana for use as a live food source for marine fish larvae. *Aquaculture* 52:87–96.

Sun, B. and Fleeger, J.W. 1995. Sustained mass culture of *Amphiascoides atopus* a marine harpacticoid copepod in a recirculating system. *Aquaculture* 136:313–321.

Thompson, P. and Hosja, W. 1996. Nutrient limitation of phytoplankton in the upper Swan River estuary, Western Australia. *Marine and Freshwater Research* 47:656–667.

Thomson, J.M. 1946. New Crustacea from the Swan River estuary. *Journal of the Royal Society of Western Australia* 30:35–53.

Tiselius, P. 1992. Behaviour of *Acartia tonsa* in patchy food environments. *Limnology and Oceanography* 37:1640–1651.

Toledo, J.D., Golez, S.N., Doi, M., and Ohno, A. 1997. Food selection of early grouper, *Epinephelus coioides*, larvae reared by the semi-intensive method. *Suisanzoshoku* 45:327–337.

Toledo, J.D., Golez, M.S., Doi, M., and Ohno, A. 1999. Use of copepod nauplii during early feeding stage of grouper *Epinephelus coioides*. *Fisheries Science* 65:390–397.

Twomey, L. and John, J. 2001. Effects of rainfall and salt-wedge movement on the phytoplankton succession in the Swan-Canning estuary, Western Australia. *Hydrological Processes* 15:2655–2669.

Uye, S. and Liang, D. 1998. Copepods attain high abundance, biomass and production in the absence of large predators but suffer cannibalistic loss. *Journal of Marine Systems* 15:495–501.

Vuorinen, I. 1987. Vertical migration of *Eurytemora* (Crustacea, Copepoda): a compromise between the risks of predation and decreased fecundity. *Journal of Plankton Research* 9:1037–1046.

Vuorinen, I., Rajasilta, M., and Salo, J. 1983. Selective predation and habitat shift in a copepod species — support for the predation hypothesis. *Oecologia (Berlin)* 59:62–64.

White, J.R. and Roman, M.R. 1992. Egg production by the calanoid copepod *Acartia tonsa* in the mesohaline Chesapeake Bay: the importance of food resources and temperature. *Marine Ecology Progress Series* 86:239–249.

Witt, U., Quantz, G., Kuhlmann, D., and Kattner, G. 1984. Survival and growth of turbot larvae *Scophthalmus maximus* L. reared on different food organisms with special regard to long-chain polyunsaturated fatty acids. *Aquacultural Engineering* 3:177–190.

10
Development of Feeding Mechanics in Marine Fish Larvae and the Swimming Behavior of Zooplankton Prey: Implications for Rearing Marine Fishes

Ralph G. Turingan, Jessica L. Beck, Justin M. Krebs, and Jason D. Licamele

ABSTRACT

A major bottleneck in the larviculture of marine fishes occurs immediately following yolk absorption when mass mortality is experienced while transitioning to exogenous prey resources. This study uses a novel approach, digital high-speed videography, to investigate the swimming velocity of zooplankton prey and link predator-induced change in prey swimming speed to the ability of fish larvae to capture prey. We test the hypotheses that (a) size of prey consumed by marine fish larvae is constrained by predator gape and (b) variation in swimming velocity among zooplankton prey is associated with their differential vulnerability to predation by marine fish larvae. *Sciaenops ocellatus* were cultured from eggs in the laboratory and used to investigate the development of fish-feeding mechanisms and feeding performance. Five zooplankton prey species, including two copepod species, were concurrently cultured and used to examine interspecific differences in swimming velocity with and without the presence of a larval-fish predator. First-feeding larvae use a simple mechanism involving the hyoid-mandible linkage (i.e., hyoid stage) to capture prey. Larvae around metamorphosis capture prey using a more complex mechanism involving the hyoid- and opercular series-mandible linkages (i.e., hyoid-opercular stage). This two-stage development of the feeding mechanism is associated with stage-specific differences observed in feeding performance: less than 50% of larvae fed during the hyoid stage while almost 100% of larvae fed at the hyoid-opercular stage. First-feeding larvae preferred small and less elusive prey; conspecifics at metamorphosis preferred larger and more elusive prey. The two copepod species had the highest swimming velocity among all zooplankton examined, especially with the presence of a larval-fish predator. Prey-capture performance in marine fish larvae is likely affected by the escape response of zooplankton prey. Thus, the authors propose that in addition to size, the swimming velocity and escape response of zooplankton be considered in the selection of an appropriate live feed for marine fish larviculture.

INTRODUCTION

Finfish aquaculture research and development have progressed tremendously over the past several decades, primarily because of the need to

Corresponding author: Ralph G. Turingan, Department of Biological Sciences, Florida Institute of Technology, 150 West University Boulevard, Melbourne, FL 32901, USA; E-mail: turingan@fit.edu; Phone: (321) 674-8037; Fax: (321) 674-7238.

culture species for food production and the aquarium trade, stock enhancement, and research. In addition, the diminishing supply of eggs and fry from the wild to meet the high demand for fish underscores the need to rely on captive breeding, spawning, and hatchery production. Numerous marine food fish (e.g., red drum *Sciaenops ocellatus*, mahi mahi *Coryphaena hippurus*, snappers and flounders) and marine ornamental fish (e.g., clownfish *Amphiprion* sp., cardinalfish *Apogon* sp., and angelfish *Pterophyllum scalare*) species have been successfully spawned in captivity. Survivorship of fish larvae, however, has been very poor. For example less than 10% of successfully hatched mahi mahi, *C. hippurus*, survive from the larval stage to the juvenile stage (R.A. Hagood personal communication). Likewise, Lazo et al. (2000) found only 4–14% survivorship through metamorphosis in red drum, *S. ocellatus*, fed different diets: live zooplankton with algae, zooplankton without algae, microparticulate diet with algae, and microparticulate diet without algae. In southern flounder, *Paralichthys lethostigma* (H.V. Daniels personal communication), and in black sea bass, *Centropristis striata* (W.O. Watanabe personal communication), survival rates to the metamorphic stage vary widely, but often fall below 15%. It is likely that a breakthrough in larviculture will underlie the success of marine finfish aquaculture by enabling more reliable and cost-effective supplies of fingerlings to support research and grow-out projects in land-based and offshore production systems.

High mortality in both wild and cultured marine fish larvae has been well documented. In the wild, high mortality of marine fish larvae has been attributed to predation on larvae by other fishes (Langsdale 1993; Fuiman and Magurran 1994) and invertebrates (Bailey and Houde 1989), starvation due to insufficient density of planktonic prey (Hunter 1976), and other biotic and abiotic factors. In aquaculture systems, where most of the above factors that influence survivorship of oceanic marine fish larvae are controlled, high mortality during larviculture remains a major obstacle to successful rearing of large numbers of finfish species (Watanabe and Kiron 1994). In particular, catastrophic mortality has been associated with first feeding (i.e., the period during which larvae switch from endogenous to exogenous feeding; Hagen 1993). This "critical period" for the larvae coincides with the time of complete yolk resorption and the onset of active foraging. This phenomenon has triggered extensive research aimed at elucidating the causes of mass mortality during the early stages of exogenous feeding in cultured marine fish larvae.

Several factors, including the development of sense organs, swimming and feeding mechanisms, and behavior, as well as the size, conspicuousness, behavior, density, and nutritional value of available prey influence feeding success, and thus, mortality in cultured marine fish larvae (Blaxter 1986; Watanabe and Kiron 1994). Because marine fish larvae are visual feeders, the development of a pure cone retina during the first-feeding stage requires light (threshold light intensity = 0.1 lux) for feeding (e.g., Bagarinao and Hunter 1983; Blaxter 1968, 1986). Water currents disperse planktonic prey organisms and food particles, as well as affect the swimming behavior of larval fish and their prey; detection of water currents and disturbances (e.g., movement of prey and predators) is facilitated by the development of free neuromasts and lateral line canals soon after hatching (Disler 1971; Iwai 1980; Blaxter et al. 1983; Fuiman and Magurran 1994). The early development of olfactory pits and taste buds enables larvae of Atlantic herring (Dempsey 1978), Mozambique tilapia (Iwai 1980), red sea bream (O'Connell 1981), and other marine fish species to respond to chemical stimuli elicited by larval fish prey. Swimming abilities (i.e., to search for and pursue prey, as well as to escape from predation) improve with the development of fins, beyond the finfold stage (Bailey and Batty 1984; Fuiman 1993). The availability of optimal larval fish prey may influence the association between mass mortality and first feeding in marine fish larvae. Size, density, ability to elicit chemical stimuli, mobility, color, and nutritional value of live prey and formulated feeds affect capture rates in marine fish larvae (Watanabe and Kiron 1994; Turingan 1999; Krebs and Turingan 2003).

At both interspecific and intraspecific levels of analysis, the size of prey consumed by marine fish larvae increases with larval body size (Houde and Schekter 1980; Kiorboe et al. 1985; Rice et al. 1987; Schael et al. 1991; Pryor and Epifanio 1993; Pepin and Penney 1997). Hunter (1981) postulated that these relationships are widespread in fishes, which may indicate that "these patterns are the result of positive size selection inherent in species or ecological groups of larvae." Selection of prey, especially based on size, has been found to correlate with mouth width or gape and gut length of larvae (e.g., Shirota 1970; Duray and Kohno 1990; Doi et al. 1997; Krebs and Turingan 2003). These studies suggest that larval mouth size limits the size of prey that could be consumed by the larva. Most larvae, however, do not consume prey as large as predicted based on the relationship between maximum prey diameter and maximum larval mouth width (Dabrowski and Bardega 1984; Krebs and Turingan 2003). This suggests that there are constraints other than mouth width or mouth gape that limit the size of prey a larval fish could successfully consume.

In fishes that capture prey by suction feeding, such as most fish larvae, a successful prey capture requires an explosive suction-feeding behavior in which rapid expansion of the oral cavity draws water and prey into the mouth (Lauder 1980). The suction-feeding mechanism in fishes involves a complex system of bones, muscles, tendons, and ligaments in the head (see Turingan and Wainwright 1993). Much, however, of what is known about the mechanism of prey capture, prey processing, and prey transport into the esophagus in fishes is based on analyses of feeding behavior and mechanics in juvenile and adult fishes (for review see Ferry-Graham and Lauder 2001). Surprisingly, relatively few studies have investigated the prey-capture mechanism of marine fish larvae (Hunt von Herbing et al. 1996; Liem 1991; Turingan 1999). Recent studies have suggested that developing marine fish larvae undergo an ontogenetic shift in prey-capture mechanics, from a simple system involving only the hyoid apparatus and head rotation to a more complex system involving the hyoid, head rotation, and the opercular series (Hunt Von Herbing et al. 1996; Turingan 1999). This functional transformation occurs after the onset of exogenous feeding; apparently, first-feeding larvae are still at the hyoid-stage of development. This first-feeding stage (hyoid stage) is associated with massive decline in population numbers, a major concern during the hatchery phase of aquaculture. Contemporary research then needs to address the question, "Is the hyoid-mandible linkage a sufficient mechanism to drive prey-capture performance during the first-feeding stage in marine fish larvae?"

The central theme of this study is to elucidate the consequences of developmental transformations of the feeding system from a simple, hyoid-mediated prey-capture mechanism to a more integrated, hyoid- and opercular-driven feeding system for feeding performance in cultured marine fish larvae. Does feeding efficiency improve with the development of a more complex feeding mechanism through early ontogeny? Does the ability to capture larger, more elusive prey organisms correlate with the development of a more complex feeding system in marine fish larvae? What are the consequences of a hyoid-mediated (first-feeding stage) versus a hyoid- and opercular series–mediated (metamorphosis) mouth-opening mechanism for prey selection in cultured marine fish larvae?

The objectives of this study are threefold. First, the authors explore the development of the prey-capture mechanism in cultured red drum larvae. Second, they compare the swimming velocity among zooplankton prey with and without the presence of a larval fish predator. And third, they examine the feeding performance of first-feeding and metamorphosing fish larvae.

MATERIALS AND METHODS

Rearing of Fish Larvae

Newly hatched (day 0) red drum larvae were obtained from the Stock Enhancement Research Facility at Florida Marine Research Institute (Port Manatee, FL) and reared in a 143-L recirculating culture tank. Water was maintained at 25.8 ± 0.8°C and 34.4 ± 0.8 ppt salinity and circulated through biological, chemical, and mechanical filtration as well as through an ultraviolet (UV) sterilizer. Beginning at 2 days after hatching, larvae were fed rotifers (*Brachionus plicatilis*), *Artemia* sp., or both, twice daily for 1 hour. During subsequent days, larvae were randomly sampled for feeding trials and to characterize the development of the feeding mechanism. Sampling of larvae for these trials coincided with specifically defined stages of development. First feeding in larval red drum occurred between 3 days and 5 days after hatching. First feeding was defined as the period during yolk sac absorption and "first exogenous feeding." Flexion of the notochord began with the onset of caudal ray formation and dorsal deflection of the notochord tip and continued through the beginning of metamorphosis (about 7–14 days after hatching). Because of the extended duration of this period (7 days), two trials were performed and were classified as "early notochord flexion" (7–10 days after hatching) and "late notochord flexion" (11–14 days after hatching). Metamorphosis began around 15 days after hatching, was marked by the appearance of distal pterygiophores in the median fins, and was completed for most larvae around 19 days after hatching when the full complement of fin rays had developed and larvae morphologically resembled the adult.

Feeding trials were conducted during each of the stages described above. To account for variability in growth, the days chosen for each feeding trial were selected such that a majority of larvae had reached the given stage of development during the day that the trial was run. Larvae for experimental trials were sampled from the hatchery tank at first feeding (3 days after hatching, early notochord flexion (day 9), late notochord flexion (day 13), and metamorphosis (day 18).

Rearing Zooplankton Prey

Brachionus rotundiformis was acquired from the Florida Institute of Technology, Vero Beach Laboratory. *Artemia franciscana* was cultured from cysts and raised to 9 days after hatching. Two copepod species (*Acartia tonsa* and *Nitokra lacustris*) were acquired from University of Texas and North Carolina State University, respectively. All prey types were transferred to closed culture systems in the aquaculture laboratory at Florida Institute of Technology, under conditions of continuous light and aeration, and fed a mixture of *Nannochloropsis* sp. algae paste, Roti-Rich supplement (Florida Aqua Farms, Dade City, FL, USA) and Culture Selco (INVE Aquaculture NV, Baasrode, Belgium), with 100% water changes performed every 3 days.

Development of the Feeding Mechanism

Every day from day 3 to day 30, three fish were randomly sampled from the hatchery, preserved in 10% formalin solution, and then cleared and stained to observe the development of cranial bones and cartilage. Only larvae that met the criteria for each stage of development (see "Larval rearing" section above) were used in the analyses. Representative larvae (freshly killed, preserved, and cleared and stained for bones and cartilage) in each stage of development were examined under an Olympus (Olympus America Inc., Melville, NY, USA) stereomicroscope for the presence of key biomechanical structures of the feeding mechanism. Standard length (SL) or notochord length (NL), maximum gape height, and maximum gape width were measured from each fish. Length was measured to the nearest 0.1 mm using an ocular micrometer mounted on an Olympus microscope. Gape measurements were made to the nearest 0.01 mm with the mouth opened such that the premaxilla and the dentary formed a 90-degree angle. This is considered to be the maximum functional mouth gape for fish larvae (Shirota 1970). Gape height was measured from the anterior-most tip of the premaxilla to the anterior-most tip of the dentary (see Krebs and Turingan 2003). Gape width was measured on the ventral surface of the larva as the distance between the left and right articular-quadrate joints (after Kiorboe et al. 1985).

Filming Fish Larvae and Zooplankton Prey

Prey swimming behaviors and larval feeding events were filmed at a rate of 250 and 500 frames per second (fps), respectively, using a low-light Redlake High-Speed MotionScope 2000S monochrome camera (Redlake MASD Inc., San Diego, CA, USA) with a Sony 45-mm low-light macro lens. Animals were placed in a 5.7 × 1.5 × 0.8–cm (approximately 6.8 mL) polyvinyl chloride chamber supported by a custom-made kinematic bench composed of a 2.5 × 37 × 90–cm aluminum plate with antivibration cushions used to suppress external vibrations. A 1 × 1–mm plastic grid was attached to the rear of the chamber to serve as a scale for kinematic measurements.

The filming chamber was mounted on an XYZ table, allowing accurate control of image focus as well as for movement of the chamber in any direction along the Y-Z plane to keep the animal in focus while filming. Lighting was provided by a Dolan-Jenner fiber optic illuminator (Model PL-180; Dolan-Jenner Industries Inc., Lawrence, MA, USA) with a 50-W, 115-V halogen bulb. Light was transmitted via a flexible fiber optic light guide and reflected through the back of the chamber using a white microscope reflector. Decreasing the amount of direct illumination minimized the amount of heat, thereby reducing an increase in water temperature that could have affected the behavior of the animals.

Filming events were limited to vertical movements of the prey in front of the camera due to the two-dimensional field of the filming setup. As a result, several criteria were used to increase accuracy of measurement of zooplankter swimming velocity. First, specimens were kept in focus throughout the entire swimming event to eliminate errors in measurement due to lateral movements of the animals. Second, the swimming velocity of each prey type was measured according to characteristic movements of the zooplankter. For example, in *A. franciscana*, one sweep of the appendages was considered one stride; in copepods, one stride consisted of one jump or hop; and in *B. rotundiformis* and the ciliate species, a steady swimming movement over a predetermined number of frames was analyzed. Prey types were acclimated to the filming chamber 2 minutes prior to filming. Swimming behavior was recorded at a rate of 250 fps under constant light and temperature regimes to minimize the effects of these environmental factors on swimming behavior. The recorded video was played back at 1% of the recorded rate and analyzed frame by frame. The swimming velocity (in millimeters per second) of at least five different individuals of each prey type, both with and without the presence of a larval fish predator, was calculated using Redlake motion analysis software (Redlake MASD Inc., San Diego, CA, USA).

Feeding Trials

Feeding trials were conducted on days 3, 9, 13, and 18, during which 150 larvae were sampled from the hatchery tank using a 1-L glass beaker and distributed evenly between three 2-L glass bowls containing 1 L of tank water. The outer surface of each bowl was covered with black plastic to prevent external interference (e.g., movement, light) and to enhance prey contrast. Larvae were allowed to acclimate for 1 hour prior to feeding. During the first-feeding trial, two 100-W incandescent lamps were used. For all subsequent trials, full-spectrum fluorescent lighting (78–81 $\mu E/m^2/sec$; 15 W) was provided.

Following the acclimation period, prey were introduced to each bowl at a density of five prey per milliliter for each prey type. Prey consisted of rotifers (0.22 mm ± 0.04 mm [S.E.] in length × 0.13 mm ± 0.02 mm in width), newly hatched *Artemia* nauplii (designated A0) (0.42 mm ± 0.04 mm × 0.14 mm ± 0.02 mm) and 1-day-old (designated A1) *Artemia* nauplii (0.60 mm ± 0.06 mm × 0.14 mm ± 0.02 mm). The prey sizes used in these experiments nearly covered the size range available to wild sciaenid larvae (Pryor and Epifanio 1993). Larvae were allowed to feed for 2 hours, after which the contents of each bowl, including larvae, were immediately siphoned through a 53-μm sieve. All contents were preserved in either 5% formalin or in a mixture of formaldehyde and glutaraldehyde (4CF-1G after McDowell and Trump 1976) for later analysis of larval morphometrics and feeding performance (feeding success and prey size selectivity).

Data Analyses

Gape height and width were regressed against length (NL or SL) for all four developmental stages to examine mouth gape through development in larval red drum. Both gape dimensions were compared for differences across developmental stages of larval fish using a nonparametric, one-way analysis of variance (ANOVA; Kruskal-Wallis). Dunn's multiple comparison test was used for *a posteriori* comparisons.

Feeding performance was measured as the percentage of larvae in each stage with at least one prey item in the gut. Feeding performance was also measured as the mean number of prey consumed by larvae in each stage. Kruskal-Wallis test was used to determine if there was a significant difference, in terms of both measures of feeding performance, between stages of larval development. Significant differences were isolated using Dunn's multiple comparison tests. To determine the effect of variability in larval length within each stage on feeding performance, mean prey number consumed was regressed against body length for each stage from first feeding to metamorphosis.

Mean prey length consumed by larvae in each stage was calculated, and these values were compared using a parametric one-way ANOVA. Mean prey width consumed by larvae in each stage was calculated and compared using Kruskal-Wallis to examine the possibility of differences in prey size selectivity between stages. Differences were isolated using Tukey's (prey length) and Dunn's pairwise (prey width) multiple comparison tests. A prey selectivity index (Jacobs 1974) was calculated for each prey type using the following formula:

$$D = (R_k - P_k) / [(R_k + P_k) - (2R_kP_k)] \quad (10.1)$$

where R_k represents the proportion of each prey type in the diet and P_k represents the proportion of each prey type in the environment. The value for D ranges from −1 to +1, with a negative value representing lesser selectivity toward a specific prey item than a positive value and a zero value representing no selectivity.

Two-way ANOVAs were used to test for significant differences in swimming velocities between the various zooplankters in trials conducted both with and without the presence of a larval fish predator. Post hoc tests were conducted to determine where significant differences in swimming velocity between different prey types were found. T-tests comparing the swimming velocities of each prey type before and after the addition of a larval fish predator were used to determine if the zooplankters in this study were able to significantly increase their swimming speed when threatened by a larval fish predator.

RESULTS

Development of the Feeding Mechanism in Red Drum Larvae

Anatomical examinations of fresh, dead, formalin-preserved, cleared and stained specimens of an age and size series of red drum larvae revealed a two-stage development of the feeding mechanism in this species. Development of the hyoid apparatus is completed on day 3 (Table 10.1). This suggests that the hyoid-mandible linkage is the first of the three musculoskeletal linkages to develop, thus the first mechanism to mediate mouth opening in first-feeding larvae (Fig. 10.1A). This stage is coincident with first feeding, thus this stage is also called the "hyoid-mandible" stage. Beginning on day 13, older larvae have a functional gill apparatus, and ossification of the entire head is nearly complete, indicating the development of the opercular-mandible linkage. This linkage is an

Table 10.1. Stage-specific morphological features of larval red drum

Larval stage	Age (days)	Key morphological features	SL (mm)
First feeding	3–6	Hyoid apparatus fully developed	4.0–6.0
Premetamorphosis	6–11	Gut coiling, notochord flexion	6.0–7.0
Metamorphosis	11–20	Gill apparatus fully developed	7.0–15.0

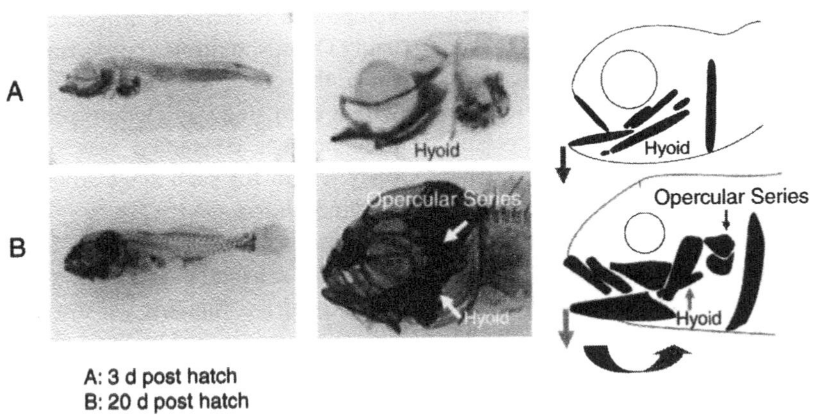

Figure 10.1. The feeding mechanism in (A) first-feeding (hyoid-stage) and (B) metamorphosing (opercular-hyoid stage) red drum (*Sciaenops ocellatus*) larvae. Note the difference in complexity of the linkages that drive the feeding mechanism between the two stages of development.

additional mechanism involved in the depression of the lower jaw and is responsible for mouth opening in postlarval teleost fishes (Fig. 10.1B). Thus, beginning at this age, a fully developed hyoid-opercular-mandible mechanism can be used for mouth opening during prey capture, gill ventilation, "coughing," and prey transport.

Relationship between Mouth Gape and Prey Size in Red Drum Larvae

Gape and prey width increased linearly with body length in red drum larvae (Fig. 10.2). However, as larvae grew in age and body length, gape width increased faster than prey width. Evidently, larvae, especially during the early stages of development, did not select for the maximum prey width as predicted by the gape width-body size relationship (Fig. 10.2). Feeding performance (i.e., number of prey consumed) varied among individuals within an age group in *S. ocellatus* larvae (Fig. 10.3). Older larvae consumed more prey than younger conspecifics of the same body size and gape (see Fig. 10.2 and Fig. 10.3).

Prey Behavior and Feeding Performance in Red Drum Larvae

Prey swimming behaviors were described according to characteristic movements of the zooplankter for both treatments (with and without a larval *S. ocellatus* predator present) (Fig. 10.4). Swimming velocity of zooplankton prey varied greatly between treatments and among prey types. Mean swimming velocity of zooplankton prey ranged from 0.66 ± 0.09 S.E. to 41.30 ± 4.79 S.E. mm/sec in trials conducted without a predator. In contrast, mean swimming velocities ranged from 0.87 ± 0.06 S.E. to 121.03 ± 16.69 S.E. mm/sec when a larval fish predator was present (Fig. 10.4). The ciliate species used in this study exhibited the slowest swimming velocity, while a copepod species, *A. tonsa*, demonstrated the fastest swimming velocity for both treatments.

A two-way ANOVA, with swimming velocity as the dependent variable and prey type and predator presence/absence as the main effects, revealed that swimming velocity varied among prey types ($F_{12, 104} = 357.28$, $P < 0.001$) and be-

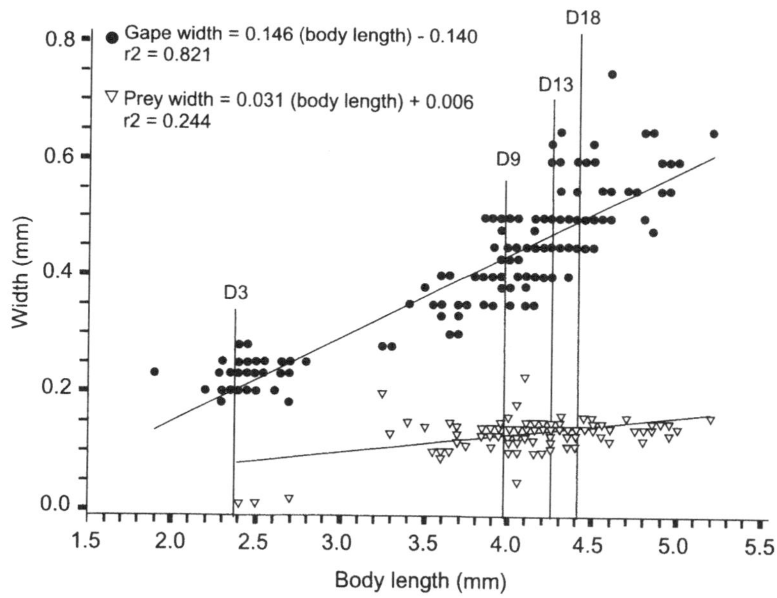

Figure 10.2. Relationship between gape size and body size (notochord or standard length) (circle) and between prey size and body size (triangle) in cultured red drum, *Sciaenops ocellatus*, from first feeding (3 days after hatch [D 3]) to metamorphosis (D 18). The mean body size at each age is indicated by a flag. Note the difference in slopes between these two relationships; prey consumed were much smaller than gape size at any age of larvae.

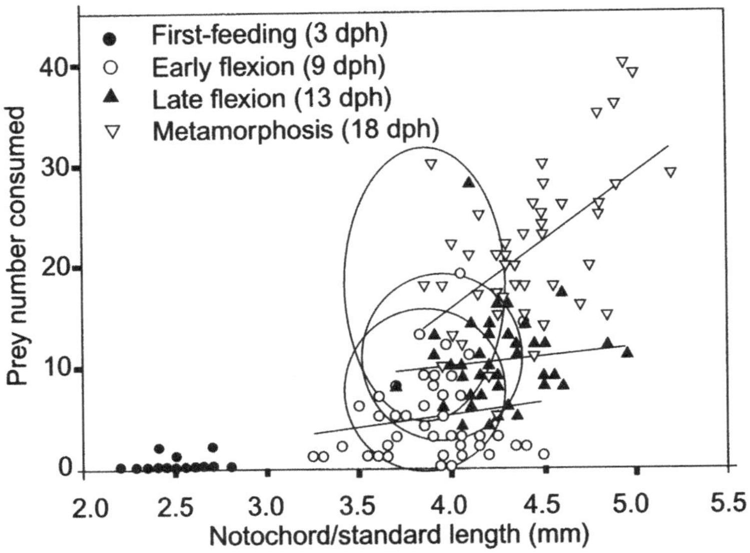

Figure 10.3. Variation in fish size (measured as notochord or standard length) and feeding performance (measured as number of prey consumed) among individual fish within each stage of development. Older red drum, *S. ocellatus,* larvae at later stages of development were more successful at capturing prey than younger larvae with similar gape width, suggesting that feeding performance is associated with stage of development rather than size of larvae.

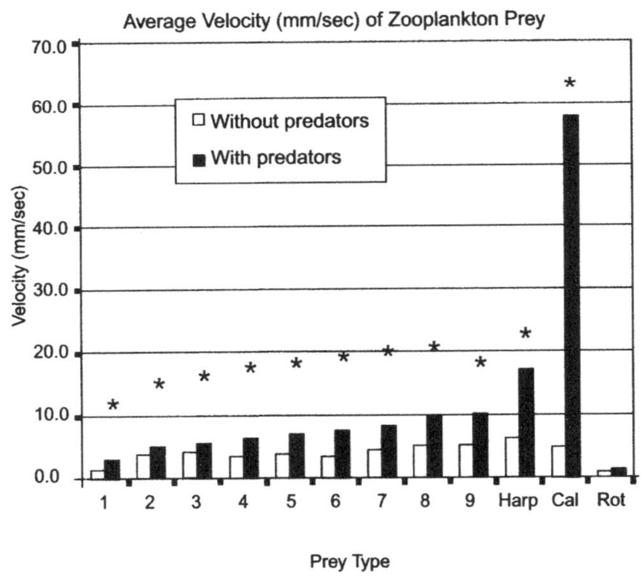

Figure 10.4. Comparison of $\log_{10}$-transformed mean swimming velocity (mm/sec) of zooplankton prey between prey without a predator (Without predators) and prey with a red drum larval predator (With predators). An asterisk indicates prey that significantly (*t* test; $P < 0.05$) increased its swimming velocity when a predator was introduced into the filming chamber. Prey 1–9 were *Artemia* sp. at 1 through 9 days after hatching. Harp, harpacticoid copepod; Cal, calanoid copepod; Rot, *Brachionus rotundiformis*.

tween the presence and absence of predator treatments ($F_{1, 104} = 363.34$, $P < 0.001$). Interaction between prey type and the presence or absence of a predator also influenced prey swimming velocity ($F_{12, 104} = 6.20$, $P < 0.001$). Results of post hoc multiple pairwise comparison tests revealed that prey swimming velocity increased significantly in all prey types with the exception of *B. rotundiformis* and the ciliate species (Fig. 10.4).

Larval red drum began feeding on small prey such as rotifers (*Brachionus* sp.) 3 days after hatching. During the first week of feeding, these larvae are very inefficient at capturing prey. Feeding performance of red drum larvae in the current study was low; less than 5% of larvae could successfully capture prey during the hyoid period (i.e., first-feeding larvae) (Fig. 10.5). In contrast, 85% to almost 100% of 13-day-old and older larvae successfully captured prey. These older larvae have a functional gill apparatus, thus a fully developed hyoid-opercular-mandible mechanism for mouth opening (see above, Development of the feeding mechanism section). Red drum larvae switched prey, from strictly small, less mobile rotifers (*Brachionus* sp., mean swimming velocity = 0.38 ± 0.06 mm/sec) around first feeding to larger, more elusive *Artemia* nauplii (mean swimming velocity = 4.69 ± 0.32 mm/sec) around metamorphosis.

The percentage of larvae ($n = 182$) with food in the gut (feeding success) was lowest during day 3 (2%), as very few ($n = 4$) first-feeding larvae successfully fed (Fig. 10.5). Of the first-feeding larvae that consumed prey, a total of 11 prey items were ingested, consisting of seven A0, one A1, and three copepods, which were incidentally introduced into the system through water changes. Mean prey number consumed by first-feeding larvae was 0.24 prey items per individual (Fig. 10.5). Larvae at early notochord flexion (day 9) were more successful at capturing prey than first-feeding larvae as 37.9% of early flexion larvae had prey in the gut. Mean prey number consumed by each larva increased as well (4.95 prey items). During late notochord flexion (day 13), larvae were more successful at capturing prey (86.2% of larvae fed). They also consumed more prey than first-feeding and early flexion-stage larvae (mean = 10.36 prey items per individual). Larvae at metamorphosis (day 18) were more efficient at capturing prey than those at the previous stages. About 97% of 150 metamorphic larvae had consumed prey. On average, these larvae consumed 21.98 prey per individual. Those larvae that did not feed during this stage appeared to be developing at a slower rate than those that were successful (J. Krebs, personal observation).

Larvae at all stages of development selected prey of significantly different sizes, except for late flexion and metamorphic larvae, which, relative to each another, exhibited no difference in prey-size selection for prey length (one-way

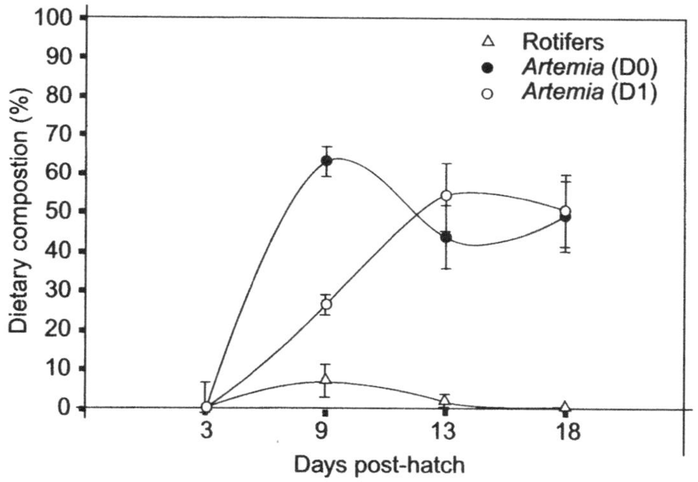

Figure 10.5. Percentage of each prey type consumed by cultured red drum larvae within each age class.

ANOVA; $P < 0.001$) or prey width (Kruskal-Wallis; $P < 0.001$; Dunn's method). The size of prey selected by larvae increased from day 3 to day 9. Because only a small number of first-feeding larvae consumed prey, it was decided that the value for mean prey size selected at this stage did not represent the actual value. Therefore, the authors decided to eliminate this fish from further analysis of stage-specific variation in size of prey consumed by larvae. Early flexion larvae selected mainly (A0) *Artemia* (63.2%) and some larger (A1) *Artemia* (26.5%), but few rotifers (7.2%; Fig. 10.5). Prey selectivity during late flexion (day 13) shifted toward larger prey items. The diet of larvae during this stage consisted mainly of larger (A1) *Artemia* (54.1%) and (A0) *Artemia* (43.8%). Rotifers were seldom selected by late flexion larvae (1.9% of the diet). Larvae at metamorphosis (day 18) consumed primarily (A0) and (A1) *Artemia* (49.2% and 50.4% of diet, respectively). Only four larvae consumed rotifers. Despite consuming larger prey through development, larvae did not exclusively select for the largest available prey. Prey selectivity values calculated using Jacobs' index of selectivity are summarized in Table 10.2.

Table 10.2. Jacobs' prey selectivity indices for cultured red drum, *Sciaenops ocellatus* larvae

	Jacob's Selectivity indices for each prey type		
Developmental stage	ROTS	ART 0	ART 1
First feeding	−1.00	0.778	−0.333
Early flexion	−0.724	0.579	−0.139
Late flexion	−0.925	0.221	0.405
Metamorphosis	−0.984	0.320	0.341

Note: This measure is based on the proportion of prey consumed by fish larvae relative to the proportion of prey available in the hatchery environment (Jacobs 1974). Values for this index range from −1 to +1: −1 = negative selectivity, 0 = no selectivity, and +1 = strong selectivity.

ROTS, Rotifer prey; ART 0, newly hatched *Artemia* nauplii; ART 1, 1-day-old *Artemia* nauplii.

DISCUSSION

Notwithstanding the magnitude of larviculture research conducted over the past four decades, performance of larvae at first-feeding and larval nutrition continue to be critical considerations before a species is selected for aquaculture. To the knowledge of the authors, very few studies have directly investigated the causal relationship between prey type and prey-capture success in cultured fish larvae. Understanding the causal link between development of the feeding mechanism and prey-capture performance is vital to understanding mortality and survivorship in cultured marine fish larvae.

In this study, the use of state-of-the-art digital high-speed videography allowed the investigators to quantify the movement of prey with and without the presence of a larval fish predator. First-feeding larvae use a simple mechanism involving the hyoid-mandible linkage (i.e., the hyoid stage) to capture prey. Larvae around metamorphosis capture prey using a more complex mechanism involving the hyoid and opercular series of mandible linkages (i.e., the hyoid-opercular stage). This two-stage development of the feeding mechanism is associated with stage-specific differences in feeding performance: less than 50% of larvae fed during the hyoid stage, while almost 100% of larvae at the hyoid-opercular stage consumed prey. First-feeding larvae selected small and less elusive prey; conspecifics at metamorphosis selected larger and more elusive prey. The two copepod species had the highest swimming velocity among all zooplankton examined, especially with the presence of a larval fish predator. Prey-capture performance in marine fish larvae is influenced by the size, escape response, and speed of zooplankton prey. Thus, the authors propose that in addition to size, the swimming velocity and escape response of zooplankton be considered in the selection of an appropriate live feed for marine fish larviculture.

Although measurements of mouth dimensions from preserved specimens is a good first step in the exploration of the possible role the feeding apparatus plays in selection of prey and feeding success in larval fishes, in nature mouth gape is a dynamic parameter that could vary depending on the size of prey encountered by the fish (Liem 1980, 1991; Wainwright and Turingan 1993). In juvenile and adult fishes, the dynamics of prey capture is driven by three musculoskeletal linkages: (a) epaxialis muscles-neurocranium, (b) levator operculi muscles-opercular bones-mandible, and (c) hypaxial muscles-sternohyoideus-hyoid apparatus-mandible (Lauder 1985). The first coupling involves the contraction of the epaxialis muscles, which causes the neurocranium to rotate dorsally, thus expanding the roof of the mouth cavity. In the second linkage system, rotation of the gill cover upon contraction of the levator operculi muscle pulls the lower jaw ventrally, thus opening the mouth. The third musculoskeletal coupling is mediated by the hyoid apparatus. In this linkage, contraction of the sternohyoideus muscle causes the hyoid bar to retract; this movement of the hyoid is transmitted to the lower jaw via the hyomandibular ligament, causing the lower jaw to rotate ventrally. In most fishes, further expansion of the buccal cavity that follows mouth opening is facilitated by the contraction of the levator arcus palatini muscle. The key feature of a prey-capture event in juvenile and adult fishes is the synchronous action of all of these linkage systems (Liem 1980; Muller and Osse 1984; Lauder 1985; Turingan and Wainwright 1993).

In sharp contrast to juvenile and adult fishes, prey capture during first feeding in larval fishes may involve only one of the suite of mouth-opening and -expanding mechanisms employed by juvenile and adult fishes: the hyoid-mandible coupling (Otten 1982; Liem 1991; Hunt von Herbing et al. 1996). Anatomical and histological study of the feeding mechanism of cultured mahi mahi and red drum have documented the complete development of the hyoid apparatus around 3 days after hatching (Turingan, unpublished data). These results suggest that the hyoid-mandible linkage is the first of the three musculoskeletal linkages to develop, thus the initial mechanism to mediate mouth opening in first-feeding larvae (Fig. 1; Turingan 1999).

The transformation from the hyoid-mandible–driven, first-feeding mechanism to the hyoid-opercular-mandible–driven feeding mechanism in older, metamorphosing larvae is associated with different feeding performance between the two larval fish-feeding stages. It is evident that this two-stage development of the feeding mechanism in cultured larvae is associated with the ability of larvae to capture prey, especially during the earlier stages of development. Larval mahi mahi and red drum have been shown to begin feeding on small prey such as rotifers (*Brachionus* sp.) the third day after hatching (Krebs and Turingan 2003). During the first week of feeding, these larvae are very inefficient at capturing prey. In contrast, 13-day-old larvae are very successful at prey capture. Larvae at metamorphosis and older juveniles have a functional gill apparatus, thus a fully developed hyoid-opercular-mandible mechanism for mouth opening. Krebs and Turingan (2003) also revealed that red drum larvae switched prey, from strictly small, less mobile rotifers (*Brachionus* sp.) to the more elusive *Artemia* nauplii around metamorphosis.

Despite the importance of studies that demonstrate the strong correlation between mouth gape and prey-size selection in fish larvae, the present study revealed that prey size selected by cultured red drum larvae is only weakly associated with gape size. This is consistent with the findings of Hunter (1981) and Bremigan and Stein (1994), who concluded that fish larvae consume prey that are at most 25% of mouth gape. These findings suggest that factors other than mouth gape (physiological and behavioral) influence prey selection in cultured fish larvae.

Size and swimming behavior of zooplankton prey appear to influence prey selection in cultured fish larvae. Zooplankton, such as *A. franciscana* nauplii, *B. rotundiformis*, *A. tonsa*, and *N. lacustris*, exhibit a wide range of swimming velocity and behavioral patterns that affect the ability of fish larvae to capture them (Buskey et al. 1993). An even more interesting characteristic of these species and possibly other zooplankton types that are used as live prey in larviculture is the ability of the prey to accelerate its swimming behavior in the presence of a larval fish predator. The ability of prey species to significantly increase swimming speed to avoid capture by a larval fish predator may be an adaptive escape response to the predatory threat (Rao 2003). Zooplankton prey that accelerate their swimming in the presence of a larval fish predator, such as both copepod species and, to a lesser degree, day 1 through day 9 *A. franciscana* used in this study, may be too elusive for first-feeding larvae to capture. Kerfoot et al. (1980) demonstrated that among zooplankton prey, calanoid copepods are one of the best at eluding predators due to a heightened number of mechanoreceptors that sense disturbances in the normal flow around them, allowing them to evade attackers with speeds greater than any other zooplankter. These highly elusive prey types may be too energetically costly to pursue during early stages of larval fish development, resulting in increased unsuccessful prey-capture attempts (Coughlin 1991) and, ultimately, starvation. Prey-specific predator avoidance behaviors are likely detected by larval fish predators, which may underlie the positive selection of slower-moving rotifers and ciliates by cultured fish larvae, especially during the time of first feeding (Sifa and Mathias 1987; Coughlin 1991; Osse and Boogaart 1999; Hernandez 2000).

CONCLUSION

Understanding predator–prey interactions is key to successful feeding and nutrition in larviculture. Mechanical properties and the functional morphology of the feeding mechanism underlie the ability of the developing larva to capture its optimal prey from those available in the hatchery environment. The feeding mechanism of the developing larva influences feeding performance in a number of ways. First, it is well known that maximum functional mouth gape limits the maximum size of prey that can be captured by a larval fish at a given size and age. Second, the muscular-skeletal linkages determine the speed and magnitude of buccal expansion and compression, key kinematic events that influence the size and mobility of prey that larvae can capture at a particular stage of development. Third, these expansive and compressive events also influence the ability of the larva to transport the captured prey into the esophagus for swallowing. These roles that the feeding mechanism play in the feeding biology of the marine fish larva underscore the need to address the development of the mechanical and architectural designs of the feeding apparatus in the formulation of stage-specific feeding protocols for cultured fish larvae.

ACKNOWLEDGMENTS

The authors thank William Halstead and Robert Wattendorf of the Florida Fish and Wildlife Conservation Commission for providing fertilized red drum eggs for this study. This study was funded by a Florida Sea Grant research development grant R/LR-A-38 to R.G. Turingan.

REFERENCES

Bagarinao, T. and Hunter, J.R. 1983. The visual feeding threshold and action spectrum of northern anchovy (*Engraulis mordax*) larvae. *California Cooperative Oceanic Fisheries Investigations. Reports* 24:245–254.

Bailey, K.M. and Batty, R.S. 1984. Laboratory study of predation by *Aurelia aurita* on larvae of cod, flounder, plaice and herring: development and vulnerability to capture. *Marine Biology* 83:287–291.

Bailey K.M. and Houde, E.D. 1989. Predation on eggs and larvae of marine fishes and the recruitment problem. *Advances in Marine Biology* 25:1–83.

Blaxter, J.H.S. 1968. Visual thresholds and spectral sensitivity of herring larvae. *Journal of Experimental Biology* 48:39–53.

Blaxter, J.H.S. 1986. Development of sense organs and behavior of teleost larvae with special reference to feeding and predator avoidance. *Transactions of the American Fisheries Society* 115:98–114.

Blaxter, J.H.S., Gray, J.A.B., and Best, A.C.G. 1983. Structure and development of the free neuromasts and lateral line system of the herring. *Journal of the Marine Biological Association of the UK* 63:247–260.

Bremigan, M.T. and Stein, R.A. 1994. Gape-dependent larval foraging and zooplankton size: Implications for fish recruitment across systems. *Canadian Journal of Fisheries and Aquatic Sciences* 51:913–922.

Buskey, E.J., Coulter, C., and Strom, S. 1993. Locomotory patterns of microzooplankton: potential effects on food selectivity of larval fish. *Bulletin of Marine Science* 53(1):29–43.

Coughlin, D.J. 1991. Ontogeny of feeding behavior of first-feeding Atlantic salmon (*Salmo salar*). *Canadian Journal of Fisheries and Aquatic Sciences* 48:1896–1904.

Dabrowski, K. and Bardega, R. 1984. Mouth size and predicted food size preferences of larvae of three cyprinid fish fishes. *Aquaculture* 40:41–46.

Dempsey, C.H. 1978. Chemical stimuli as a factor in feeding and intraspecific behaviour of herring larvae. *Journal of the Marine Biological Association of the UK* 58:739–747.

Disler, N.N. 1971. Lateral line sense organs and their importance in fish behaviour. Israel Program for Scientific Translations, Jerusalem, Israel.

Doi, M., Ohno, A., Kohno, H., Taki, Y., and Singhagraiwan, T. 1997. Development of feeding ability in red snapper (*Lutjanus argentimaculatus*) early larvae. *Fisheries Science* 63:845–853.

Duray, M. and Kohno, H. 1990. Development of mouth width and larval growth in larvae of three marine fish species. *Philippine Journal of Science* 119:247–255.

Ferry-Graham, L. and Lauder, G.V. 2001. Aquatic prey capture in ray-finned fishes: a century of progress and new directions. *Journal of Morphology* 248:99–119.

Fuiman, L.A. 1993. Development of predator evasion in Atlantic herring, *Clupea harengus* L. *Animal Behavior* 45:1101–1116.

Fuiman, L.A. and Magurran, A.E. 1994. Development of predator defenses in fishes. *Reviews in Fish Biology and Fisheries* 4:145–183.

Hagen, N. 1993. Effect of different prey and larval densities on the gut content of plaice (*Pleuronectes platessa* L.) at initial feeding. pp. 180–182. In *Physiological and Biochemical Aspects of Fish Development*, edited by B.T. Walther and H.J. Fyhn, pp. 180–182. Bergen, Norway: University of Bergen.

Hernandez, LP. 2000. Intraspecific scaling of feeding mechanics in an ontogenetic series of zebrafish, *Danio rerio*. *Journal of Experimental Biology* 203:3033–3043.

Houde, E.D. and Schekter, R.C. 1980. Feeding by marine fish larvae: developmental and functional responses. *Environmental Biology of Fishes* 5:315–334.

Hunt von Herbing, I., Miyake, T., Hall, B.K., and Boutilier, R.G. 1996. Ontogeny of feeding and respiration in larval Atlantic cod (*Gadus morhua*) (Teleostei, Gadiformes): I. Morphology. *Journal of Morphology* 227:15–35.

Hunter, J.R., editor. 1976. Report of a colloquium on larval fish mortality studies and their relation of fishery research, January 1975. NOAA (National Oceanic and Atmospheric Administration) Tech. Rep. NMFS Cir. 395:1–5. U.S. Dept. of Commerce, Washington, D.C., USA.

Hunter, J.R. 1981. Feeding ecology and predation of marine fish larvae. In *Marine Fish Larvae: Morphology, Ecology and Relation to Fisheries*, edited by R. Lasker, pp. 33–77. Seattle, Washington, USA: Washington Sea Grant Program.

Iwai, K. 1980. Sensory anatomy and feeding of fish larvae. In *Fish Behavior and Its Use in the Capture and Culture of Fishes*, edited by J.E. Bardach, J.J. Magnuson, R.C. May, and J.M. Reinhart, pp. 124–145. Manila, Philippines: International Center for Living Aquatic Resources Management (ICLARM).

Jacobs, J. 1974. Quantitative measurement of food selection. *Oecologia* 14:413–417.

Kerfoot, K.C., Kellogg, D.L., Jr., and Strickler, J.R. 1980. Visual observations of live zooplankters: evasion, escape and chemical defenses. In *Evolution and Ecology of Zooplankton Communities*, edited by W.C. Kerfoot, pp. Hanover, New Hampshire, USA: The University Press of New England.

Kiorboe, T.K., Munk, P., and Stottrup, J.G. 1985. First feeding by larval herring (*Clupea harengus L.*) *Dana* 5:95–107.

Krebs, J.M. and Turingan, R.G. 2003. Intraspecific variation in gape-prey size relationships and feeding success during early ontogeny in red drum, *Sciaenops ocellatus*. *Environmental Biology of Fishes* 66:75–84.

Langsdale, J.R.M. 1993. Developmental changes in the opacity of larval herring, *Clupea harengus*, and their implications for vulnerability to predation. *Journal of the Marine Biological Association of the UK* 73:225–232.

Lazo, J.P., Dinis, M.T., Holt, G.J., Faulk, C., and Arnold, C.R. 2000. Co-feeding microparticulate diets with algae: toward eliminating the need of zooplankton at first feeding in larval red drum (*Sciaenops ocellatus*). *Aquaculture* 188:339–351.

Lauder, G.V. 1980. Hydrodynamics of prey capture by teleost fishes. In *Biofluid Mechanics II*, edited by D. Schneck, pp. 161–181. New York, New York, USA: Plenum Press.

Lauder, G.V. 1985. Aquatic feeding in lower vertebrates. In *Functional Vertebrate Morphology*, edited by M. Hildebrand, D. Bramble, K. Liem, and D. Wake, pp. 210–229. Cambridge, Massachusetts, USA: Harvard University Press.

Liem, K.F. 1980. Adaptive significance of intra- and interspecific differences in the feeding repertoires or cichlid fishes. *American Zoologist* 20:295–314.

Liem, K.F. 1991. A functional approach to the development of the head of teleosts: implication on the constructional morphology and constrains. In *Constructional Morphology and Evolution*, edited by N. Schmidt-Kittler and K. Vogel, pp. 231–249. Berlin, Germany: Springer-Verlag.

McDowell, E.M. and B.F. Trump. 1976. Histologic fixatives suitable for diagnostic light and electron microscopy. *Archives of Pathology and Laboratory Medicine* 100:405–414.

Muller, M. and Osse, J.W.M. 1984. Hydrodynamics of suction feeding in fish. *Transactions of the Zoological Society of London* 37:51–135.

O'Connell, C.P. 1981. Development of organ systems in the northern anchovy *Engraulis mordax* and other teleosts. *American Zoologist* 21:429–446.

Osse, J.W.M. and Boogaart, J.G.M. 1999. Dynamic morphology of fish larvae, structural implications of friction forces in swimming, feeding and ventilation. *Journal of Fish Biology* 55(A):156–174.

Otten, E. 1982. The development of the mouth-opening mechanism in a generalized *Haplochromis species: H. elegans* Trewavas, 1933 (Pisces, Cichlidae). *Netherlands Journal of Zoology* 32:31–48.

Pepin, P. and Penney, R.W. 1997. Patterns of prey size and taxonomic composition in larval fish: are there general size dependent models? *Journal of Fish Biology* (Supp. A) 51:84–100.

Pryor, V.K. and Epifanio, C.E. 1993. Prey selection by larval weakfish (*Cynoscion regalis*): the effects of prey size, speed and abundance. *Marine Biology* 116:31–37.

Rao, T.R. 2003. Ecological and ethological perspectives in larval fish feeding. *Journal of Applied Aquaculture* 13 (1–2):145–178.

Rice, J.A., Crowder, L.B., and Holey, M.E. 1987. Exploration of mechanisms regulating larval survival in Lake Michigan bloater: a recruitment analysis based on characteristics of individual larvae. *Transactions of the American Fisheries Society* 116:703–718.

Schael, D.M., Rudstam, L.G., and Post, J.R. 1991. Gape limitation and prey selection in larval yellow perch (*Perca flavescens*), freshwater drum (*Aplodinotus grunniens*) and black crappie (*Pomoxis nigromaculatus*). *Canadian Journal of Fisheries and Aquatic Sciences* 48:1919–1925.

Shirota, A. 1970. Studies on the mouth size of fish in the larval and fry stages. *Bulletin of the Japanese Society of Scientific Fisheries* 36:353–368. (Translation of the Fisheries Research Board of Canada No. 1978).

Sifa, L. and Mathias, J.A. 1987. The critical period of high mortality of larvae fish — a discussion based on current research. *Chinese Journal of Oceanology and Limnology* 5(1):80–96.

Turingan, R.G. 1999. Two-stage development of the feeding mechanism in larval marine fishes: consequences for feeding performance. *American Zoologist* 39(5):85A.

Turingan, R.G. and Wainwright, P.C. 1993. Functional and morphological bases of durophagy in the queen triggerfish, *Balistes vetula* (Pisces, Tetraodontiformes). *Journal of Morphology* 215:101–118.

Wainwright, P.C. and Turingan, R.G. 1993. Coupled versus uncoupled functional systems: motor plasticity in the queen triggerfish *Balistes vetula*. *Journal of Experimental Biology* 180:209–227.

Watanabe, T. and Kiron, V. 1994. Prospects in larval fish dietetics. *Aquaculture* 124:223–251.

11
Copepods as Live Prey: A Review of Factors That Influence the Feeding Success of Marine Fish Larvae

Edward J. Chesney

ABSTRACT

The critical importance of fish larvae successfully capturing live prey during their first few days of life has been a concept central to aquaculture production and fisheries recruitment for many years. In the marine environment, copepods and their nauplii are the dominant prey of many first-feeding fish larvae, often making up as much as 90–100% of their diet. Although their abundance in marine environments and size dictates the importance of copepods as prey, fish larvae exhibit some degree of selection for different types of copepods and other prey. A number of factors may contribute to prey selectivity. Marine fish larvae vary in their ability to capture different sizes and types of zooplankton prey at first feeding because of associated differences in larval size, visual acuity, swimming patterns and abilities of the larvae, and the size, swimming behavior, and other attributes of their prey. Environmental factors such as light (quality and quantity), turbidity, and temperature also affect capture rates. In the culture of fish larvae, tank color is also known to affect capture rates of live prey. Prey size (width) is one of the most significant factors affecting capture success and is strongly related to mouth size of a larva. Vital rates (respiration rates, assimilation) that determine how captured energy is used by larvae can also vary depending on the species of larvae, prey type and quality, temperature, and stage of larval development. The consequence of this variability is that fish larvae are likely to vary significantly in their energy requirements and efficiency in using ingested prey at a given stage of development early in their life cycle. Several size-related attributes can affect the ability of the larva to feed at a given food density, including perceptive or reactive distance, swimming speed, and mouth size. Foraging behavior and competence can also have significant effects on the ability of a larva to capture prey of a specific size and prey density. Mean reactive distances of fish larvae typically range from 0.70–1.43 body lengths (BL) at first feeding. Swimming speeds while foraging range from 0.41–2.69 BL per second at first feeding in species studied to date, but increase with size and stage of larval development. Feeding studies and direct observations of foraging behavior show that fish larvae gain competence in prey capture quickly. Capture success can be low at first feeding (3–30%), but increases as larvae become more competent. Competence includes gaining size, swimming ability, maneuverability, and experience in aiming toward and capturing prey. First-feeding larvae also face bioenergetic constraints because of low assimilation efficiencies and high weight-specific metabolism. The combination of bioenergetic and foraging challenges may be especially problematic for small first-feeding larvae growing at high temperatures. The diversity of prey sizes offered by copepods provides an advantage over traditional types of cultured live prey for small first-feeding larvae because the size spectrum of life stages allows larvae to capture prey within an optimal size range and capture efficiency that facilitates faster growth.

INTRODUCTION

The culture of marine fishes now stands at only 781,000 metric tons, totaling less than 1% of world fisheries production (FAO 2002). This is despite the fact that marine fishes dominate fishery landings and aquaculture currently makes up a significant proportion of global fish production. Many teleost fishes, including those now in production or under aquaculture development, produce relatively small planktonic larvae that require live zooplankton prey at first feeding (Tucker 1998). In nature, fish larvae prey upon a wide variety of different types and sizes of zooplankton. In the marine environment, the diet of fish larvae typically includes protozoans, mollusk larvae, appendicularians, rotifers, copepods, marine clodocerans, and phytoplankton (Arthur 1976; Hunter 1981; Turner 1984). All life stages of copepods are important prey for all life stages of fish. Many marine fish larvae continue to feed on the larger stages of copepods until they reach metamorphosis while planktivorous fishes such as anchovies and sardines spend their entire life cycle consuming zooplankton (Lazzaro 1987). Estuarine and coastal marine environments provide a fertile habitat that serve as primary nursery areas for many marine fishes of interest for aquaculture (Tucker 1998). Within this habitat, copepods are abundant and serve as an important prey resource for wild fish and invertebrate larvae of many species.

Copepods as Prey for Fish Larvae

The critical importance of fish larvae successfully capturing live prey during their first few days of life has been a concept central to fisheries recruitment and aquaculture production for many years (Hjort 1914; Houde 1987; Cushing and Horwood 1994; Avault 1996). In the marine environment copepod life stages typically make up the bulk of the zooplankton biomass in the size range suitable as prey for most larval and planktivorous fishes (Turner 1984; Dagg and Govoni 1996). Numerous diet studies in the field show that the nauplius through adult stages of copepods are important prey items of many marine fish larvae, typically making up greater than 50% or more of their stomach contents (Hunter 1981; Setzler-Hamilton et al. 1982; Govoni et al. 1983; Houde and Lovdal 1984; Turner 1984; Economou 1991; Munk and Nielsen 1994). Although the abundance of copepods in marine environments and the size of copepod life stages dictates that they are likely to be important prey for fish larvae, some degree of apparent selection has been observed for different types of copepods and other prey. Without knowing precise encounter rates between fish larvae and their potential prey, it is impossible to determine the degree of selectivity with certainty. From laboratory studies it is known that in a mixed prey field, both adult and juvenile planktivorous fish (Zaret 1980) and fish larvae (Buskey et al. 1993) select certain prey over others, including selection for copepods over other prey types (Margulies et al. 2001). A number of factors are known to contribute to prey selectivity. Fish larvae vary in their ability to capture different sizes and types of zooplankton prey at first feeding because of differences in larval size, visual acuity, swimming patterns and abilities of the larvae, and the size, swimming behavior, visibility, and other attributes of their zooplankton prey (see Buskey, 2005).

Observed differences in capture rates for different types of copepod nauplii suggest that understanding the ability of fish larvae to capture a particular kind of copepod nauplii should be a consideration in the development of copepods as live feeds. The ability of a larva to capture a particular type of prey might be influenced by larval perception of the prey, "catchability" (size, shape, and predator avoidance), and palatability. Perception is influenced by visibility of the prey (morphology, pigmentation, behavior) and the distribution and movements of a larva versus its prey. Another factor is movement of the prey relative to the movement of a fish larva in search of prey and the foraging ability of a particular species of larva. This chapter offers an overview of the complex set of factors that affect the interaction between fish larvae and their zooplankton prey, how these interactions help us understand the subtleties of larval feeding, and how the prey size requirements are likely to change for fish larvae as they develop.

PREY SIZE SPECTRA AND COPEPODS AS PREY

Although the behavior of the prey is an important consideration in the ability of cultured fish larva to effectively capture zooplankton, prey size (width) is equally important since it affects the ability of the fish larva to capture and swallow prey. Clearly, mouth size or gape limits what a larva can capture (Shiroto 1970; Pearre 1980). The size of the esophagus, however, sets an upper limit on the ability of a larva to swallow or ingest live prey (Busch 1996). Relationships between mouth size (gape) and prey capture, extensively studied in the field and laboratory, have focused on understanding fish recruitment and prey size requirements of fish larvae (Shiroto 1970; Hunter 1981; Dabrowski and Bardega 1984; Munk 1992; Bremigan and Stein 1994; Busch 1996). Body and mouth size of fish larvae are highly variable at first feeding. This indicates that the ability of larvae to capture a particular size of prey also varies significantly (Fig. 11.1). In general smaller larvae have been reported to eat smaller prey at first feeding, although larger larvae often capture prey of widely varying sizes at first feeding. Some studies have shown the efficiency of prey capture for different sizes of prey is determined by the ratio of the size of the gape of the larva to the size of the prey (Munk 1992). Larvae capture prey with greater efficiency when attacking prey of relatively smaller sizes (Munk 1992). Nevertheless, as they grow, fish larvae tend to attack and capture larger and larger prey and begin to ignore prey that is apparently too small.

Perhaps the single greatest advantage copepods offer fish larvae as a prey is the diversity and the size spectra afforded by their numerous developmental stages (Fig. 11.2 and Fig. 11.3). Prey width is generally considered the most important metric for the ability of the larva to ingest a particular type of prey (Shiroto 1970; Pearre 1980). It may also be a better estimator of weight of copepod life stage than either prosome length or total length (Pearre 1980). Size–weight relationships from the literature (Chisholm and Roff 1990) were used to estimate the spectra of widths and weights offered by five different species of tropical neritic copepods. Figures 11.2 and 11.3

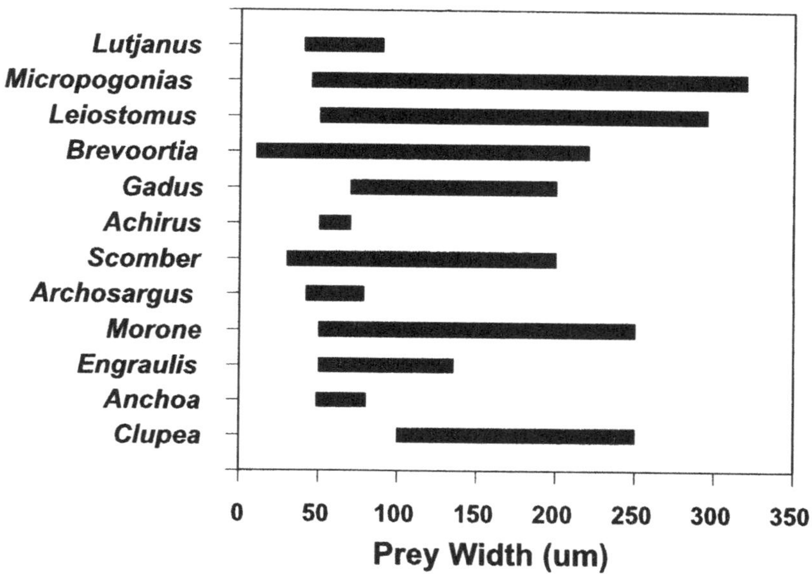

Figure 11.1. A comparison of prey size spectra eaten by 12 species of marine fish larvae derived from field and laboratory studies. Data are for red snapper, *Lutjanus campechanus* (E.J. Chesney, personal observation); croaker, *Micropogonius undulatus*; spot, *Leiostomus xanthurus*; menhaden, *Brevoortia tyrannus* (Govoni et al. 1983); cod, *Gadus morhua* (Kane 1984; Economou 1991); lined sole, *Achirus lineatus* (Stepien 1976); sea bream, *Archosargus rhomboidalis*; Bay anchovy, *Anchoa mitchilli* (Houde and Lovdal 1984); Pacific mackerel, *Scomber japonicus* (Hunter and Kimbrell 1980); striped bass, *Morone saxatilis* (Setzler-Hamilton et al. 1982); Northern anchovy, *Engraulis mordax* (Arthur 1976); and Atlantic herring, *Clupea harengus* (Checkley 1982).

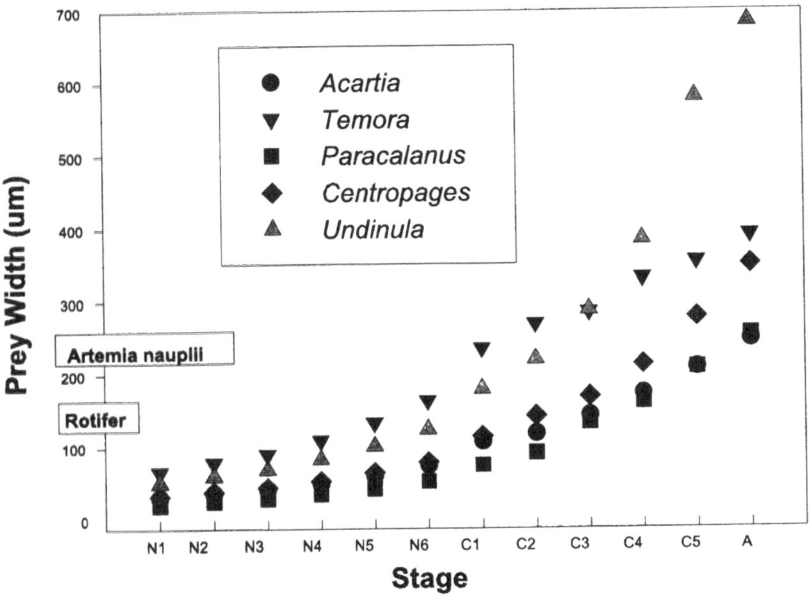

Figure 11.2. Zooplankton size spectra (width) for five species of copepods, *Acartia clausi, Temora turbinata, Paracalanus aculeatus, Centropages velificatus,* and *Undinula vulgaris,* for nauplius through copepodid and adult stages. Sizes are estimated based upon regressions in Durbin and Durbin (1978) and Chisolm and Roff (1990). Widths of the rotifer *Brachionus plicatilis* (L), and *Artemia salina* are plotted for comparison.

show prey size spectra (as widths) based upon those regressions for a variety of different types of copepod from the nauplius through the copepodid stages. The average size of the rotifer *Brachionus plicatilis* (L-strain) and *Artemia* nauplii are plotted for comparison.

These relationships illustrate two important points. First, copepods offer a significantly broader range of prey sizes for first-feeding larvae than the zooplankton typically used as live feed such as rotifers and *Artemia.* Second, not all of the copepods in this example produce nauplii small enough to be suitable as prey for the smallest fish larvae at first feeding. For example the nauplius I widths of *Acartia* sp. and *Paracalanus* sp. average 30 μm and 21 μm, respectively, compared with the much larger nauplius I of *Undinula* sp. (50 μm) and *Temora* sp. (66 μm). The latter species are perhaps a little too big for some of the smallest marine fish larvae. This suggests that the most useful species for aquaculture may be those small enough to offer suitable size spectra for larvae requiring sizes smaller than rotifers and having characteristics that make them suitable for mass culture (Hernandez Molejon and Alvarez-Lajoncheres 2003).

CHARACTERISTICS OF FISH LARVAE FROM DIFFERENT ENVIRONMENTS

Researchers attempting to unravel the role of the early life stage in the recruitment of fishes have noted that differences in attributes of larvae from contrasting environments offer variable potentials for fish larvae to be food limited (Houde 1989; Pepin 1989; Houde and Zastrow 1993; Houde 1994; Fuiman 2002). These variable potentials will have a direct impact on the effort and the techniques required to culture fish larvae from these contrasting habitats. Based upon a comparison of 42 freshwater species from 21 families and 42 species of marine fishes from 34 families, Houde and Zastrow (1993) concluded that, on average, freshwater fishes produce larger eggs and larvae than marine fishes and their eggs tend to be demersal. The marine fishes they compared generally had smaller planktonic eggs (median = 1.02

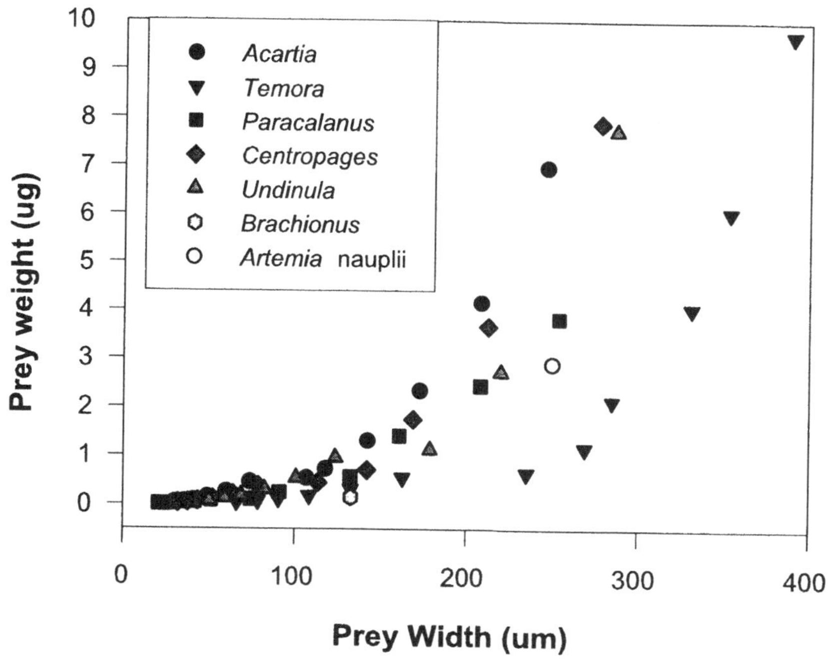

Figure 11.3. Zooplankton size (width) and dry weight (μg) spectra for five species of copepods, *Acartia clausi, Temora turbinata, Paracalanus aculeatus, Centropages velificatus,* and *Undinula vulgaris,* for nauplius through copepodid and adult stages. Sizes and weights are estimated based upon regressions in Durbin and Durbin (1978) and Chisolm and Roff (1990). Widths and dry weights of the rotifer *Brachionus plicatilis* (L), and *Artemia salina* are plotted for comparison.

mm) and shorter incubation periods ($I = 10.9 \pm 0.27$ days) than similar-sized freshwater fish eggs ($I = 7.0 \pm 0.33$ days) (Houde 1994; Fuiman 2002). These characteristics have important implications and consequences for the duration of the larval period of fishes from different habitats (Houde 1989, 1994). Houde (1989) calculated temperature-adjusted growth rates and estimated that, on average, duration of the larval period based upon temperature-adjusted growth rates is 36 days for marine fishes compared with only 21 days for freshwater fishes.

Another important characteristic of fish larvae with relevance to aquaculture is the interaction between larval size and developmental temperature. This has been emphasized as a factor that can influence the likelihood for a given species of larva to be food limited in the sea (Houde 1989). A comparison of early life-cycle characteristics of fishes from different environments demonstrated that, on average, fishes from low latitudes (tropical and subtropical environments) have significantly smaller eggs and larvae and higher temperature-adjusted metabolic rates (Houde 1989; Houde and Zastrow 1993). Although they are generalizations, these insights have obvious consequences for the culture of some tropical and subtropical marine larvae, especially with regard to prey-size requirements and the overall effort required to culture marine species compared with the freshwater species, which presently dominate global aquaculture production. Overall, the three principal conclusions that can be drawn from Houde's analyses (that are significant for the rearing of larval fishes) are that (a) small tropical and subtropical fish larvae have higher temperature-adjusted weight-specific metabolic rates; (b) this will make them more likely to be vulnerable to food limitation; and (c) their generally smaller size indicates that, on average, they will require smaller prey at first feeding.

THE BIOENERGETICS OF LARVAL LIFE

Bioenergetics can play a major role during the early development of fish larvae, especially from an energy accumulation and allocation standpoint

Table 11.1. A hypothetical comparison of energy allocation in larval and juvenile fish

Life stage	Ingested energy (I)	Assimilation rate (A)	Assimilated energy ($I*A$)	Percent allocated to metabolism ($M = M_B + M_A + M_{SDA}$)	Energy allocated to M ($M*A$)	Energy units available for growth	Percent of ingested energy for G
Larval fish	100	0.60	60	65	39	21	21%
Juvenile fish	100	0.80	80	60	48	32	32%

(Pederson 1997). It was pointed out in the previous section that, in general, tropical and subtropical marine fish larvae are at a disadvantage energetically because of their comparatively low assimilation efficiencies and higher temperature-adjusted metabolic rates (Houde 1989, 1994). Larvae respire cutaneously prior to development of branchial respiration and have substantially higher weight-specific metabolic rates (Rombough 1988) and lower assimilation efficiencies than their juvenile counterparts. Until fish larvae fully develop their digestive and circulatory systems, they continue to face substantial bioenergetic disadvantages. Despite these constraints, fish larvae often have growth rates comparable to or higher than their juvenile and adult counterparts (Pederson 1997). Low assimilation rates and higher weight-specific metabolism alone should significantly affect their relative ability to accumulate biomass, especially during the first few days of life. However, they face the additional challenges of limited visual acuity, poorly developed sensory systems, small size, and poor swimming ability.

The relative energy-gathering challenges faced by larvae can be illustrated with a simple bioenergetic comparison (Table 11.1). The energy available for growth (G) is equal to the energy ingested (I) minus the unassimilated energy ($F + U$ = feces and urine) and the energy required for metabolism (M). Metabolism can be partitioned into the energy required for basal metabolism (M_B), energy used for activity (M_A), and the energy required for protein synthesis or specific dynamic action (M_{SDA}). This simple analysis shows that from an efficiency standpoint, larvae are at a disadvantage compared with their juvenile counterparts in terms of accumulating energy (Table 11.1). At first feeding, all these factors are less efficient (on a weight-specific basis) for fish larvae and limit their ability to effectively accumulate energy relative to their juvenile counterpart, except possibly the energy required for protein synthesis (Pederson 1997). So how do fish larva manage to grow at higher rates? The only plausible answer is that they must eat a significantly higher proportion of their body weight per day. They may also require less energy for protein synthesis as food consumption increases (M_{SDA}), and this could allow them to allocate more assimilated energy to growth (Pederson 1997).

FORAGING BEHAVIOR OF FISH LARVAE

Fish larvae undergo an array of developmental changes during the first few weeks of life that dramatically affect their ability to locate and capture prey (Table 11.2). Variability in development among species affects individual foraging abilities.

Size is critically important, because it affects the ability of larvae to capture prey of appropriate size and effectively forage. For small marine larvae reared at high temperatures, feeding success in the first few days of life plays a major role in their overall likelihood of survival.

Foraging behavior can be broken down and characterized by five fundamental elements: search, perception, pursuit, attack, and capture (Holling 1959, O'Brien 1987). Foraging behavior has been observed and quantified for several species of freshwater and marine fish larvae and has recently been modeled for a marine fish larva (Fiksen and MacKenzie 2002). Several key patterns in the foraging behavior of fish larvae have emerged from those observations (Table 11.3): (a) larvae typically perceive and attack prey in the hemisphere in front of them and above their horizontal axis; (b) both reactive distances and swimming speeds of foraging fish larvae tend to scale to their body length; (c) first-feeding larvae encounter many more prey than they actually attack

Table 11.2. Building blocks of development that can affect feeding success and growth of fish larvae

Developmental attributes	Status at first feeding	Consequences for first-feeding larvae
Sensory capability		
Visual acuity	Limited visual angle; rods typically absent	Reduced prey perception
Chemoreception	Olfactory pits present	Reduced prey detection
Mechanoreception	Free neuromasts present; inner ear undeveloped	Reduced prey encounter
Foraging ability		
Swimming ability	Weak swimmers, limited musculature; viscous forces dominate	Limited ability to glide
Capture success	Poorly developed fins: limited maneuverability	Low capture success
Bioenergetic factors		
Digestive enzymes	Poorly developed	Low assimilation efficiency
Swim bladder inflation	Typically undeveloped or uninflated	Increased metabolic costs
Metabolism	High weight-specific respiration rate; cutaneous respiration	High metabolic costs
Digestive flora	Largely absent	Poor disease resistance

or capture; (d)capture rates increase with size and age of the larva as their swimming ability, maneuverability, and visual acuity increases; and (e) fish larvae vary in their patterns of search, pursuit, and attack. This can have consequences for their relative abilities to encounter and capture prey.

Larvae typically perceive and attack prey above their horizontal plane and at angles typically in the range of 0–45 degrees to the right and left of their central axis and from 0 to 90 degrees above their horizontal axis. Larvae have been observed to perceive prey better within certain areas of their prey-location volume (Luecke and O'Brien 1981). For example, early in development, white crappie larvae (*Pomoxis annularis*) tended to locate prey more frequently at angles between 10 and 70 degrees from their body position (Browman and O'Brien 1992b). Juveniles located most prey between 0 and 20 degrees from their body position. Prey size also affects the ability of larvae to locate prey within the location volume. Wazenbock and Shiemer (1989) compared reactive distances of three species of larvae for two prey sizes. They found that for each species, the larger of the two prey were perceived at a greater distance. Similar results have been documented for juvenile and adult planktivorous fishes (Dunbrack and Dill 1984; O'Brien 1987).

Reactive distance and swimming speed of a larva can vary significantly for any given prey encounter and between individual larvae. Reported mean reactive distances of first-feeding fish larvae range from 0.70 to 1.43 BL (Table 11.3), although many prey are encountered at greater distances and in close proximity. Not all prey within the prey-location volume are approached or attacked. Even when prey are perceived and approached, the pursuit often ends in either a failed or aborted attack for first-feeding fish larvae (Drost 1987). As larvae grow, the proportion of successful attacks increases dramatically, and aborted attacks become less frequent. Capture success can be low at first feeding (3–30%) but increases as larvae become more competent (Drost 1987). Competence includes gaining size, swimming ability, maneuverability, and experience in aiming toward and capturing prey (Drost 1987). Capture success increases dramatically during early development and can approach 100% after several days of growth for fast-growing larvae reared at high temperatures, up to a few weeks for slower growing larvae.

Swimming speed is an important attribute of first-feeding fish larvae because it accounts for a major metabolic expenditure (Pedersen 1997). Swimming faster or swimming in a turbulent environment can increase prey encounter, but with a metabolic cost that must be balanced with energy intake in order to grow quickly.

Swimming speeds of foraging larvae can vary

Table 11.3. A comparison of size, swimming speed, and reactive distance of 12 marine and 8 freshwater species of fish larvae at first feeding

Marine Species	Dry wt. at first feeding (μg)	Length at first feeding TL (mm)	Swimming speeds (BL/sec)	Reactive distance (BLs)	Reactive distance (cm)	Reaction volume (cc)	Source
Clupea harengus	80	9.5	0.48	0.70	0.67	0.154	Rosenthal and Hempel (1970); Batty (1987)
Anchoa mitchilli	10	3.5	0.90	0.84	0.26	0.013	Chesney, personal observation
Engraulis mordax	20	4.2	0.41	0.80	0.33	0.020	Hunter (1972)
Morone saxatilis	200	5.8	0.71	0.90	0.52	0.064	Chesney, personal observation
Archosargus rhomboidalis	10	2.7	—	0.80[a]	0.14	0.005	—
Scomber japonicus	30	3.6	0.8	0.80[a]	0.27	0.013	Hunter and Kimbrell (1980)
Pseudopleuronectes americanus	10	3.9	—	0.80[a]	0.29	0.016	—
Achirus lineatus	10	2.1	—	0.80[a]	0.11	0.002	—
Gadus morhua	60	4.6	2.69	0.83[b]	0.38	0.029	MacKenzie and Kiørboe (1995); Hunt von Herbing et al. (2001)
Pleuronectes platessa	105	6.5	—	0.80[b]	0.52	0.074	Blaxter and Staines (1971); Arnold and Holford (1989)
Pagrus auratus	—	3.6	—	0.80[a]	0.29	0.013	—
Lutjanus campechanus	8	2.1	—	0.80[a]	0.17	0.002	Chesney, personal observation
Freshwater species							
Alosa pseudoharengus	40	3.8	—	0.92	0.35	0.022	Miller et al. (1993)
Perca flavescens	200	5.5	—	1.43	0.79	0.255	Miller et al. (1993)
Coregonus hoyi	650	9.5	—	0.5	0.49	0.062	Miller et al. (1993)
Pomoxis annularis	—	4.5	0.80	1.33	0.6	0.112	Browman and O'Brien (1992a, 1992b)
Notemigonus crysoleucus	—	4.8	2.18	0.44	0.21	0.005	Browman and O'Brien (1992a, 1992b)
Rutilis rutilis	400	8.3	1.43	0.77	0.64	0.137	Wazenbock and Schiemer (1989); Keckeis and Schiemer (1992)
Alburnus alburnus	340	8.4	2.62	0.98	0.82	0.292	Wazenbock and Schiemer (1989)
Abramis ballerus	1500	13	0.80	0.79	1.03	0.567	Wazenbock and Schiemer (1989)

Note: Reaction volumes were estimated assuming a volume equal to a one-eighth sphere with r equal to reactive distance. BL, body length; RD, reactive distance; TL mm, total length in millimeters.

[a]Projected based upon mean RDs for other marine species.

[b]Based upon RDs from two studies.

widely but typically average around 1–1.5 BL/sec and range from 0.41 to 2.69 BL/sec (Table 11.3). These speeds are relatively slow compared to burst or sustainable swimming speeds observed when larvae are forced to swim in a flume or a current (Houde 1969; Blaxter 1986; Meng 1993) or react to a real or simulated predator (Webb and Corrolla 1981; Bailey and Batty 1984). Swimming speed of foraging larvae can be affected by temperature, prey availability, and larval size (Hunter 1981). Increasing temperature increased swimming speeds in older northern anchovy larvae (Hunter 1981). On average, jack mackerel tended to swim faster than northern anchovy larvae at all life stages at similar temperatures (Hunter 1981). Both species tended to swim significantly faster when prey were present than when not, indicating that larvae modify their behavior to increase their probability of encountering prey (Hunter 1981). Prey encounter also decreases average swimming speeds because of the predation cycle, the time it takes to approach and capture prey. For fish larvae this typically averages a few seconds of slower, more deliberate movement as a larva approaches, attacks, and then swallows its prey before continuing to search (Browman and O'Brien 1992a, 1992b).

First-feeding fish larvae are intensely focused on finding and capturing prey. When disturbed, fish larvae immediately return to their search pattern. The s-start is the typical attack posture of most fish larvae at first feeding (Batty and Morley 1994), although some species of fish larvae have been reported to use other postures for attacking prey, including a c-start or a modified s-start (MacKenzie and Kiørboe 1995). Fish larvae also use a c-start to respond to an attack by a predator or other external stimuli (Batty 1989; Batty and Morley 1994). An s-start consists of the larvae coiling its body into an s-shape as it approaches within a few millimeters of its prey and then rapidly striking toward the prey in a single motion. Some larvae such as cod do not coil into a full s, but instead bend the caudal part of their body while using their pectoral fins for positioning (Munk 1995).

Fish larvae use two principal search patterns: pause-travel or "saltatory" search and cruise search. In pause-travel search, the forage pattern is punctuated by frequent stops where the larva scans for prey while remaining motionless (Browman and O'Brien 1992a, 1992b). Cruise searchers are able to locate prey while they are moving, thus scanning a greater volume of water, but at a greater energetic cost. Very small larvae may be restricted to pause-travel search because of the energetic constraints imposed by being small. Viscous forces (i.e., high Reynold's numbers) limit the ability of small larvae to glide, thus increasing the metabolic costs of a cruise-type search pattern for a small larva (Vlymen 1974; Webb 1975; Weihs 1980). Larvae less than approximately 5 mm total length (TL) are dominated by viscous forces, and intermittent or beat and glide swimming is inefficient (Weihs 1980). Beat and glide becomes a more efficient mode of swimming for northern anchovy (*Engraulis mordax*) once a larva reaches 15 mm TL (Weihs 1980).

Swimming speed is also affected by the foraging pattern that a larva employs and the physical conditions within its environment. The more frequently a pause-travel searcher stops, and the longer the duration of those stops, the slower its average swimming speed. Light intensity and the abundance of prey have been shown to influence swimming speed of larvae (Hunter 1981; Batty 1987). Reduced light often slows the average speed of larvae while low food density may induce them to swim faster to encounter prey. Of course prey encounters and attacks by themselves will reduce the average rate of travel of the larva for both pause-travel and cruise-search patterns.

An example of how the forage pattern of a larva might affect their rate of travel and efficiency of searching a given volume of water is illustrated in Fig. 11.4 and Fig. 11.5. Three search patterns are compared with regard to the distance traveled and their relative rates of searching a given volume of water. For each search pattern, three sizes of larvae are compared: 3.5, 6, and 9.5 mm TL. In this comparison, all larvae swim at 1.0 BL/sec and have a reactive distance of 0.8 BL. For the search strategies that employ a pause, each pause is 0.25 second. Search strategy 1 consists of a swim, a pause, and a "flick," where the larva flicks its body and rotates to scan a new volume of water after a short pause. This strategy has been observed for first-feeding bay anchovy (Chesney, personal observation). Search strategy 2 employs a swim of 1.0 BL followed by a short pause to scan for prey, followed by another swim. Search strategies 1 and 2 are variations of pause-

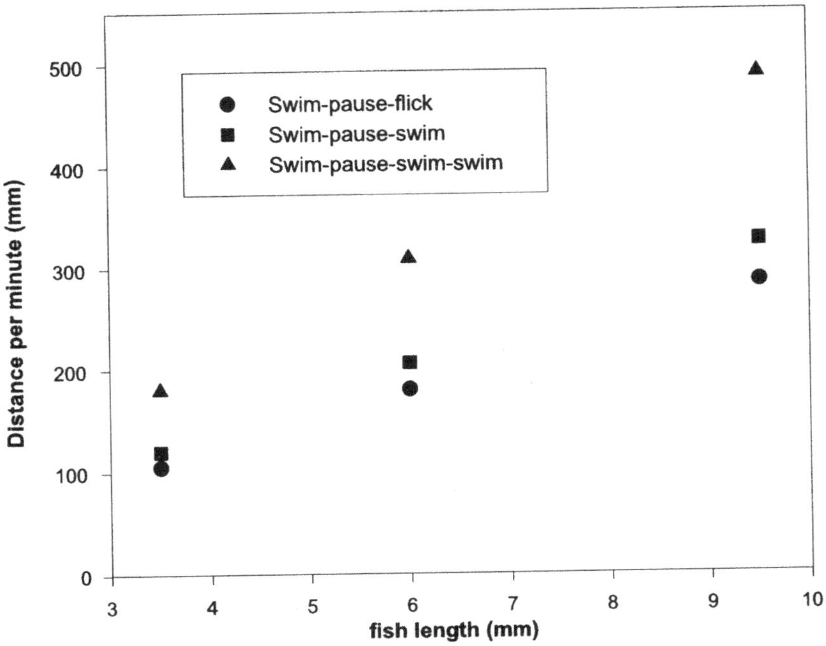

Figure 11.4. Distance traveled per minute by foraging fish larvae of three sizes (3.5, 6, 9.5 mm TL). Three search strategies are compared: two pause-travel search patterns (swim-pause-flick and swim-pause-swim) and a cruise-search strategy (swim-pause-swim-swim). The fish larvae were assumed to swim at 1.0 BL/sec, with a pause duration of 0.25 seconds.

travel search. Strategy 3 is a cruise-search strategy where the larva searches while swimming. Again, the larva swims at 1.0 BL/sec and has a reactive distance of 0.8 BL. As may be expected, the cruise searcher covers more ground than either of the pause-travel searchers. Relative distances traveled are similar for small larvae but differ substantially for larger larvae (Fig. 11.4). While the volume searched still favors the cruise searcher, the swim-pause-flick version of the pause-travel search strategy effectively searches a greater volume than the swim-pause-swim strategy, presumably while expending less energy (Fig. 11.4 and Fig. 11.5). The disadvantage of the swim-pause-flick search strategy is that it is only effective if a larva is located in a volume with an adequate density of prey for frequent encounters.

As sight feeders, the development of visual acuity of fish larvae is a critical part of their sensory development that affects prey-capture success (Table 11.3). Visual acuity changes dramatically as larvae develop, and this affects both their prey encounter and capture efficiency. The eyes of most fish larvae lack pigment at hatching, but they soon develop pigmented eyes capable of detecting movement (Blaxter 1986). Changes of visual acuity in fish larvae have been estimated based upon both behavioral studies (reviewed in the section on forage behavior) and histological methods. The rates and patterns of development vary widely depending on the type of larvae and its developmental patterns and rates (Blaxter and Staines 1971; Blaxter 1988; Neave 1984a, 1984b; Fuiman and Delbos 1998; Higgs and Fuiman 1998; Miller et al. 1993; Pankhurst 1994; Fuiman 2002). Both behavior and histological assessments show a progressive pattern of improved visual acuity as larval development proceeds. Visual sensitivity increases, the lens begins to develop, visual angles improve, and cones begin to differentiate, all leading to increased perceptive abilities as larvae develop into juvenile fishes. The main point is that development of visual acuity during early life reduces some of the constraints on the ability of the larva to find food or avoid predators.

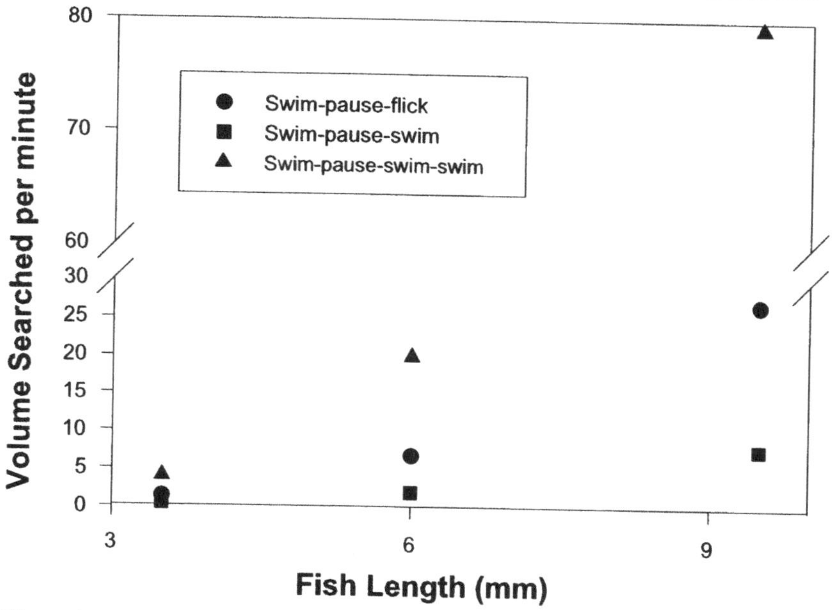

Figure 11.5. Volume searched per minute by foraging fish larvae of three sizes (3.5, 6, 9.5 mm TL). Three search strategies are compared: two pause-travel search patterns (swim-pause-flick and swim-pause-swim) and a cruise-search strategy (swim-pause-swim-swim). The fish larvae were assumed to swim at 1.0 BL/sec with pause duration of 0.25 seconds.

ENVIRONMENTAL FACTORS AFFECTING PREY CAPTURE

Light, Turbidity, Turbulence

Fish larvae are sight feeders requiring adequate levels of light and the appropriate contrasting background to effectively capture prey (Blaxter 1986; Boeuf and Le Bail 1999). Numerous feeding studies of fish larvae have reported reduced feeding success at light intensities below some optimum level. A light intensity of about 0.1 lux is generally considered to be the threshold for feeding (Blaxter 1986), although there are some reports of fish larvae feeding in the dark for flatfish, herring, striped bass, and weakfish (Blaxter 1966, 1969a, 1969b; Batty et al. 1986; Chesney 1989; Connaughton et al. 1994). Such low light levels should be expected to severely limit the effectiveness of prey capture by most fish larvae, especially at first feeding when visual acuity is developing and is relatively poor compared to the juvenile life stage. Light levels of a few hundred lux are considered adequate for most larvae to effectively feed. Optimal light levels reported for various species cover a range of intensities, but in general, light levels below 450–500 lux have been reported to reduce growth or feeding success in a number of different species (Blaxter 1966, 1969a, 1969b; Dabrowski 1982; Batty et al. 1986; Batty 1987; Chesney 1989; Miner and Stein 1993; Connaughton et al. 1994; Downing and Litvak 2001). Several studies have documented reduced feeding of fish larvae at light intensities higher than the optimal range (Blaxter 1966, 1969a, 1969b; Connaughton et al. 1994; Downing and Litvak 2001; Trotter et al 2003), while some larvae such as southern flounder, *Paralichthys lethostigma*, appear to feed more effectively at somewhat lower light intensities (Denson and Smith 1997; Henne and Watanabe 2003). Light quality (spectra) is also known to influence feeding success of fish larvae (Downing and Litvak 2001). Downing and Litvak (2001) reported higher incidents of first feeding and prey capture when haddock (*Melanogramus aeglefinus*) larvae were exposed to blue light (470 nm) rather than full-spectrum white or green light (530 nm).

Both photoperiod and light intensity affected initial swim bladder inflation, growth, survival, and postinflation viability of striped trumpeter larvae (Trotter et al 2003). Optimal photoperiods for swim bladder inflation (18 hours light/6 hours dark) and survival (24 hours light/0 hours dark) lowered postinflation viability in striped trumpeter.

In larval culture, the color of the tank walls of the culture vessel can interact with light in ways that significantly affect the ability of the larva to capture prey and develop. Several studies have investigated the effect of tank color on the feeding success of fish larvae. In general, better survival and growth occurs in tanks with darker walls as opposed to light-colored walls. Black is the color most reported to have a positive effect, but in some cases tan and light green have shown promise for larval culture. Ostrowski (1989) investigated the effect of tank color on the dolphin larvae, *Coryphaena hippurus*. He compared black-painted fiberglass tanks to unpainted (tan) tanks with dolphin larvae feeding upon rotifers at several prey densities. He reported a 130% better survival rate in black versus unpainted (tan) tanks. Ostrowski attributed some of the better survival to the background contrast that the black provided the larvae to better see their prey. Similar results have been reported for other species. Black tanks were superior to white tanks in promoting swim bladder inflation in striped bass (Martin-Tichaud and Peterson 1997). White bass, *Morone chrysops*, reared in clear glass aquaria died after only 6 days, while those in black-walled tanks had a survival rate of 48.7% and grew to 17.2 mm TL in 24 days (Denson and Smith 1996). Milkfish larvae, *Chanos chanos*, fed better in black tanks than in unpainted tan tanks and had higher survival and growth (Duray 1995). Studies with invertebrate larvae showed similar results. Rodrigues et al. (1998) found that black tanks were superior to white tanks for the production of *Macrobrachium rosenbergii* larvae. They found that the incidence of sunlight improved growth and survival of larvae and that black tanks were superior to white tanks. However, neither sunlight nor tank color influenced the food capture or ingestion capabilities of the larval prawns.

Contrary to most other studies, Duray et al. (1996) found that grouper (*Epinephelus suillus*) larvae captured more rotifers in tan tanks than in black tanks when green water was present, although growth and survival rates were similar in both treatments. Better survival of Tambaqui larvae (*Colossoma macropomum*) was reported in light green tanks compared with dark tan tanks. No difference in the size and weight of the larvae was observed. These researchers postulated that the light green tank color made it easier for the larvae to visualize their prey, resulting in higher survival rates (Pedreira and Sipauba-Taveres 2001). The presence or absence of phytoplankton undoubtedly plays a role in the relative effects of tank color on feeding (Tamaru et al. 1994), as well as the likely nutritional benefit for the larvae (Reitan et al.1997). Reports of higher larval survival in black tanks may be partly because of better contrast and feeding success, but dark tank walls may also reduce the impact of stray light on the distribution of larvae in tanks, providing a better environment for feeding.

Turbidity can affect the interaction between a fish and its zooplankton prey. Turbidity scatters and reflects light, reducing light penetration in water. It can also reduce the reactive distance and prey field of juvenile and adult fishes by limiting their visual perception of prey (Vinyard and O'Brien 1976; Benfield and Minello 1996). In contrast to juveniles and adult fishes, turbidity has generally been reported to have a neutral or positive effect (Boehlert and Morgan 1985) on the feeding of fish larvae, especially under culture conditions. Research with cultured walleye larvae, *Stizostedion vitreum*, showed that turbidity significantly increased survival, growth, and swim bladder inflation over those reared in clear water (Bristow and Summerfelt 1994; Bristow et al.1996; Rieger and Summerfelt 1997). Turbidity may also reduce cannibalism under some circumstances by reducing the tendency of the larvae to cling to the sidewalls (Bristow et al.1996; Rieger and Summerfelt 1997). Turbidity combined with turbulence generated by heavy aeration had a neutral effect on growth rates of striped bass (*Morone saxatilis*) larvae (Chesney 1989).

Studies with striped trumpeter (*Latris lineata*) larvae in tanks and bluegill (*Lepomis macrochirus*) in field mesocosms indicated that higher turbidity levels increased larval feeding success (Miner and Stein 1993; Cobcroft et al. 2001). A number of factors are likely to contribute to the improved prey capture by larvae under turbid conditions, including improved prey contrast and better dispersion of prey and larvae within a tank

(Rieger and Summerfelt 1997). Turbidity disperses light more evenly and thus reduces phototaxis by larvae toward bright spots at the surface and on the sidewalls of the tanks (Bristow et al. 1996; Rieger and Summerfelt 1997). Better dispersion of light means that the phototaxis of the prey and the fish larvae will be reduced, leading to better dispersion of both within a culture system. This should result in more effective feeding and more uniform growth of the larvae. Turbidity from the addition of algae to a culture system or the growth of algae within a system is likely to have a dual effect that favors larval feeding and growth. The "greenwater" effect should provide a better visual environment for the larvae to detect and capture prey and provide the added benefit provided by algae by bolstering the nutritional value of the zooplankton.

Turbidity has been reported to reduce the feeding success of juvenile fishes by reducing their visual field (Vinyard and O'Brien 1976; Benfield and Minello 1996). Although some studies have reported reduced feeding of fish larvae under turbid conditions, results may have been confounded by not accounting for the attenuation of light that turbidity causes. It may be that the generally short reactive distances of a few millimeters (see Table 11.3) required by a larva to perceive and attack small zooplankton would rule out any dramatic effect of turbidity on prey visibility. This would especially be true for small first-feeding larvae, as long as light levels are adequate for a given species to see its prey.

Turbulence is an important physical factor that is rarely considered in the culture of fish larvae. It has been modeled and studied by fisheries scientists because of the potential role it might play in affecting the relationship between prey contact and capture by fish larvae in the sea (Rothschild and Osborn 1988; MacKenzie and Kiørboe 1995). Models have demonstrated that turbulence is likely to influence feeding success of larvae in the sea by changing the rates of contact between a larva and their zooplankton prey (Rothschild and Osborn 1988). A dome-shaped feeding response of fish larvae to turbulence has been postulated, where intermediate levels of turbulence resulted in the highest rate of prey contact and capture (MacKenzie et al. 1994). Low turbulence presumably lowered contact rates, and very high turbulence apparently interfered with prey capture.

Laboratory studies and models with cod and herring larvae showed that turbulence potentially increases encounter rates of fish larvae with their prey and feeding success and that turbulence is more likely to favor increased prey encounter for small rather than larger larvae (Kiørboe and MacKenzie 1995; MacKenzie and Kiørboe 1995). Search behavior also played a role in encounter rates of cod, *Gadus morhua,* and herring larvae, *Clupea harengus*, with turbulence favoring encounter frequency for a pause-travel-search pattern in cod larvae over a cruise-search behavior in herring larvae (Fiksen and Mackenzie 2002). Search behavior may also play a role in prey size selection (Fiksen and Mackenzie 2002).

STRATEGIES FOR USING COPEPODS AS LIVE PREY

The greatest benefits from the production of copepods are likely to come from establishing high rates of feeding and growth in the earliest stages of the smallest fish larvae. Small larvae grown at high temperatures hatch within a day and begin to feed within a few days of being spawned. Many tropical and subtropical larvae require very small prey. Despite the need for a fast start, it also seems likely that larvae will not necessarily require small copepod nauplii once they can grow to sizes where traditional live feeds, such as rotifers and *Artemia* nauplii, can be eaten. The only other consideration is that some benefits not strictly related to size could come from the use of copepods as live feeds. Intermediate copepod life stages could provide benefits for survival of some species that may not require the smallest stages of copepods. For example, fish larvae whose initial prey size requirements are in a size range greater than rotifers but smaller than *Artemia* nauplii might benefit by supplementation with large copepod nauplii or copepodids as live feed.

CONCLUSION

The future of commercial fish culture will include a diversity of fish species from the marine environment. First feeding is the critical stage where the proper size and type of prey of adequate nutritional quality must be available to promote rapid growth and high survival of fish larvae. In addition, providing the proper environmental conditions for growth will also improve the success rate for species currently considered difficult to rear. Some of the most promising high-value species that are currently under development by aquaculturists have small larvae that grow at high temperatures and require small prey (< 80 μm width) at first feeding. Copepod nauplii offer a diverse size spectra and nutritious prey that can meet the specialized needs of small fast-growing fish larvae (Stoettrup et al.1998). Rearing methods are needed for the mass propagation of suitably small copepods as live prey that can meet the needs of these species and be practical for broad-scale application by commercial aquaculture businesses (Hernandez Molejon and Alvarez-Lajoncheres 2003).

REFERENCES

Arnold, G.P. and Holford, B.H. 1989. The reactive perceptive field of the larval plaice (*Pleuronectes platessa* L.): a three-dimensional analysis of visual feeding (Abstract). In *The Early Life History of Fish. The 3rd International Council for the Exploration of the Seas (ICES) Symposium*. Bergen, Norway, 3–5 October 1988. *Rapports el Proces Verbaux des Conseil International Pour l'Exploration de la Mer* 191:474.

Arthur, D.K. 1976. Food and feeding of larvae of three fishes occurring in the California current, *Sardinops sagax, Engraulis mordax, and Trachurus symmetricus. Fishery Bulletin* 74:517–530.

Avault, J.W., Jr. 1996. *Fundamentals of Aquaculture*. Baton Rouge, Louisiana, USA: AVA Press. 889 pp.

Bailey, K.M. and Batty, R.S. 1984. Laboratory study of predation by *Aurelia aurita* on larvae of flounder, plaice, and herring: development and vulnerability to capture. *Marine Biology* 83:287–291.

Batty, R.S. 1987. Effect of light intensity on activity and food-searching of larval herring, *Clupea harengus*: a laboratory study. *Marine Biology* 94:323–327.

Batty, R.S. 1989. Escape response of herring larvae to visual stimuli. *Journal of the Marine Biological Association of the United Kingdom* 69(3):647–654.

Batty, R. S., Blaxter, J.H.S., and Libby, D.A. 1986. Herring (*Clupea harengus*) filter-feeding in the dark. *Marine Biology* 91:371–375.

Batty, R.S. and Morley, S.A. 1994. Temperature and s-start feeding strikes by larval herring. *Journal of Fish Biology* 45 (Supplement A):11–15.

Benfield, M.C. and Minello, T.J. 1996. Relative effects of turbidity on reactive distance and feeding behavior of an estuarine fish. *Environmental Biology of Fishes* 46(2):211–216.

Blaxter, J.H.S. 1966. The effect of light intensity on the feeding ecology of herring. In *Light as an Ecological Factor, the 6th Symposium of the British Ecological Society*, edited by R. Bainbridge, pp. 393–409. Oxford, U.K.: Blackwell Science.

Blaxter, J.H.S. 1968. Light intensity, vision and feeding in young plaice. *Journal of Experimental Marine Biology and Ecology* 2:293–307.

Blaxter, J.H.S. 1969a. Visual thresholds and spectral sensitivities of herring larvae. *Journal of Experimental Biology* 48:39–53.

Blaxter, J.H.S. 1969b. Visual thresholds and spectral sensitivities of flatfish larvae. *Journal of Experimental Biology* 51:221–230.

Blaxter, J.H.S. 1986. Development of sense organs and behaviour of teleost larvae with special reference to feeding and predator avoidance. *Transactions of the American Fisheries Society* 115:98–114.

Blaxter, J.H.S. 1988. Pattern and variety in development. In *Fish Physiology, Vol. II. The Physiology of Developing Fish, Part A: Eggs and Larvae*, pp. 1–48. San Diego, California, USA: Academic Press.

Blaxter, J H.S. and Staines, M.E. 1971. Food searching potential in marine fish larvae. In *4th European Marine Biology Symposium*, edited by D.J. Crisp, pp. 467–485. Cambridge, U.K.: Cambridge University Press.

Boehlert, G.W. and Morgan, J.B. 1985. Turbidity enhances feeding abilities of larval Pacific herring, *Clupea harengus pallasi. Hydrobiologia* 123:161–170.

Boeuf, G. and Le Bail, P.Y. 1999. Does light have an influence on fish growth? *Aquaculture* 177:129–152.

Bremigan, M.T. and Stein, R.A. 1994. Gape dependent larval foraging and zooplankton size: Implications for fish recruitment across systems. *Canadian Journal of Fisheries and Aquatic Sciences* 51:913–922.

Bristow, B.T. and Summerfelt, R.S. 1994. Performance of larval walleye cultured intensively in clear and turbid water. *Journal of the World Aquaculture Society* 25(3):454–464.

Bristow, B.T., Summerfelt, R.C., Clayton, R.D. 1996. Comparative performance of intensively cultured larval walleye in clear, turbid and colored water. *Progressive Fish-Culturist* 58(1):1–10.

Browman, H.I. and O'Brien, W.J. 1992a. Foraging and prey search behaviour of golden shiner

(*Notemigonus crysoleucas*) larvae. *Canadian Journal of Fisheries and Aquatic Sciences* 49:813–819.

Browman, H.I. and O'Brien, W.J. 1992b. The ontogeny of prey search behaviour in the white crappie, *Pomoxis annularis. Environmental Biology of Fishes* 34:181–195.

Busch, A. 1996. Transition from endogenous to exogenous nutrition: larval size parameters determining the start of external feeding and size of prey ingested by Ruegen spring herring *Clupea harengus. Marine Ecology Progress Series* 130:39–46.

Buskey, E.J. (2005). Behavioral characteristics of copepods that affect their suitability as food for larval fishes. In *Culture of Copepods and Applications to Marine Finfish Larval Rearing*, edited by C.-S. Lee, P.J. O'Bryen, and N.M. Marcus, Ames, Iowa, USA: Blackwell Publishing Professional.

Buskey, E.J., Coulter, C., and Strom, S. 1993. Locomotor patterns of microzooplankton: potential effects on food selectivity of larval fish. *Bulletin of Marine Science* 53(1):29–43.

Checkley, D.M., Jr. 1982. Selective feeding of Atlantic herring (*Clupea herengus*) larvae on zooplankton in natural assemblages. *Marine Ecology Progress Series* 9:245–253.

Chesney, E.J., Jr. 1989. Estimating the food requirements of striped bass larvae, *Morone saxatilis*: effects of light, turbidity and turbulence. *Marine Ecology Progress Series* 53:191–200.

Chesney, E.J., Jr. 1993. A model of the survival and growth of striped bass larvae, *Morone saxatilis*, in the Potomac River, 1987. *Marine Ecology Progress Series* 92:15–25.

Chisholm, L.A. and Roff, J.C. 1990. Size-weight relationships and biomass of tropical neritic copepods off Kingston, Jamaica. *Marine Biology* 106:71–77.

Cobcroft, J.M., Pankhurst, P.M., Hart, P.R., and Battaglene, S.C. 2001. The effects of light intensity and algae induced turbidity on feeding behaviour of larval striped trumpeter. *Journal of Fish Biology* 59(5):1181–1197.

Connaughton, V.P., Epifanio, C.E., and Thomas, R. 1994. Effects of varied irradiance on feeding in larval weakfish. *Journal of Experimental Marine Biology and Ecology* 180:151–163.

Cushing, D.H. and Horwood, J.W. 1994. The growth and death of fish larvae. *Journal of Plankton Research* 16(3):291–300.

Dabrowski, K.R. 1982. The influence of light intensity on feeding of fish larvae and fry. I. *Coregonus pollan* (Thompson) and *Esox luscius* (L.). *Zoologische Jahrbuecher Abteilung fuer Allgemeine Zoologie* 86:341–351.

Dabrowski, K.R. and Bardega, R. 1984. Mouth size and predicted food size preferences of larvae of three cyprinid fish species. *Aquaculture* 40:41–46.

Dagg, M.J. and Govoni, J.J. 1996. Is ichthyoplankton predation an important source of copepod mortality in subtropical waters? *Marine and Freshwater Research* 47:137–144.

Denson, M.R. and Smith, T.I.J. 1996. Larval rearing and weaning techniques for white bass, *Morone chrysops. Journal of the World Aquaculture Society* 27(2):194–201.

Denson, M.R. and Smith, T.I.J. 1997. Diet and light intensity effects on survival, growth and pigmentation of southern flounder, *Paralichthys lethostigma. Journal of the World Aquaculture Society* 28(4):366–373.

Downing, G. and Litvak, M.K. 2001. The effect of light intensity and spectrum on the incidence of first feeding by larval haddock. *Journal of Fish Biology* 59(6):1566–1578.

Drost, M.R. 1987. The relationship between aiming and catch success in larval fishes. *Canadian Journal of Fisheries and Aquatic Sciences* 44:304–315.

Dunbrack, R.L. and Dill, L.M. 1984. Three-dimensional prey reaction field of the juvenile coho salmon (*Oncorhynchus kisutch*). *Canadian Journal of Fisheries and Aquatic Sciences* 41:1176–1182.

Duray, M.N. 1995. The effect of tank color and rotifer density on rotifer ingestion, growth and survival of milkfish (*Chanos chanos*) larvae. *Philippine Scientist* 32:18–26.

Duray, M.N., Estudillo, C.B., and Alpasan, L.G. 1996. The effect of background color and rotifer density on rotifer intake, growth and survival of the grouper (*Epinephalus suillus*) larvae. *Aquaculture* 146 (3-4):217–224.

Durbin, E.G. and Durbin, A.G. 1978. Length and weight relationships of *Acartia clausi* from Narragansett Bay, Rhode Island. *Limnology and Oceanography* 23:958–969.

Economou, A.N. 1991. Food and feeding ecology of five gadoid larvae in the northern North Sea. *International Council for the Exploration of the Sea (ICES) Journal of Marine Science* 47:339–351.

FAO 2002. The state of world fisheries and aquaculture (SOFIA 2002). Food and Agriculture Organization, Rome, Italy. 150 pp.

Fiksen, O. and MacKenzie, B.R. 2002. Process based models of feeding and prey selection in larval fish. *Marine Ecology Progress Series* 243:151–164.

Fuiman, L.A. 2002. Special considerations of fish eggs and larvae. In *Fishery Science: The Unique Contributions of Early Life History Stages*, edited by L.A. Fuiman and R.G. Werner. Oxford, U.K.: Blackwell Publishing. 326 pp.

Fuiman, L.A. and Delbos, B.C. 1998. Developmental changes in visual sensitivity of red drum, *Sciaenops ocellatus*. *Copeia* 4:936–943.

Govoni, J.J., Hoss, D.E., and Chester, A.J. 1983. Comparative feeding of three species of larval fishes in the northern Gulf of Mexico: *Brevoortia patronus*, *Leistomus xanthurus* and *Micropogonius undulatus*. *Marine Ecology Progress Series* 13:189–199.

Henne, J.P. and W.O. Watanabe 2003. Effects of light intensity and salinity on growth, survival and whole body osmolality of larval southern flounder, *Paralichthys lethostigma*. *Journal of the World Aquaculture Society* 34(4):450–465.

Hernandez Molejon, O.G. and Alvarez-Lajonchere, L. 2003. Culture experiments with *Oithona oculata* Farran, 1913 (copepoda: Cyclopoida), and its advantages as food for marine fish larvae. *Aquaculture* 219(1-4):471–483.

Higgs, D.M. and Fuiman, L.A. 1998. Associations between sensory development and ecology in three species of clupeoid fish. *Copeia* 1:133–144.

Holling, C.S. 1959. The components of predation as revealed by a study of small mammal predation of the European pine sawfly. *The Canadian Entomologist* 91:293–320.

Hjort, J. 1914. Fluctuations in the great fisheries of northern Europe. *Rapports el Proces Verbaux des Conseil International Pour l'Exploration de la Mer* 20:1–228.

Houde, E.D. 1969. Sustained swimming ability of larvae of walleye (*Stizostedion vitreum vitreum*) and yellow perch (*Perca flavescens*). *Journal of the Fisheries Research Board of Canada* 26:1647–1659.

Houde, E.D. 1987. Fish early life history dynamics and recruitment variability. *Transactions of the American Fisheries Society Symposium* 2:17–29.

Houde, E.D. 1989. Comparative growth, mortality and energetics of marine fish larvae: temperature and implied latitudinal effects. *Fishery Bulletin* 87(3):471–495.

Houde, E.D. 1994. Differences between marine and freshwater fish larvae: implications for recruitment. *International Council for the Exploration of the Sea (ICES) Journal of Marine Science* 51(1):91–97.

Houde, E.D. and Lovdal, J.A. 1984. Seasonality of occurrence, foods and food preferences of ichthyoplankton in Biscayne Bay, Florida. Estuarine Coastal and Shelf Science, 18(4):403–419.

Houde, E.D. and Shekter, R.C. 1981. Growth rates, rations and cohort consumption of marine fish larvae in relation to prey concentrations. *Rapports el Proces Verbaux des Conseil International Pour l'Exploration de la Mer* 178:441–453.

Houde, E.D. and Zastrow, C.E. 1993. Ecosystem and taxon-specific dynamic and energetic properties of larval fish assemblages. In *Advances in the Early Life History of Fishes*, edited by H.G. Moser, P.E. Smith, and L.A. Fuiman. *Bulletin of Marine Science* 53(2):290–335.

Hunt von Herbing, I., Gallagher S.M., and Halteman, W. 2001. Metabolic costs of pursuit and attack in the early larval Atlantic cod. *Marine Ecology Progress Series*, 216:201–212.

Hunter, J.R. 1972. Swimming and feeding behavior of larval anchovy, *Engraulis mordax*. *Fishery Bulletin* 70:821–838.

Hunter, J.R. 1981. Feeding ecology and predation of marine fish larvae. In *Marine Fish Larvae*, edited by R. Lasker, pp.33–77. Seattle, Washington, USA: University of Washington Press.

Hunter, J.R. and Kimbrell, C.A. 1980. Early life history of Pacific mackerel, *Scomber japonicus*. *Fishery Bulletin* 78:89–101.

Kane, J. 1984. The feeding habits of co-occurring cod and haddock larvae from Georges Banks. *Marine Ecology Progress Series* 16:9–20.

Keckeis, H. and Schiemer, F. 1992. Food consumption and growth of larvae and juveniles of three cyprinid species at different food levels. *Environmental Biology of Fishes* 33(1-2):33–45.

Kiørboe, T. and MacKenzie, B.R. 1995. Turbulence enhanced encounter rates in larval fish: effects of spatial scale, larval behavior and size. *Journal of Plankton Research* 17(2):2319–2331.

Lazzaro, X. 1987. A review of planktivorous fishes: their evolution, feeding behaviors, selectivities and impacts. *Hydrobiologia* 146:97–167.

Luecke, C. and O'Brien, W.J. 1981. Prey location volume of a planktivorous fish: a new measure of prey vulnerability. *Canadian Journal of Fisheries and Aquatic Sciences* 38:1264–1270.

MacKenzie, B.R. and Kiørboe, T. 1995. Encounter rates and swimming behavior of pause travel and cruise larval fish predators in calm and turbulent environments. *Limnology and Oceanography* 40(7):1278–1289.

MacKenzie, B.R., Miller, T.J., Cyr, S., and Leggett, W.C. 1994. Evidence for a dome shaped relationship between turbulence and larval fish ingestion rates. *Limnology and Oceanography* 39:1790–1799.

Margulies, D., Wexler, J.B., Bentler, K.T., Suter, J.M., Masuma, S., Tezuka, N., Teruya, K., Oka, M., Kanematsu, M., and Nikaido, H. 2001. Food selection of yellowfin tuna, *Thunnus albacares*, larvae reared in the laboratory. *Bulletin. Inter-American Tropical Tuna Commission/Boletin. Comision Interamericana del Atun Tropical. La Jolla, California* 22(1):9–33.

Martin-Tichaud, D.J. and Peterson, R.H. 1997. Factors affecting swim bladder inflation success in larval striped bass (*Morone saxatilis*). In *Bulletin of the Aquaculture Association of Canada, Special Publication* 2:111.

Meng, L. 1993. Sustainable swimming speeds of striped bass larvae. *Transactions of the American Fisheries Society* 122: 702–708.

Miller, T.J., Crowder, L.B., and Rice, J.A. 1993. Ontogenetic changes in behavioural and histological measures of visual acuity in three species of fish. *Environmental Biology of Fishes* 37:1–8.

Miner, J.G. and Stein, R.A. 1993. Interactive influence of turbidity and light on larval bluegill (*Lepomis macrochirus*) foraging. *Canadian Journal of Fisheries and Aquatic Sciences* 50(4):781–788.

Munk, P. 1992. Foraging behaviour and prey size spectra of larval herring *Clupea harengus*. *Marine Ecology Progress Series* 80:149–158.

Munk, P. 1995. Foraging behaviour of larval cod (*Gadus morhua*) influenced by prey density and hunger. Marine Biology. 122(2):205–212.

Munk, P. and Nielsen, T.G. 1994. Trophodynamics of the plankton community at Dogger Bank: predatory impact by larval fish. *Journal of Plankton Research* 16(9):1225–1245.

Neave, D.A. 1984a. The development of retinomotor reactions in larval plaice, (*Pleuronectes platessa*, L.) and turbot (*Scapthalmus maximus*, L.). *Journal of Experimental Marine Biology and Ecology* 76:167–175.

Neave, D.A. 1984b. The development of visual acuity in larval plaice, (*Pleuronectes platessa*, L.) and turbot (*Scapthalmus maximus*, L.). *Journal of Experimental Marine Biology and Ecology* 78:167–175.

O'Brien, W.J. 1987. Planktivory by freshwater fish: thrust and parry in the pelagia. In *Predation, Direct and Indirect Impacts on Aquatic Communities*, edited by W.C. Kerfoot and A. Sih, Hanover, New Hampshire, USA: University Press of New England. 386 pp.

Ostrowski, A.C. 1989. Effect of rearing tank background color on early survival of dolphin larvae. *Progressive Fish-Culturist* 51(3):161–163.

Pankhurst, P.M. 1994. Age related changes in the visual acuity of larvae of the New Zealand snapper, *Pagrus auratus*. *Journal of the Marine Biological Association of the United Kingdom* 74:337–349.

Pearre, S., Jr. 1980. The copepod width-weight relation and its utility in foodchain research. *Canadian Journal of Zoology* 58:1884–1891.

Pederson, B.H. 1997. The cost of growth in young fish, a review of new hypotheses. *Aquaculture* 155:259–269.

Pedreira, M.M. and Sipauba-Taveres, L.H. 2001. Effect of light green and dark brown colored tanks on survival and development of Tambiqui larvae, *Colossoma macropomum* (Osteichthyes, Serrasalmidae). *Acta Scientiarum* 23(2):521–525.

Pepin, P. 1989. Predation and starvation of larval fish: a numerical experiment of size and growth dependent survival. *Biological Oceanography* 6(1):23–44.

Rieger, P.W. and Summerfelt, R.C. 1997. The influence of turbidity on larval walleye, *Stizostedion vitreum*, behavior and development in tank culture. *Aquaculture* 159(1-2):19–32.

Reitan, K.I., Rainuzzo, J.R., Øie, G., and Olsen, Y. 1997. A review of the nutritional effects of algae in marine fish larvae. *Aquaculture* 155:207–221.

Rodrigues, C.C.B., Rodrigues, J.B.R., Moschen, S. 1998. Effects of sunlight and tank coloration on the larval development and post-larvae production of *Macrobrachium rosenbergii* (De Man, 1879). In *Aquaculture Brazil '98. Sustainable development, Vol. II*, November 2–6 1998. pp. 423–435. Recife, P.E., Brazil.

Rombough, P.J. 1988. Respiratory gas exchange, aerobic metabolism and effects of hypoxia during early life. In *Fish Physiology, Vol. II. The Physiology of Developing Fish, Part A: Eggs and Larvae*, pp. 59–144. San Diego, California, USA: Academic Press.

Rothschild, B.R. and Osborn, T.R. 1988. Small-scale turbulence and plankton contact rates. *Journal of Plankton Research* 10:465–474.

Rosenthal, H. and Hempel, G. 1970. Experimental studies in feeding and food requirements of herring larvae (*Clupea harengus* L.). In *Marine Foodchains*, edited by J.H. Steele, pp. 344–364. Edinburgh, U.K.: Oliver and Boyd.

Setzler-Hamilton, E., Jones, P.W., Martin, F.D., Ripple, K., Mihursky, J.A., Drewry, G.E., and Beaven, M. 1982. Comparative feeding habits of white perch and striped bass larvae in the Potomac estuary. In *Proceedings of the Fifth Annual Meeting of the Potomac Chapter of the American Fisheries Society, May 1981*, edited by R. A. Richkus, pp. 139–157. UM SG RS-82-08. College Park, Maryland, USA: Maryland Sea Grant.

Shiroto, A. 1970. Studies on the mouth size of fish in the larval and fry stages. *Bulletin of the Japanese Society of Scientific Fisheries* 36(4):353–368 (Translated into English by Fisheries Research Board of Canada Translation Services 1978).

Stepien, W.P., Jr. 1976. Feeding of laboratory reared larvae of the sea bream, *Archosargus rhomboidalis* (Sparidae). *Marine Biology* 38:1–16.

Stoettrup, J.G., Shields, R., Gillespie, M., Gara, M.B., Sargent, J.R., Bell, J.G., Henderson, R.J., Tocher,

D.R., Sutherland, R., Noess, T., Mangor Jensen, A., Naas, K., and van der Meeren, T. 1998. The production and use of copepods in larval rearing of halibut, turbot and cod. Proceedings of the Live Feeds Session, Aquaculture Canada 1998. *Bulletin of the Aquaculture Association of Canada* 98(4):41–45.

Tamaru, C.S., Murashige, R., and Lee, C.S. 1994. The paradox of using background phytoplankton during the larval culture of striped mullet, *Mugil cephalus* L. *Aquaculture* 119:167–174.

Trotter, A.J., Battaglene, S.C., and Pankhurst, P.M. 2003. Effect of photoperiod and light intensity on initial swim bladder inflation, growth and post inflation viability in cultured striped trumpeter (*Latris lineata*) larvae. *Aquaculture* 224:141–158.

Tucker, J.W., Jr. 1998. *Marine Fish Culture*. Norwell, Massachusetts, USA: Kluwer Academic Publishers. 750 pp.

Turner, J.T. 1984. The feeding ecology of some zooplankton that are important prey items of larval fish. NOAA (National Oceanic and Atmospheric Administration) Tech. Report NMFS 7, 1–28, U.S. Dept. of Commerce, Washington, D.C., USA.

Vinyard, G.L. and O'Brien, W.J. 1976. Effects of light and turbidity on the reactive distance of bluegill (*Lepomis macrochirus*). *Journal of the Fisheries Research Board of Canada* 33:2845–2849.

Vlymen, W.J., III. 1974. Swimming energetics of larval anchovy, *Engraulis mordax*. *Fishery Bulletin* 72(4):885–899.

Wazenbock, J. and Schiemer, F. 1989. Prey detection in cyprinids during early development. *Canadian Journal of Fisheries and Aquatic Sciences* 46:995–1001.

Webb, P.W. 1975. Hydrodynamics and energetics of fish propulsion. Fisheries Research Board of Canada Special Bulletin no. 190, 158 pp.

Webb, P.W. and Corrolla, R.T. 1981. Burst swimming performance of northern anchovy, *Engraulis mordax*, larvae. *Fishery Bulletin* 79:143–150.

Weihs, D. 1980. Energetic significance of changes in swimming modes during growth of larval anchovy, *Engraulis mordax*. *Fishery Bulletin* 77(3):597–604.

Zaret, T.M. 1980. The effect of prey motion on planktivore choice. In *Evolution and Ecology of Zooplankton Communities*, edited by W. C. Kerfoot, pp. 5994–6603. Hanover, New Hampshire, USA: University Press of New England.

12
Intensive and Extensive Production Techniques to Provide Copepod Nauplii for Feeding Larval Red Snapper *Lutjanus campechanus*

Ronald P. Phelps, Gede S. Sumiarsa, Emily E. Lipman, Hsiang-Pin Lan, Komarey Kao Moss, and Allen D. Davis

ABSTRACT

A series of studies were conducted at the Claude Peteet Mariculture Center, Gulf Shores, Alabama, to develop protocols for obtaining large quantities of copepod nauplii from organically fertilized ponds or from laboratory-based systems. Use of alfalfa meal, fishmeal, rice bran, and wheat bran as organic fertilizers to stimulate copepod production were compared. The use of aeration to improve copepod production in wheat bran–fertilized ponds was also studied. Fatty acid profiles of pond-produced nauplii were described, and the effects of acclimation and enrichment were determined.

Maximum densities of > 5,000 nauplii/L were common in most pond management protocols and averaged > 1,000/L. Aerated wheat bran–fertilized ponds averaged 5,079 nauplii/L. Zooplankton production was cyclic with rotifer abundance peaking at 6–14 days after filling and nauplii abundance at 10–14 days after filling. An average of 4.9 million nauplii were collected per trap set. When nauplii were acclimated from 10–15 ppt to 32 ppt over 6 hours, there were reductions of 44.2% in total lipids, 44.2 and 55.5% in polar and nonpolar docosahexaenoic acid (DHA), respectively, and 23.9 and 64.9% in polar and nonpolar eicosapentaenoic acid (EPA), respectively. A 3-hour enrichment in A-1 Super Selco returned total fatty acid, DHA, EPA, and arachidonic acid (ARA) levels of nauplii to values similar to that of preacclimation.

The intensive production of the cyclopoid copepod *Apocyclops panamensis* was studied. Adult stocking rates of 320, 640, 1,280, 2,560, and 5,120/L were evaluated in a series of trials. Wild-caught adult copepods were stocked into 40-L clear plastic bags filled with 30 ppt artificial seawater and *Isochrysis galbana*, at a density of 500,000 cells/mL. A mean yield of 16,942 ± 794 nauplii/L was obtained when adult *A. panamensis* were stocked 5,120/L. Production per adult female ranged from 8.1 to 8.5 nauplii per female at adult stocking densities of 320–2,560/L and 4.7 nauplii per female at 5,120 adults/L.

A closed batch system was evaluated, tracking two populations over time. A total of 10 trials, ranging from 4 to 9 days each, were conducted over a period of 77 days, restocking adults into new bags at each harvest. Production of nauplii and copepodids in a closed system varied depending on the adult density stocked. Mean pro-

---

Corresponding author: Ronald P. Phelps, Department of Fisheries and Allied Aquacultures, Auburn University, Auburn, AL 36849,USA; Phone: (334) 844-9317; Fax: (334) 844-9208; E-mail: RPPhelps@acesag.auburn.edu.

duction in population A was 10,939 ± 3,200/L, but ranged from 144 to 31,532/L. Mean production in population B was 14,366 ± 3,095/L and ranged from 250 to 36,800/L. Females in the closed system averaged 14.13 nauplii and copepodids per female.

INTRODUCTION

The small mouth gape of red snapper *Lutjanus campechanus* larvae requires that a small food organism be available at first feeding. Copepod eggs and nauplii are an important food for many marine fish larvae in their natural habitat (Hunter and Lasker 1981; Houde 1978; Toledo et al. 1999). In hatcheries copepods have been used as the first feed for larvae of halibut *Hippoglossus hippoglossus*, turbot *Scophthalmus maximus*, and cod *Gadus morhua* L. (Støttrup et al. 1998), mangrove red snapper, *Lutjanus argentimaculatus* (Singhagraiwan and Doi 1993), sea bream *Archosargus rhomboidalis*, bay anchovy *Anchoa mitchilli*, and lined sole *Achirus lineatus* (Houde 1975, 1978). Larval mangrove snapper raised on rotifers failed to survive after 5 days after hatching, but when the larvae were fed copepod nauplii, survival was 40% after 21 days after hatching (Doi and Singhagraiwan 1993). Daily specific growth rates of cod have been shown to increase from 2.8 to 21% with an increase in availability of copepod nauplii (van der Meeren and Naess 1993). Survival of turbot larvae was 73% when fed *Artemia* nauplii and 93% when fed copepod nauplii (Kuhlmann et al. 1979).

Hatchery techniques are not well established for the mass production of copepod nauplii. Producing copepods under a controlled environment requires complex facilities and sophisticated skill (Støttrup and Norsker 1997). In addition, culturing live food organisms in a controlled environment may not be economical for certain hatchery applications (Graves and Morrow 1988).

An alternative way to obtain live food organisms is to collect zooplankton from fertilized ponds (Graves and Morrow 1988; Misra and Phelps 1992). Species abundance, however, may vary from day to day and the act of trapping may damage organisms and affect their nutritional quality. Salinity at which copepods are produced and at which they are being given as food may differ and also reduce copepod survival and nutrient quality. Changes in salinity stimulate crustaceans to adjust their osmoregulation. It has been observed in the adult calanoid copepod *Eurytemora affinis* and zoea 1 of several decapod crustacean larvae that extracellular osmoregulation is associated with energy expenditure as part of active ion transport, involving degradation of energy-rich compounds such as lipid and protein (Gonzales and Bradley 1994; Kimmel and Bradley 2001; Torres et al. 2002). When copepod nauplii are to be used as feed for larval fish, changes in salinity may affect their nutrient quality.

Nutrient quality of several types of organisms can be improved through supplemental enrichment. Various enrichment regimes have been evaluated to improve the nutrient quality of rotifers (Støttrup and Attramadal 1992; Fernandez-Reiriz et al. 1993) and *Artemia* (van Ballaer et al. 1985; Mourente et al. 1993; Craig et al. 1994; Rainuzzo et al. 1994). Additional enrichment of copepods is also possible to ensure adequate nutrition for fish larvae (Craig et al. 1994; Salhi et al. 1997; Planas and Cunha 1999; Toledo et al. 1999).

Extensive production of copepods in ponds can vary seasonally, and not all marine hatcheries have access to ponds that can be dedicated to copepod production. Intensive production of copepods in an indoor environment offers better control over species composition and nutritional quality. The feasibility of culturing copepods in intensive indoor systems has been investigated, and maximum densities of 26,000 harpacticoid copepod *Tigriopus brevicornis* nauplii/L (Vilela 1992) and 33,000 cyclopoid copepod *Apocyclops royi* nauplii/L (Cheng et al. 2001) have been reported. Intensive copepod production systems, however, have been difficult to maintain over extended periods of time (Davis 1983; Støttrup and Norsker 1997).

The following reports on a series of trials conducted at the Claude Peteet Mariculture Center, Gulf Shores, AL, USA, to provide copepod nauplii to feed larvae of red snapper *Lutjanus campechanus*. Topics addressed include: pond production techniques, trapping, acclimation and enrichment considerations, as well as intensive production trials with *Apocyclops panamensis*.

MATERIALS AND METHODS

Extensive Production Techniques

Pond Production

Three pond fertilization studies were conducted to optimize the production of copepod nauplii using outdoor 1,100-m^2 ponds having an average depth of 1.0 m. Details of each study are given in Table 12.1. In each study organic fertilizers were given at rate of 250 kg/ha, followed by weekly applications of 125 kg/ha using three replicate ponds per treatment. Liquid inorganic fertilizer (38-8-0) was added to all ponds initially at 20 L/ha and at 10 L/ha at each week thereafter. In Study III, one of the treatments with wheat bran was provided with aeration for 8 hours (1000–0600) per day using a 1-hp Aero-2 aspirator pump aerator.

Zooplankton sampling began 7 days after adding water to the pond. In Studies I and II, zooplankton was sampled every 2 days, while in Study III sampling was weekly. A 25-L composite sample, 5 L taken from the top 30 cm of water from each corner of the pond and 5 L near the drain, was taken, concentrated to 1 L over 21-μm mesh netting, and fixed in Lugol's fixative. Zooplankton was enumerated as to the abundance per liter of copepod nauplii, adult copepods, rotifers, and other organisms. Representative samples were preserved in 4% buffered formalin for later identification.

Dissolved oxygen (DO) and temperature were monitored daily between 0500 and 0600 hours using a YSI 55 DO meter. Total ammonia, pH, and salinity were monitored once a week using a LaMotte Ammonia Nitrogen Test Kit (± 0.02 ppm; LaMotte Co., Chestertown, MA, USA), a Fisher Scientific Accumet meter (Fisher Scientific International Inc., Hampton, NH, USA), and a Fisher Scientific refractometer, respectively. Pond transparency was measured weekly with a secchi disc to the nearest centimeter.

In all three studies, data collected over time by treatment were analyzed as repeated measures analysis of variance (ANOVA) using SAS Version 6.12 for Windows (SAS Institute Inc. 1996). When significant differences among treatments were detected, a least significant difference (LSD) multiple-range test was applied (Sokal and Rohlf 1981). Correlations of zooplankton abundance to other environmental parameters were analyzed and considered significant where $r^2 > 0.50$ and $P < 0.05$.

Table 12.1. Pond fertilization protocols used in Studies I, II, and III

Study	Organic fertilizer	Study period	Salinity
I	Alfalfa meal Wheat bran Fish meal	37 days	14 ppt
II	Alfalfa meal Wheat bran	30 days	12 ppt
III	Rice bran Wheat bran Wheat bran[a]	50 days	21 ppt

Note: Organic fertilizers were added at an initial rate of 250 kg/ha, followed weekly by 125 kg/ha applications. Inorganic fertilizer (38-8-0) was added at an initial rate of 20 L/ha, followed weekly by 10 L/ha applications.

[a]Aeration was provided nightly to three wheat bran–fertilized ponds.

Nutritional Quality Considerations

Fatty acid profiles were determined for unenriched copepod nauplii *Apocyclops panamensis* collected as part of fertilization trials from alfalfa meal–fertilized ponds in Study II, and from rice bran– and wheat bran–fertilized ponds in Study III. Ponds were sampled every morning (0700–0900 hours) to identify ponds in which the number of copepod nauplii was at least 90% of the total zooplankton organisms (≥ 30 μm). Copepods were trapped from such ponds using the trap shown in Fig. 12.1. The trap consisted of three cylindrical containers (140 L each) connected in series by 5.08-cm pipe. A set of 40-μm and 100-μm nylon filter bags with drawstrings around the mouth, 80-cm long and 17.5-cm diameter, were tied to each water outlet. The 100-μm mesh bag was placed inside the 40-μm mesh bag. A sump pump (1/3 HP-115 V) with a mean pumping rate of 133.81 ± 8.5 L/min was connected to a water supply pipe and suspended in the water column of the pond. Water passed through the 100-μm bag then the 40-μm bag into the surrounding container, where it then flowed back to the pond.

Nauplii were harvested from the trap and held in brackish water at 200/mL and acclimated in a plastic container with a maximum volume of 120 L. Acclimation was accomplished by dilution of the initial holding water (8~12 ppt) with 32 ppt

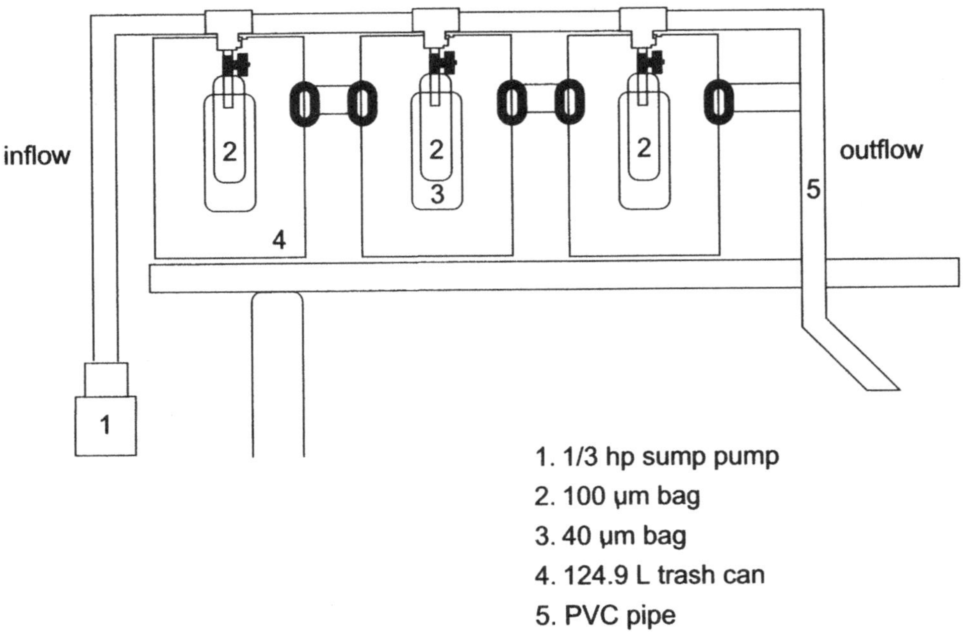

Figure 12.1. Design of trapping apparatus used to capture wild copepods from ponds, consisting of a 1/3-horsepower (hp) sump pump, 100-μm bags, polyvinyl chloride (PVC) pipe, and three 124.9-L trash cans.

seawater to achieve an end point of 32 ppt. Aeration was added at 250 mL/min to both the acclimator and the seawater supply. Copepod nauplii were not fed during the experiments. Salinity was measured using a refractometer (± 1 ppt). Water flow and salinity was monitored hourly to maintain the salinity increase evenly. Temperature and DO were measured hourly using a YSI DO meter (Model 8510; YSI Inc., Yellow Springs, OH, USA). Aeration was adjusted to maintain the DO level above 4 mg/L.

In each trial, before acclimation, at least 50 million of the collected copepod nauplii (approximately 5.0 g wet weight) were sieved through 21-μm Nitex plankton net, rinsed with approximately 1 L of distilled water, placed in screwed-cap glass jars, and stored at −20°C as described by Christie (1982). After acclimation, a minimum sample of 5.0 g (wet weight) of copepod nauplii were collected and frozen following the same procedures. All samples were held for lipid extraction and fatty acid analysis.

Enrichment of Copepod Nauplii

A series of protocols to enrich the nutrient quality of the copepod nauplii were evaluated using a commercial enrichment product, A1 Super Selco (INVE Americas Inc., Salt Lake City, UT, USA), marketed for rotifers and *Artemia*. Copepod nauplii were enriched for 3-, 6-, and 12-hour periods following the manufacturer's recommendation of 0.6 mg enrichment material pre liter of seawater. Enrichment trials were conducted indoors in 20-L plastic buckets (18 L working volume) filled with 32 ppt seawater. Density of copepod nauplii in every trial was 200 individuals/mL (3.6×10^6 copepod nauplii in each experimental unit). Two replicates were used for each treatment. Each bucket was provided with aeration of 0.4–0.6 L/min. Water quality was monitored at the beginning and end of each trial. Water temperature was measured to the nearest 0.1°C, DO to the nearest 0.01 mg/L, and total ammonia-nitrogen (TAN) to the nearest 0.01 mg/L.

Nauplii for these studies were obtained from one pond initially fertilized with 50 kg/ha of each of alfalfa meal, wheat bran, menhaden fish meal, and rice bran. In addition, two liquid fertilizers of 32-2-0 (N:P:K) and 10-34-0 were combined to give a 38-8-0, and the combination was given into each pond at an initial dose of 20 L/ha. This pond was subsequently given half the initial dose of both organic and inorganic fertilizers weekly. Nauplii were trapped and handled as described previously.

Survival of copepod nauplii for each enrichment protocol was determined by counting the number of live and dead nauplii per milliliter in a Sedgwick-Rafter counting cell, from three 50-mL samples taken at the beginning and the end of each enrichment period. After each enrichment trial, copepod nauplii were sieved and frozen following the same procedure as for copepod nauplii collected prior to acclimation. Wet weight of copepod nauplii in each trial was determined separately by weighing known numbers of copepod nauplii, sieved in a 21-μm Nitex plankton net (10 $\times$ 10 cm^2), in a Mettler H10 balance to the nearest 0.01 mg. Dry weight of copepod nauplii was determined following heat drying (AOAC 1970).

Lipid Extraction and Fatty Acid Analysis

Lipid extraction was conducted at the Fish Nutrition Laboratory of the Department of Fisheries and Allied Aquacultures, Auburn University, AL, USA, according to a modified method of Folch et al. (1957), where the extracted lipid was separated into polar and neutral lipid fractions using a Whatman solid-phase extraction cartridge (catalog #6804-1705). Both lipid fractions were diluted in 200 μL methylene chloride (DCM) put in a 0.3-mL insert (National Scientific Company, Atlanta, GA, USA) for fatty acid analysis. Each sample had two replicates.

The relative fatty acid methyl ester (FAME) fatty acid composition of these solutions was determined by a gas chromatograph (GC) Shimadzu GC-17A, with helium as a carrier gas. Fatty acid values were expressed as area percentage of the total fatty acids.

Statistical Analyses

Values are presented in mean ± standard error. If required, some data were transformed into square root for count data and arc sine square root for proportions before analyses. Data were analyzed using a *t* test and one-way ANOVA model procedure of SAS Version 6.12 for Windows (SAS Institute 1996). When significant differences among treatment means were detected, a LSD multiple-range test was applied (Sokal and Rohlf 1981). The differences were considered to be significant at a probability level of $P = 0.05$.

Intensive Production

General Procedures

Trials were conducted in an indoor laboratory using adult copepods collected from brackish water (10–27 ppt salinity) ponds using the trap design described earlier. The collected zooplankton were brought back to the lab and sieved through a series of different mesh sizes to separate copepods from rotifers, aquatic insects, and other unwanted organisms. The copepods were placed in a 19-L bucket of 1 μm filtered pond water and provided with slight aeration. Salinity of the water was increased to 30 ppt through gradual addition of 40 ppt artificial seawater. The acclimation process took approximately 30 minutes. After acclimation, the density of adult copepods in the 19-L bucket was estimated using a Sedgewick-Rafter counting cell. The number of males and females and the number of females carrying eggs were counted. Culture units were stocked based on the total number of adults required.

The culture units were 100-L, 6-mm-thick, clear, plastic bags suspended from a wooden frame. The bags were filled with 40 L of 1-μm filtered pond water and commercial sea salts to give a salinity of 30 ppt. A low-pressure blower provided aeration through weighted airstones. Light was provided by four Sylvania GTE Gro-Lux 75-W light bulbs on a 24-hour light cycle. At stocking, *Isochrysis galbana* was added to each bag at a density of 5 $\times$ 10^5 cells/mL. Water quality analyses for DO, pH, temperature, salinity, and total ammonia were measured at the beginning and end of each trial.

Experimental Trials

A series of progressively higher adult copepod stocking densities were evaluated in four trials to determine the effect of adult stocking density on nauplii production using wild-caught adults. Each culture bag was inoculated with *I. galbana* at a density of 5 $\times$ 10^5 cells/mL. Trial 1 was stocked with 320 adults/L with 10 replicates and was run for 8 days. Trial 2 was stocked at 640 and 1,280 adults/L, with four replicates per treatment. In subsequent trials the high density from the previous trial was repeated, and the second density was twice that rate. In Trial 3, adults were stocked at 1,280 adults/L and 2,560 adults/L. Trial 4 tested the stocking densities of 2,560 adults/L and 5,120 adults/L. Trials 2–4 each lasted 4 days, with four replicates per treatment.

At the end of the culture period, the bags were harvested by siphoning the water through a 150-μm sieve to capture adults and a 35-μm sieve to capture the copepodids and nauplii. The sieves were washed into 230-mL sample containers. Two samples from each sample container, four samples per bag, were counted using a Sedgewick-Rafter slide using the same method previously described, but nauplii and copepodid stages were also counted. Where rotifers and other contaminates were found, they were also enumerated.

Data were analyzed using ANOVA and multiple comparisons to find significant differences at $P < 0.05$. Multiple regressions were also run to see what factors affected production. All analyses were run on SAS, Version 6.12 (SAS Institute Inc. 1988).

Closed-System Production

An initial stock of *A. panamensis* was captured from ponds and handled as described earlier. Two 100-L, clear, 6-mm-thick plastic bags were filled with 40 L of 1-μm filtered water and commercial sea salts to give a salinity of 30 ppt. Aeration and lighting was provided as described earlier. *Isochrysis galbana* was added to each bag at a density of 5×10^5 cells/mL. Pond-caught adults were stocked into the initial bags at 320/L to establish populations A and B.

At the end of the culture period, the bags were harvested by siphoning the water through a 150-μm sieve to capture adults and a 35-μm sieve to capture the copepodids and nauplii. The sieves were washed into 230-mL sample containers to hold the retained animals. Two samples from each sample container, four samples per bag, were counted using the same method previously described, but nauplii and copepodid stages were also counted. Where rotifers and other contaminates were found, they were also enumerated.

After counting, new culture bags were prepared as described above. The adults harvested from the first trial were then stocked into bags to start a new culture, Trial 2. Population A was stocked into a new bag and population B was also stocked into a new bag, so each population was kept separate throughout the 12 trials. For each subsequent trial, the harvested adults from population A were restocked into a new bag, and the same was done for population B. Restocking density was dependent on the number of adults harvested from the previous trial. Adult stocking density ranged from 160.0 to 6,498.0 adults/L. A total of 12 trials, ranging from 4 to 9 days each, were conducted over a period of 77 days. Trials were harvested when visually the algae was cleared from the bags. This usually happened after 4 to 5 days, depending on the stocking density. When the adult stocking densities were low, < 1,000 adults/L, the cultures were run for a longer period of time (6 to 7 days) to reestablish an adult population.

Data were analyzed using ANOVA analysis and multiple comparisons to find significant differences at $P < 0.05$. Multiple regressions were also run to see what factors affected production. All analyses were run on SAS, Version 6.12 (SAS Institute Inc. 1988).

RESULTS

Extensive Production Techniques

Pond Production of Copepod Nauplii

Abundance of copepod nauplii was highly variable, and no one organic fertilizer showed a consistent advantage (Table 12.2). In Study I the average copepod nauplii densities were 887 ± 172, 1,000 ± 361, and 1,071 ± 415/L for alfalfa-, wheat bran–, and fishmeal-fertilized ponds, respectively. In Study II the average copepod nauplii densities were 2,112 ± 594 and 2,520 ± 706/L for alfalfa- and wheat bran–fertilized ponds, respectively. The highest levels of nauplii production were achieved in Study III, where the mean densities were 3,125 ± 912, 3,391 ± 523, and 5,080 ± 1,224/L, respectively, for ponds fertilized with rice bran, wheat bran, and wheat bran with supplemental aeration.

Nauplii abundance varied over the course of each study. In Study I nauplii abundance peaked on day 10 at mean densities per fertilization regime of 4,484 ± 1,184, 3,893 ± 1,220, and 4,533 ± 1,193/L for alfalfa-, fishmeal-, and wheat bran–fertilized ponds, respectively. A second peak of 1,301 ± 203 occurred on day 32 in alfalfa-fertilized ponds and on day 34 of 3,285 ± 811 and 1,984 ± 743/L in fishmeal and wheat bran–fertilized ponds. *Acartia* sp. was the predominate copepod in the first peak. It later became infested with the parasitic protozoan *Epistylis* sp., and both adult copepod and nauplii abundance declined. Up to 79% of the adult copepods and 12.6% of the nauplii were infested. *Apocyclops panamensis,* the dominant copepod in the second wave, was not infested with *Epistylis* sp. In Study II, peak abundance of nauplii per treatment occurred on day 20 and day 26 at 3,684 ± 1,259 and 2,518 ± 571/L for alfalfa- and wheat bran–fertilized ponds. *Apocyclops panamensis* was the predominate copepod. In Study III mean peak densities per treatment were 4,566 ± 1,334, 6,070 ± 1.327, and 9,534 ± 775/L on day 35, day 49, and day 14 for the rice bran, wheat bran without aeration, and wheat bran with aeration, respectively.

Environmental parameters varied across studies and in some cases, among fertilization treatments within a study (Table 12.3). Mean early morning water temperatures were 28.1 ± 0.2, 29.8 ± 0.2, and 28.6 ± 0.2°C, respectively, for Studies I, II, and III. Mean salinities were 14.5 ± 0.1, 12.7 ± 0.2, and 19.0 ± 0.2 ppt, respectively, for Studies I, II, and III. Early morning DO in Study I averaged 3.8 ± 0.2, 3.2 ± 0.1, and 3.1 ± 0.2 mg/L for alfalfa-, wheat bran–, and fishmeal-fertilized ponds, respectively. In Study II early morning DO averaged 3.1 ± 0.0 and 2.8 ± 0.2 mg/L for alfalfa- and wheat bran–fertilized ponds, respectively. The use of aeration in Study III significantly increased early morning DO ($P = 0.029$). Aerated ponds averaged 4.0 ± 0.0, while rice bran– and wheat bran–fertilized ponds without aeration averaged 2.0 ± 0.0 and 2.7 ± 0.0 mg/L. In Studies I and II alfalfa-fertilized ponds were more transpar-

Table 12.2. Summary of results of zooplankton production in Studies I, II, and III

Item	Study I	Study II	Study III
Mean rotifer density (no./L)	AM: 3,193 ± 587[a]	AM: 8,800 ± 798[a]	RB: 3,895 ± 1,150[a]
	WB: 4,387 ± 598[a]	WB: 7,876 ± 1,483[a]	WB: 2,226 ± 599[a]
	FM: 5,671 ± 421[a]	—	WBa: 7,881 ± 2,163[b]
Mean adult copepod density (no./L)	AM: 219 ± 131[a]	AM: 293 ± 116[a]	RB: 752 ± 288[ab]
	WB: 247 ± 66[a]	WB: 826 ± 144[b]	WB: 403 ± 221[a]
	FM: 142 ± 85[a]	—	WBa: 1,028 ± 746[b]
Mean copepod nauplii density (no./L)	AM: 887 ± 172[a]	AM: 2,112 ± 594[a]	RB: 3,125 ± 912[a]
	WB: 1,000 ± 361[a]	WB: 2,520 ± 706[a]	WB: 3,391 ± 523[a]
	FM: 1,070 ± 415[a]	—	WBa: 5,080 ± 1,224[a]
Ectocommensal fouling disease	Yes	No	No

Note: Alfalfa meal (AM), fish meal (FM), rice bran (RB), and wheat bran with aeration (WBa) or without aeration (WB) were applied initially at 250 kg/ha.

[a and b]Values sharing the same letter in the same item column are not significantly different ($P > 0.05$).

Table 12.3. Summary of water quality parameters in Studies I, II, and III

Item	Study I	Study II	Study III
Average early morning dissolved oxygen (mg/L)	AM: 3.8 ± 0.2[a]	AM: 3.1 ± 0.0[a]	RB: 2.0 ± 0.0[a]
	WB: 3.2 ± 0.1[a]	WB: 2.8 ± 0.2[a]	WB: 2.7 ± 0.0[a]
	FM: 3.1 ± 0.2[a]	—	WBa: 4.0 ± 0.0[b]
Average secchi disk transparency (cm)	AM: 123.0 ± 3.8[a]	AM: 83.8 ± 5.4[a]	RB: 63.2 ± 3.9[a]
	WB: 82.4 ± 6.6[b]	WB: 58.1 ± 4.3[b]	WB: 57.1 ± 2.3[a]
	FM: 79.4 ± 5.0[b]	—	WBa: 55.6 ± 2.2[a]
Average salinity (ppt)	14.5 ± 0.1	12.7 ± 0.2	19.0 ± 0.2
Median pond water temperature (°C)	29.1 ± 0.2	31.7 ± 0.2	28.6 ± 0.2
Average of water temperature range (°C)	4.2 ± 0.2	3.5 ± 0.2	5.5 ± 0.3

Note: Alfalfa meal (AM), fish meal (FM), rice bran (RB), and wheat bran with (WBa) or without aeration (WB) were applied initially at 250 kg/ha.

[a and b]Values sharing the same letter in the same item column are not significantly different ($P > 0.05$).

Table 12.4. Polar and nonpolar fatty acid profiles of arachidonic acid (ARA, 20:4*n*-6), EPA (20:5*n*-3), DHA (22:6*n*-3), total omega-3 and omega 6, omega-3/omega-6, and DHA/EPA ratios, total unsaturated and saturated fatty acid (% total fatty acid) and its ratio, polar and nonpolar lipid content (% dry weight and % polar or nonpolar lipid), total lipid content (% dry weight), and individual naupliar weight (μg) of postharvest *Apocyclops panamensis* nauplii trapped from alfalfa meal (Study II) and rice bran, wheat bran, and mixed (rice and wheat bran) meals (Study III) fertilization in brackish water ponds

	Alfalfa meal (Study II)	Rice bran (Study III)	Wheat bran (Study III)	Mixed meals (Study III)
Polar fatty acids				
ARA	0.5 ± 0.0	1.5 ± 0.3	1.0 ± 0.1	1.4 ± 0.0
EPA	11.4 ± 0.7	9.8 ± 0.4[a]	12.6 ± 0.8[b]	13.8 ± 0.8[b]
DHA	23.8 ± 2.7	23.0 ± 2.2	22.9 ± 0.3	22.4 ± 0.4
Σ omega-3	36.7 ± 2.0	33.4 ± 2.6	36.2 ± 1.0	38.1 ± 1.4
Σ omega-6	1.8 ± 0.1	2.5 ± 0.3	3.1 ± 0.1	3.6 ± 0.2
Omega-3/omega-6	19.7 ± 1.4	13.3 ± 0.7	11.8 ± 0.6	10.8 ± 1.0[b]
DHA/EPA	2.1 ± 0.4	2.4 ± 0.1[a]	1.8 ± 0.1[b]	1.6 ± 0.1[b]
Σ Fatty acids	89.5 ± 3.5	86.7 ± 3.5	83.0 ± 2.9	84.1 ± 1.7
Polar lipid	0.41 ± 0.07	0.51 ± 0.09	0.36 ± 0.19	0.82 ± 0.48
% Polar lipid	7.2 ± 1.1	6.5 ± 1.0	21.3 ± 1.8	11.4 ± 6.0
Nonpolar fatty acids				
ARA	1.2 ± 0.1	1.3 ± 0.0	1.3 ± 0.2	1.4 ± 0.0
EPA	16.4 ± 2.9	14.0 ± 0.1	15.6 ± 3.2	14.8 ± 0.4
DHA	26.4 ± 1.4	27.4 ± 1.6	27.0 ± 1.8	28.1 ± 2.3
Σ omega-3	43.9 ± 1.5	42.3 ± 1.6	43.7 ± 4.9	44.1 ± 1.8
Σ omega-6	2.0 ± 0.1	1.5 ± 0.1	1.9 ± 0.3	2.1 ± 0.1
Omega-3/omega-6	22.0 ± 0.4	28.2 ± 0.1[a]	23.1 ± 0.5[b]	21.5 ± 0.4[b]
DHA:EPA	1.7 ± 0.4	2.0 ± 0.1	1.8 ± 0.2	1.9 ± 0.2
Σ Fatty acids	82.9 ± 1.8	82.0 ± 0.8	80.6 ± 3.6	84.5 ± 3.8
Nonpolar lipid	5.25 ± 0.07	7.26 ± 0.18[a]	4.99 ± 0.15[b]	6.08 ± 0.05[a,b]
% Nonpolar lipid	92.8 ± 1.1	93.5 ± 1.0	78.7 ± 1.8	88.6 ± 6.0
Total lipid	5.66 ± 0.15	7.76 ± 0.27	6.35 ± 0.34	6.90 ± 0.52
Naupliar dry weight (μg)	0.08 ± 0.00	0.16 ± 0.01	0.13 ± 0.02	0.13 ± 0.01

Note: Values are mean ± S.E. Only parameters with significant differences ($P < 0.05$) are noted.

ent than the other fertilization regimes. In Study III, rice bran–fertilized ponds were more transparent than wheat bran ponds.

Unenriched Copepod Nauplii from Various Fertilization Regimes

Given in Table 12.4 are the fatty acid profile (polar and nonpolar), DHA/EPA ratio, lipid content (percentage of dry weight), and individual dry weight (micrograms) of copepod nauplii trapped from ponds of the fertilizer studies and sampled before acclimation. There was no difference in mean lipid content in copepod nauplii trapped from rice bran–, wheat bran–, and rice and wheat bran mixed meals–fertilized ponds ($P = 0.18$). Mean nonpolar lipid content of copepod nauplii trapped from wheat bran–fertilized ponds in Study III was lower than those from other fertilization regimes ($P = 0.003$). There were no significant differences in mean contents and ratios of other nonpolar fatty acid profiles in copepod nauplii trapped from three fertilization regimes in Study III ($P > 0.05$) (Table 12.4). There was also no significant difference in mean individual copepod nauplii dry weight ($P = 0.29$). Except for mean content of EPA and the DHA/EPA ratio, there were no significant differences in mean abundance and ratios of other polar fatty acid profiles in copepod nauplii trapped from three fertilization regimes in Study III ($P > 0.05$).

Table 12.5. Polar and non-polar fatty acid profile of ARA, EPA, DHA, total omega-3 and omega-6, omega-3/omega-6 and DHA/EPA ratios, total unsaturated and saturated fatty acid (% total fatty acid) and its ratio, polar and nonpolar lipid content (% dry weight and % polar or nonpolar lipid), total lipid content (% dry weight), and individual naupliar weight (µg), survival rate (%), and change (%) of postharvest and 6-hour after acclimation of *Apocyclops panamensis* nauplii trapped from mixed meals fertilization in brackish-water ponds from Study III

	After harvest	6 hours after acclimation	Change (%)
Polar fatty acids			
ARA	1.4 ± 0.0[a]	1.0 ± 0.0[b]	−28.6
EPA	13.8 ± 0.8[a]	10.5 ± 0.4[b]	−23.9
DHA	22.4 ± 0.4[a]	12.5 ± 0.7[b]	−44.2
Σ omega-3	38.1 ± 1.4[a]	24.2 ± 1.1[b]	−36.5
Σ omega-6	3.6 ± 0.2[a]	3.0 ± 0.1[a]	−16.7
Omega-3/omega-6	10.8 ± 1.0[a]	8.1 ± 0.4[a]	−25
DHA/EPA	1.6 ± 0.1[a]	1.2 ± 0.0[b]	−25
Σ Fatty acids	84.1 ± 1.7[a]	68.3 ± 1.4[b]	−18.8
Polar lipid	0.82 ± 0.48[a]	0.62 ± 0.21[a]	−24.4
% Polar lipid	11.4 ± 6.0[a]	15.8 ± 4.4[a]	38.6
Nonpolar fatty acids			
ARA	1.4 ± 0.0[a]	1.0 ± 0.1[b]	−28.6
EPA	14.8 ± 0.4[a]	5.2 ± 0.3[b]	−64.9
DHA	28.1 ± 2.3[a]	12.5 ± 1.1[b]	−55.5
Σ omega-3	44.1 ± 1.8[a]	18.3 ± 0.7[b]	−58.5
Σ omega-6	2.1 ± 0.1[a]	1.4 ± 0.0[b]	−33.3
Omega-3/omega-6	21.5 ± 0.4[a]	12.9 ± 0.3[b]	−40
DHA/EPA	1.9 ± 0.2[a]	2.4 ± 0.4[a]	26.3
Σ Fatty acids	84.5 ± 3.8[a]	48.6 ± 2.0[b]	−42.5
Nonpolar lipid	6.08 ± 0.05[a]	3.23 ± 0.06[b]	−46.9
% Nonpolar lipid	88.6 ± 6.0[a]	84.2 ± 4.4[a]	−5
Total lipid	6.90 ± 0.52[a]	3.85 ± 0.26[b]	−44.2
Naupliar weight	0.13 ± 0.01[a]	0.08 ± 0.00[b]	−38.5
Survival rate		90.4 ± 3.4	

Note: Values are mean ± S.E. Only parameters with significant differences ($P < 0.05$) are noted.

Unenriched Copepod Nauplii after Acclimation

Results of fatty acid profiles of nonpolar and polar lipid classes, survival rate (percentage), individual dry weight (microgram), and change values (percentage) of after harvest and 6-h after acclimation of copepod nauplii *A. panamensis* trapped from the mixed meals–fertilized pond in Study III are shown in Table 12.5. Acclimation from an average initial salinity of 19.7 ± 0.2 ppt up to full seawater salinity of 32.0 ± 0.0 ppt was conducted over 6 hours at a copepod nauplii density of 200 individuals per liter. Mean total lipid content was significantly lower ($P = 0.029$) after acclimation, decreasing from 6.90 ± 0.52 to 3.85 ± 0.26% dry weight after 6 hours, a 44.2% decrease. Likewise, nonpolar lipid content was significantly lower after acclimation ($P < 0.001$), with a 46.9% decrease. Mean contents of ARA, EPA, DHA, total omega-3, omega-6, omega-3/omega-6, and total fatty acids in copepod nauplii showed significant decreases after acclimation ($P < 0.05$), while other components did not decrease significantly ($P > 0.05$). There were significant decreases in mean content of polar ARA, EPA, DHA, total omega-3, DHA/EPA ratio, and total fatty acids in copepod nauplii ($P < 0.05$), while other components were not significantly

Table 12.6. Polar and nonpolar fatty acid profile of ARA, EPA, DHA, total omega-3 and omega-6, omega-3/omega-6 and DHA/EPA ratios, total unsaturated and saturated fatty acid (% total fatty acid) and its ratio, nonpolar lipid content (% dry weight and % nonpolar lipid), and total lipid content (% dry weight), naupliar dry weight (DW, μg) and survival rate (SR, %) of *Apocyclops panamensis* nauplii enriched with A-1 Super Selco for 3 hours, 6 hours, and 12 hours in normal dose

	A-1 Super Selco		
	3 hours	6 hours	12 hours
Polar fatty acid			
ARA	1.0 ± 0.2	1.3 ± 0.0	1.6 ± 0.3
EPA	10.5 ± 0.7	11.3 ± 1.2	11.9 ± 0.8
DHA	16.3 ± 0.2	20.8 ± 1.0	21.2 ± 0.6
Σ omega-3	28.2 ± 0.4	34.0 ± 2.1	36.6 ± 2.6
Σ omega-6	3.0 ± 0.3	3.3 ± 0.1	3.8 ± 0.3
Omega-3/omega-6	9.6 ± 0.7	10.3 ± 0.3	9.7 ± 0.1
DHA/EPA	1.6 ± 0.1	1.8 ± 0.1	1.8 ± 0.1
Σ Fatty acids	73.9 ± 0.7[a]	82.9 ± 1.6[b]	86.4 ± 5.5[b]
Polar lipid	1.36 ± 0.01[a]	1.94 ± 0.12[a,b]	2.30 ± 0.21[b]
% Polar lipid	18.7 ± 0.2	20.7 ± 0.8	21.4 ± 1.2
Nonpolar fatty acid			
ARA	1.5 ± 0.0[a]	1.6 ± 0.0[a]	1.9 ± 0.1[b]
EPA	12.5 ± 1.3	15.8 ± 0.3	16.2 ± 0.5
DHA	22.3 ± 1.8	23.1 ± 1.4	24.5 ± 0.4
Σ omega-3	36.5 ± 2.9	41.1 ± 1.1	43.1 ± 1.1
Σ omega-6	2.5 ± 0.0[a]	2.7 ± 0.0[a]	3.1 ± 0.1[b]
Omega-3/omega-6	14.9 ± 1.5	15.2 ± 0.4	13.9 ± 0.8
DHA/EPA	1.8 ± 0.0	1.5 ± 0.1	1.5 ± 0.0
Σ Fatty acids	71.8 ± 3.6[a]	80.1 ± 2.3[b]	83.8 ± 1.2[b]
Nonpolar lipid	5.89 ± 0.14[a]	7.42 ± 0.09[b]	8.44 ± 0.19[b]
% Nonpolar lipid	81.3 ± 0.2	79.3 ± 0.8	78.6 ± 1.2
Total lipid	7.25 ± 0.15[a]	9.36 ± 0.20[ab]	10.74 ± 0.40[b]
Survival rate	89.0 ± 3.5[a]	58.6 ± 2.2[ab]	28.1 ± 4.1[b]
Dry weight	0.10 ± 0.01[a]	0.27 ± 0.03[b]	0.30 ± 0.06[b]

Note: Values are mean ± S.E. Only parameters with significant differences ($P < 0.05$) are noted.

lower after acclimation ($P > 0.05$). Mean survival rate after acclimation was 90.4 ± 3.4%.

Enrichment of Copepod Nauplii

Nonpolar and polar fatty acid profiles of ARA, EPA, DHA, total omega-3 and omega-6, omega-3/omega-6, and DHA/EPA ratios, total fatty acids, total lipid content, copepod naupliar dry weight, and survival rate of *A. panamensis* nauplii enriched with A-1 Super Selco for 3 hours, 6 hours, and 12 hours are shown in Table 12.6. Mean concentration of nonpolar fatty acid profiles of nauplii enriched with A-1 Super Selco for 3 hours, 6 hours, and 12 hours were not significantly different ($P > 0.05$; Table 12.6), but mean content of ARA, omega-6, total fatty acids, total lipid, and neutral lipid content were significantly different ($P < 0.05$). Except for mean total fatty acids and polar lipid content, there were no significant differences in all other fatty acid profiles of polar lipid class ($P > 0.05$; Table 12.6). Means of survival rates of copepod nauplii were significantly lower after the 12-hour enrichment period ($P = 0.006$) and averaged 28.1 ± 4.1% (Table 12.6).

DO was progressively lower the longer the enrichment period, dropping from 6.32 mg/L to 3.6 mg/L in a 12-hour period even though the holding container was aerated. Ammonia was below detectable levels at the end of the enrichment period.

Table 12.7. Evaluation of *Apocyclops panamensis* when wild-caught adults are stocked at varying densities in bags containing *I. galbana* at 5 × 10⁵ cells/mL and harvested after 4 days

Adults/L	Production/L	Change in adult abundance (%)	Production/female
320	1,630 ± 357.0[d]	−117 ± 57.2[a]	2.6 ± 0.7[b]
640	3,044 ± 117.2[d]	2.0 ± 10.9[b]	8.1 ± 1.1[a]
1,280	6,198 ± 466.4[c]	12 ± 2.3[b]	8.6 ± 0.9[a]
2,560	3,616 ± 682.3[b]	27 ± 9.1[b]	8.2 ± 1.2[a]
5,120	17,873 ± 654.0[a]	39 ± 5.5[b]	4.7 ± 0.6[b]
P value	< 0.01	< 0.01	< 0.01

Note: Mean ± standard errors of the production of nauplii and copepodids per liter, the change in density of adults in the culture, and nauplii and copepodid production per female at harvest. Means in columns with the same letter are not significantly different.

Table 12.8. Evaluation of *Apocyclops panamensis* culture when wild-caught adults are stocked at varying densities in bags containing *I. galbana* at 5 × 10⁵ cells/mL and harvested after 4 days

Adults/L	Males	Females	Females with eggs	Males (%)	Females (%)	Females with eggs (%)
320	86.0 ±20.2	143.0 ± 73.6	64.0 ± 12.0	44.3 ± 5.8[a]	56.7 ± 5.8[a]	72.2 ± 17.3[a]
640	288.3 ± 40.2	394.1 ± 50.2	216.4 ± 26.8	42.2 ± 1.0[a]	58.8 ± 1.0[a]	55.0 ± 2.0[a]
1,280	712.6 ± 54.6	744.4 ± 45.4	456.6 ± 34.4	48.7 ± 3.1[a]	51.3 ± 3.1[a]	61.3 ± 2.5[a]
2,560	1,968.6 ± 229.8	1,999.4 ± 352.0	1,105.3 ± 160.2	51.4 ± 2.8[a]	48.6 ± 2.8[a]	58.8 ± 3.8[a]
5,120	4,633.6 ± 455.9	4,010.6 ± 543.2	2,031.8 ± 237.2	53.9 ± 2.4[a]	46.1 ± 2.4[a]	51.1 ± 1.0[a]
P value	—	—	—	0.23	0.23	0.43

[a]Mean ± S.E. for the number of males, females, and egg-bearing females per liter in the culture at harvest. Mean ± S.E. for the percentage of males, females in the population, and the percentage of females with eggs in the female population. Means in columns with the same letter are not significantly different.

Intensive Production

Evaluation of Adult Stocking Density

Mean abundance of nauplii and copepodids per liter reached a peak of 17,873 ± 654 at an adult density of 5,120/L. Production was significantly ($P < 0.001$) different for each stocking density, except for the two lower stocking densities, 320 and 640 adults/L, which were similarly low (Table 12.7). Production of nauplii and copepodids per female ranged from 2.6 ± 0.7 to 8.6 ± 0.9. The highest stocking density, 5,120 adults/L, and the lowest, 320 adults/L, had a significantly ($P < 0.01$) lower mean production of nauplii and copepodids per female in comparison with the other three stocking densities of 640, 1,280, and 2,560 adults/L (Table 12.8). The change in the size of the adult population at harvest, in terms of the stocking density, was significantly ($P < 0.001$) higher for the lowest stocking density, 320 adults/L, compared with the other adult stocking densities (Table 12.8). At harvest, there was no significant difference in the percentage of males ($P = 0.23$) or females ($P = 0.23$) or the percentage of the female population with eggs ($P = 0.43$) in the culture population for the different treatments (Table 12.8). Overall, DO was positively correlated with production of nauplii and copepodids per female ($R^2 = 0.47$, $P = 0.02$).

Closed Batch Production

Net production (nauplii and copepodids) varied considerably over the 12 trials (Table 12.9 and Table 12.10). Mean production in population A was 10,939 ± 3,200/L, but ranged from 144.0 to 31,532/L. Mean production in population B was 14,366 ± 3,095/L, and ranged from 250 to 36,800/L. This variability was due in part to adult stocking density. Over the 12 trials, the adults were harvested from one trial and restocked to begin the next trial. As a result, the number of

Table 12.9. Closed system production of *Apocyclops panamensis*

Trial	Culture period (days)	Initial adult density/L	Final adult density/L	Change in abundance of the adult population (%)	Yield. (No./L)	Production No./Female
1	7	320	160	−50.0	3,640	31.7
2	9	160	3,705	2,215.6	9,670	3.5
3	4	3,705	2,202	−40.6	31,531	22.6
4	4	2,202	1,908	−13.4	26,625	26.1
5	7	1,908	4,370	129.0	25,465	8.1
6	5	4,370	2,081	−52.4	24,575	19.9
7	5	2,081	1,605	−22.9	4,575	4.7
8	4	1,605	488	−69.6	5,150	17.9
9	6	488	250	−48.8	1,500	6.0
10	6	125	1,007	705.6	383	0.8
10	7	125	115	−8.0	144	5.0
11	7	1,007	690	−31.5	4,000	10.7
12	6	690	900	30.4	4,950	7.3

Note: Initial adult density, final adult density, the percentage of change in density of adults in the culture, yield of nauplii and copepodids, and nauplii and copepodid production per female at harvest for population A. There were no significant differences at $P < 0.05$.

Table 12.10. Closed system production of *Apocyclops panamensis*

Trial	Culture period (days)	Initial adult density/L	Final adult density/L	Change in adult population abundance (%)	Yield. (No./L)	Production No./Female
1	7	320	170	−46.9	3,480	38.7
2	9	170	1,290	658.8	4,255	4.0
3	4	1,290	1,115	−13.5	11,550	15.8
4	4	1,115	688	−38.3	18,350	42.4
5	7	688	6,498	844.5	15,895	3.0
6	5	6,498	2,392	−63.2	28,395	16.0
7	5	2,392	4,227	76.7	24,150	8.8
8	4	4,227	1,985	−53.0	9,320	10.2
9	6	1,985	2,250	13.4	250	0.2
10	6	750	4,492	559.1	16,033	2.0
11	7	4,710	4,974	5.6	36,800	12.0
12	6	4,974	6,025	21.1	27,900	8.9

Note: Initial adult density, final adult density, the percentage of change in density of adults in the culture, yield of nauplii and copepodids, and nauplii and copepodid production per female at harvest for population B. There were no significant differences at $P < 0.05$.

adults stocked per trial varied over time (Table 12.9 and Table 12.10). In general there was a significant relationship between adult stocking density and nauplii and copepodid production for both populations A and B (A: $R^2 = 0.65$, $P < 0.01$; B: $R^2 = 0.52$, $P = 0.01$).

Production in population A was positively correlated to water temperature over the range of 25.8 to 31.1°C ($R^2 = 0.47$, $P = 0.01$). This was not as evident in population B (25.7 to 31.0°C, $R^2 < 0.01$, $P = 0.96$). All other regressions of water quality parameters were not significant. Production per female ranged from 0.8 to 31.7 nauplii and copepodids per female in population A and 0.2 to 42.4 per female in population B (Table 12.9 and Table 12.10). The change in abundance of the adult population, from stocking to harvest, was a weak predictor (negative) of the production of nauplii and copepodids per female in population B ($R^2 = 0.31$, $P = 0.04$). When the adult population increased 844.5%, production of nauplii and copepodids per female was 3.0. When the population of adults decreased −46.9%, production of nauplii and copepodids per female was 38.7. This relationship was not evident in population A.

In population A, the percentage of males in the population averaged 37.3% over the course of the 12 trials and ranged from 0 to 74.78% ($P = 0.59$). The percentage of females averaged 62.7% with a range of 25.2 to 100% ($P = 0.60$). The percentage of the females carrying eggs ranged from 0 to 53.7% ($P = 0.87$) and with an average of 27.1%. In population B, the average population composition was 25.4% males and 74.7% females, with 41.2% of the females carrying eggs. The percentage of males over the course of the 12 trials ranged from 10.6 to 43.9% ($P = 0.03$) and from 56.1 to 89.4% ($P = 0.03$) for females. The percentage of females carrying eggs in population B ranged from 31.7 to 50.0% ($P = 0.26$). Though there was a significant difference in the composition of adult population in population B, there appeared to be no relationship between the percentage of the males and females in the population and the productivity of the culture.

DISCUSSION

Copepod Nauplii Production in Ponds

Pond production of copepods is a successful way to produce high densities of adults and nauplii. All but one of the fertilization regimes in this study produced mean nauplii densities of *A. panamensis* greater than 1,000 nauplii/L. In natural temperate and tropical brackish water, densities of 0.6–100 calanoid, harpacticoid, and cyclopoid copepod nauplii per liter are common (Fukusho 1991; Mishra and Panigrahy 1996; Shansudin et al. 1997; Watanabe et al. 1998).

Pond production of copepods has the advantage in that large quantities of nauplii can be obtained using ponds that are often available at a hatchery, while to obtain similar quantities of nauplii from a laboratory-based system requires additional dedicated floor space, increased labor, and capital (Støttrup and Norsker 1997). Pond production of zooplankton, using organic fertilizer and inorganic fertilizers (Geiger 1983; Geiger et al. 1985; Colura et al. 1987; Doi et al. 1994; Kurten et al. 1999), has yielded average densities of 19 *Acartia* sp. nauplii per liter (Kurten et al. 1999) to 374 *Acartia* sp. and *Oithona* sp. nauplii per liter (Colura et al. 1987). Average nauplii densities produced in outdoor ponds with various fertilization regimes in this study ranged from 887.2 ± 171.8 individuals/L (both *Acartia* sp. and *A. panamensis* in alfalfa-fertilized ponds in Study I) to 5,079.7 ± 1,224.3 *A. panamensis* nauplii/L (wheat bran with aeration–fertilized ponds in Study III), with a mean peak of 9,533.7 ± 774.6 individuals/L, respectively. The average density of *A. panamensis* nauplii in wheat bran with aeration–fertilized ponds in Study III was comparable to the average density (5,150 nauplii/L) reported by Støttrup et al. (1986) for nauplii of the harpacticoid copepod *Tisbe holothuriae* in a laboratory system.

The type of organic fertilizer has been show to influence the zooplankton species composition. Geiger et al. (1985) found a significantly higher average copepod (adult and nauplii) abundance in cottonseed meal compared with chicken manure– and chicken litter–fertilized ponds. Bootes (1998) found that average copepod nauplii density in outdoor tanks was significantly higher in those with a high C/N fertilizer ratio (hog feed and cottonseed meal) compared with those with a lower C/N ratio (fishmeal). In the three studies reported here, one animal-origin (fishmeal) and three plant-origin (alfalfa meal, rice bran, and wheat bran) organic fertilizers, which varied in the quantity and quality of nutrients, were evaluated and added to the ponds, but no one fertilizer produced significantly more nauplii.

The relationship between copepod abundance, water quality, and other environmental parameters is complex. In Study III, DO in wheat bran with aeration was significantly higher than those of the other fertilization regimes ($P = 0.018$), and both rotifer and adult copepod densities for this fertilization regime were significantly higher than those of the other regimes ($P < 0.05$), and to a lesser extent, a higher density of copepod nauplii ($P = 0.072$). Geiger (1983) found that aeration and circulation of pond water seems to accelerate primary production, which can support a large zooplankton population. Ito and Satomi (1975) found that a vertical pump reduced biochemical stratification, but increased primary productivity and the zooplankton and phytoplankton population, compared with the control ponds.

There were relatively clear waves of abundance for rotifers, adult copepods, and copepod nauplii in Study I. Such waves were not as distinct in Studies II or III. Waves in zooplankton abundance were reported by Harrell and Bukowski (1990) in brackish water ponds fertilized with inorganic and organic (alfalfa cubes and soy meal) fertilizers. Allen (1976) reported that populations of copepods, with their longer life cycles, greater capacity for selective feeding, and greater ability to avoid predators, tend to reach their peak abundance following the rotifer peak.

Zooplankton production is highly variable even among similarly managed ponds. Standard errors of average densities of zooplankton in wheat bran–fertilized ponds across the studies were between 14 and 55%. High pond-to-pond variances in term of standard deviation or standard error have been commonly encountered in other fertilization studies in outdoor ponds. Geiger (1983) and Geiger et al. (1985) encountered 63–225% and 98–322% standard deviation values of average zooplankton densities, respectively. Kurten et al. (1999) had 17–162% of standard error values of average zooplankton densities in their study. Producing copepod nauplii in ponds can be highly variable, and several ponds need to be in production at one time to ensure one that has adequate numbers of nauplii for trapping.

Nutritional Characteristics

It has been suggested that variation in lipid content of copepods is greatly affected by available diets in natural waters (Fraser et al. 1989; Norrbin et al. 1990). In this study mean total lipid content of copepod nauplii was similar across fertilization regimes. Other nonpolar lipid characteristics of nauplii, however, differed among the fertilization regimes. Copepod nauplii trapped from wheat bran–fertilized ponds had a significantly lower mean content of ARA than the other regimes ($P = 0.022$). In addition, the omega-3/omega-6 ratio in the rice bran regime had significantly higher values than those of other regimes ($P = 0.02$). The DHA/EPA ratios of neutral lipid class were similar for the three fertilizer protocols in Study III, ranging from 1.6 ± 0.1 to 2.1 ± 0.4. Lokman (1994) and Shansudin et al. (1997) reported a DHA/EPA ratio of natural zooplankton (> 60% was *Oithona* sp.) between 1.0 and 2.3. Fraser et al. (1989) reported the DHA/EPA ratio in three copepod species from natural waters to be between 0.7 and 2.4 in nonpolar fatty acids. There were no significant differences in other fatty acid profiles of nonpolar lipids ($P > 0.05$), suggesting that *A. panamensis* nauplii propagated from any of the three fertilization regimes in this study were similar in nutritional value.

Among the polar lipids of copepod nauplii from the different fertilization regimes, there were only significant differences in EPA and the DHA/EPA ratio. Mean content of EPA in copepod nauplii trapped from the rice bran regime was at a significantly lower level than those of other regimes ($P = 0.004$). Ranges in the polar lipid DHA/EPA ratio from copepod nauplii were 1.6 ± 0.1 to 2.4 ± 0.1, with the ratio in the rice bran regime being the highest among the regimes ($P = 0.035$). Fraser et al. (1989) and Norrbin et al. (1990) reported DHA/EPA ratios in polar lipid fatty acids in adult copepods of 1.3–1.4 and 0.8–1.2, respectively. In general, Sargent et al. (1999) suggested that the optimal DHA/EPA ratio in lipids for marine finfish larval diets be about 2.0. The DHA/EPA ratio in both lipid classes was approximately 2.0.

Acclimation reduced the neutral lipid level of copepod nauplii (−46.9%) significantly more than that of the polar lipid (−24.4%; $P = 0.043$). It is suggested that energy reserves from triglycerides and wax ester (neutral lipids) are used in greater proportion due to unfed and crowding conditions during the acclimation. Bourdier and Amblard (1989) reported that almost entirely neutral lipid in the calanoid copepod *Acanthodiaptomus denticornis* was used up after 20 days of starvation. This study demonstrated that a salinity acclimation over 6 hours, from average salinities of 12.7 ± 0.1 to 19.7 ± 0.2 to full seawater salinity of 32–34 ppt clearly decreased nutritional quality of *A. panamensis* nauplii.

Enrichment of Copepod Nauplii

Copepod nauplii are considered to be of prime nutritional quality as live feed for marine fish larvae (Watanabe 1993), but as demonstrated above, the harvesting and acclimation of nauplii from brackish to full-strength seawater significantly reduces the nutritional quality. Enrichment of rotifers (Fernandez-Reiriz et al. 1993; Tamaru et al. 1993; Craig et al. 1994; Robin 1998) and *Artemia* (van Ballaer et al. 1985; Webster and Lovell

1990; Takeuchi et al. 1992; Han et al. 2000) with various species of phytoplankton and enrichment oil for 3–72 hours is common. Copepod nauplii also respond to enrichment. Mean content of ARA, total omega-6 fatty acids, total fatty acids, nonpolar lipid, and total lipids showed significant increase ($P < 0.05$) after prolonged enrichment with A-1 Super Selco (Table 12.6). Payne and Rippingale (2001) enriched nauplii of the calanoid copepod *Gladioferens imparipes* with *I. galbana* and *Nannochloropsis oculata* for 30 minutes up to 6 hours at 23°C, and they found a significant increase in total fatty acid content, with DHA/EPA ratio ranges of 4.9–7.0. In another study, Payne et al. (2001) compared the nutritional quality of enriched *G. imparipes* nauplii versus rotifers and *Artemia* using Super Selco and microalgae *I. galbana* and *N. oculata* for 6 hours. They found a higher mean content of DHA and a higher DHA/EPA ratio in enriched copepod nauplii.

The extensive approach for obtaining copepod nauplii is effective for producing large quantities of nauplii, but additional steps are required before they are suitable for feeding copepod nauplii. Acclimation to full-strength seawater reduces nauplii nutrient quality, but that can be restored by enrichment. Acclimation and enrichment with A-1 Super Selco, however, significantly reduced nauplii survival ($P = 0.006$). The extensive approach used at the Claude Peteet Mariculture Center is a tradeoff of high-volume production coupled with additional effort to ensure the nauplii are suitable for feeding red snapper larvae.

Intensive Production

Apocyclops panamensis appears to be suitable for intensive culture. Nauplii yields increased as stocking rates of wild-caught adults increased, with a peak of 17,872.5 ± 654.0 at an adult density of 5,120/L. In a closed system higher yields were possible. A yield of 36,800/L was obtained from a 6-day culture period when domesticated adults were stocked at 4,710/L. Maximum densities of 26,000 harpacticoid copepod *Tigriopus brevicornis* nauplii per liter (Vilela 1992) and 33,000 cyclopoid copepod *Apocyclops royi* nauplii per liter (Cheng et al. 2001) have been reported. Production trials with *Acartia* spp. have yielded up to 5,150 nauplii/L (Schipp et al. 1999). Shirgur (1989) produced *Apocyclops dengizicus* at densities averaging 16,000/L.

Total yield and production per female *A. panamensis* was related to change in the adult copepod populations. The greater the number of adults present at harvest relative to the number stocked resulted in greater total production of nauplii and copepodids per liter ($R^2 = 0.41$, $P < 0.0001$), and a decline in the production per females ($R^2 = 0.48$, $P < 0.0001$). This suggests that during a 4-day production cycle, it was possible for some of the nauplii to reach maturity and reproduce, thus increasing the total number of females (mature and immature) and, in turn, resulting in a decrease of production per female.

The optimum stocking density depends on the availability of adult copepods and other production-related resources. In these trials overall nauplii and copepodid yield per liter continued to increase as the stocking density increased ($R^2 = 0.87$, $P < 0.001$). At the wild-adult stocking densities of 640, 1,280, and 2,560, the production of nauplii and copepodids per female were similar. Production per female, however, declined when the adult stocking density was increased to 5,120/L. If space for production were available, a greater number of nauplii and copepodids could be obtained by stocking two production units at 2,560 adults/L rather than one at 5,120/L. In this study, space and adult copepod availability were limiting factors. Therefore, a stocking density of 2,560/L would give a high yield while also maintaining a high reproduction rate per female. Maintaining a closed population of *A. panamensis* giving *I. galbana* as the sole food source appears to be practical. Higher total yields of nauplii and nauplii per female were obtained using domesticated adults rather than wild-caught adults.

CONCLUSION

Copepod nauplii for feeding marine fish larvae can be provided through extensive or intensive production systems. Extensive production requires dedicated ponds prepared to optimize nauplii production. Harvesting and acclimating nauplii can impact their nutritional quality, but that can be restored through fatty acid enrichment. Species such as *A. panamensis* can be produced intensively with yields of 20,000–30,000 nauplii/L possible in a 4- to 6-day production period.

ACKNOWLEDGMENTS

The authors would like to thank the staff of the Claude Peteet Mariculture Center for their help in facilitating the field activities. The studies reported here were supported by the Alabama Department of Conservation Marine Resources Division, National Fisheries Institute and U.S. Department of Agriculture, Cooperative State Research, Education and Extension Service.

REFERENCES

Allen, J.D. 1976. Life history patterns in zooplankton. *American Naturalist* 110:165–180.

AOAC (Association of Official Analytical Chemists). 1970. Official methods of analysis. AOAC, Washington, D.C., USA.

Bootes, K.L. 1998. Culture and description of larval red snapper *Lutjanus campechanus*. Master's Thesis. Department of Fisheries and Allied Aquacultures, Auburn University, Auburn, Alabama, USA.

Bourdier, G.G., and Amblard, C.A. 1989. Lipids in *Acanthodiaptomus denticornis* during starvation and fed on three different algae. *Journal of Plankton Research* 11:1201–1212.

Cheng, S.-H., Chen, H.-C., Chang, S.-L., Chen, T.-I., and Liao, I.C. 2001. Study on the optimal density of mass culture in copepod *Apocyclops royi*. In *Sixth Asian Fisheries Forum Book of Abstracts*, p. 58. Kaohsiung, Taiwan, 25–30, November 2001. Manila, Philippines: Asian Fisheries Society.

Christie, W.W. 1982. *Lipid Analysis: Isolation, Separation, Identification and Structural Analysis of Lipids*, 2nd ed. Oxford, UK: Pergamon Press. 207 pp.

Colura, R.L., Matlock, G.C., and Maciorowski, A.F. 1987. Zooplankton abundance in unstocked mariculture ponds at three salinities. *Progressive Fish-Culturist* 49:253–259.

Craig, S.R., Arnold, C.R., and Holt, G.J. 1994. The effects of enriching live foods with highly unsaturated fatty acids on the growth and fatty acid composition of larval red drum *Sciaenops ocellatus*. *Journal of World Aquaculture Society* 25(3):424–431.

Davis, C.S. 1983. Laboratory rearing of marine calanoid copepods. *Journal of Experimental Marine Biology and Ecology* 17:119–133.

Doi, M. and Singhagraiwan, S. 1993. *Biology and Culture of the Red Snapper, Lutjanus argentimaculatus*. The Research Project of Fishery Resource Development in the Kingdom of Thailand, The Eastern Marine Fisheries Development Center, Rayong, Thailand.

Doi, M., Singhagraiwan, S., Singhagraiwan, T., Kitade, M., and Ohno, A. 1994. Culture of the calanoid copepod, *Acartia sinjiensis*, in outdoor tanks in the tropics. *Thai Marine Fisheries Research Bulletin* 5:27–36.

Fernandez-Reiriz, M.J., Labarta, U., and Ferreiro, M.J. 1993. Effects of commercial enrichment diets on the nutritional value of the rotifer (*Brachionus plicatilis*). *Aquaculture* 112:195–206.

Folch, J., Lees, M., and Stanley, G.H.S. 1957. A simple method for the isolation and purification of total lipids from animal tissues. *Journal Biological Chemistry* 226:497–509.

Fraser, A.J., Sargent, J.R., and Gamble, J.C. 1989. Lipid class and fatty acid composition of *Calanus finmarchicus* (Gunnerus), *Pseudocalanus* sp. and *Temora longicornis* Muller from a nutrient-enriched seawater enclosure. *Marine Biology and Ecology* 130:81–92.

Fukusho, K. 1991. Review of the research status of zooplankton production in Japan. In *Rotifer and Microalgae Culture Systems. Proceedings of a U.S.-Asia Workshop*, edited by W. Fulks and K.L. Main, pp. 55–60. Honolulu, Hawaii, USA: The Oceanic Institute.

Geiger, J.G. 1983. A review of pond zooplankton production and fertilization for the culture of larval and fingerling striped bass. *Aquaculture* 35:353–369.

Geiger, J.G., Turner, C.J., Fitzmayer, K., and Nichols, W.C. 1985. Feeding habits of larval and fingerling striped bass and zooplankton dynamics in fertilized rearing ponds. *Progressive Fish-Culturist* 47(4):213–223.

Gonzales, C.R.M. and Bradley, B.P. 1994. Salinity stress protein in *Eurytemora affinis*. *Hydrobiology* 292/293:461–468.

Graves, K.G. and Morrow, J.C. 1988. Method for harvesting large quantities of zooplankton from hatchery ponds. *Progressive Fish-Culturist* 50(3): 184–186.

Han, K., Geurden, I., and Sorgeloos, P. 2000. Enrichment strategies for *Artemia* using emulsions providing different levels of n-3 highly unsaturated fatty acids. *Aquaculture* 183:335–347.

Harrell, R.M. and Bukowski, J. 1990. The culture, zooplankton dynamics and predator-prey interactions of Chesapeake Bay striped bass, *Morone saxatilis* (Walbaum), in estuarine ponds. *Aquaculture and Fisheries Management* 21:195–212.

Houde, E.D. 1975. Effects of stocking density and food density on survival, growth and yield of laboratory-reared larvae of sea bream *Archosargus rhomboidalis* (L.) (Sparidae). *Journal of Fish Biology* 7:115–127.

Houde, E.D. 1978. Critical food concentrations for larvae of three species of subtropical marine fishes. *Bulletin of Marine Science* 28(3):395–411.

Hunter, J.R. and Lasker, R. 1981. Feeding Ecology and predation of marine fish larvae. In *Marine Fish Larvae-Morphology, Ecology, and Relations to Fisheries*, pp. 33–77. Seattle, Washington, USA: Washington Sea Grant Program.

Ito, Ttoi J. and Satomi, Y. 1975. The effect of artificial pond water circulation and feeding on productivity in carp ponds. *Bulletin of the Freshwater Research Laboratory, Tokio* 24:65–78.

Kimmel, D.G. and Bradley, B.P. 2001. Specific protein responses in the calanoid copepod *Eurytemora affinis* (Poppe, 1880) to salinity and temperature variation. *Journal of Experimental Marine Biology and Ecology* 266:135–149.

Kuhlmann, D., Quantz, G., and Witt, U. 1979. Rearing of turbot *Scophthalmus maximus* L. larvae: the importance of dietary long chain polyunsaturated fatty acids. *Aquaculture* 18:227–235.

Kurten, G., Hall, L., and Thompson, N. 1999. Evaluation of cottonseed meal supplementation of inorganically fertilized Florida largemouth bass spawning ponds. *North American Journal of Aquaculture* 61:115–125.

Lokman, H.S. 1994. Lipid and fatty acid composition of indigenous zooplankton from Trengganu waters of the South China Sea. *Journal of Aquaculture in the Tropics* 9:291–298.

Mishra, S. and Panighrahy, R.C. 1996. Copepods of Bahuda estuary (Orissa), east coast of India. *Indian Journal of Marine Science* 25:98–102.

Misra, S.K. and Phelps, R.P. 1992. A zooplankton harvester designed to collect rotifers. *Progressive Fish–Culturist* 54(4):267–269.

Mourente, G., Rodriguez, A., Tocher, D.R., and Sargent, J.R. 1993. Effect of dietary docosahaexaenoic acid (DHA; 22:6n-3) on lipid and fatty acid compositions and growth in gilthead seabream (*Sparus aurata* L.) larvae during first feeding. *Aquaculture* 112:79–98.

Norrbin, M.F., Olsen, R.E., and Tande, K.S. 1990. Seasonal variation in lipid class and fatty acid composition of two small copepods in Balsfjorden, northern Norway. *Marine Biology* 105:205–211.

Payne, M.F. and Rippingale, R.J. 2001. Effects of salinity, cold storage and enrichment on the calanoid copepod *Gladioferens imparipes*. *Aquaculture* 201:251–262.

Payne, M.F., Rippingale, R.J., and Cleary, J.J. 2001. Cultured copepods as food for West Australian dhufish (*Glaucosoma hebraicum*) and pink snapper (*Pagrus auratus*) larvae. *Aquaculture* 194:137–150.

Planas, M. and Cunha, I. 1999. Larviculture of marine fish: problems and perspectives. *Aquaculture* 177:171–190.

Rainuzzo, J.R., Reitan, K.I., Jorgensen, L., and Olsen, Y. 1994. Lipid composition in turbot larvae fed live feed cultured by emulsions of different lipid classes. *Comparative Biochemical Physiology* 107A:699–710.

Robin, J.H. 1998. Use of borage oil in rotifer production and *Artemia* enrichment: effect on growth and survival of turbot (*Scophthalmus maximus*) larvae. *Aquaculture* 161:323–331.

Salhi, M., Izquierdo, M.S., Hernandez-Cruz, C.M., Socorro, J., and Fernandez-Palacios, H. 1997. The improved incorporation of polyunsaturated fatty acids and changes in liver structure in larval gilthead sea bream fed on microdiets. *Journal of Fish Biology* 51:869–879.

Sargent, J., McEvoy, L., Estevez, A., Bell, G., Bell, M., Henderson, R.J., and Tocher, D.R. 1999. Lipid nutrition of marine fish during early development: current status and future directions. *Aquaculture* 179:217–229.

Schipp, G.R., Bosmans, J.M.P., and Marshall, A.J. 1999. A method for hatchery culture of tropical calanoid copepods, *Acartia* spp. *Aquaculture* 174:81–88.

Shansudin, L., Yusof, M., Azis, A., and Shukri, Y. 1997. The potential of certain indigenous copepod species as live food for commercial fish larval rearing. *Aquaculture* 151:351–356.

Shirgur, G.A. 1989. Phased fertilisation for culturing copepods. *Crustaceana* 56:113–126.

Singhagraiwan, T. and Doi, M. 1993. Induced spawning and larval rearing of the red snapper, *Lutjanus argentimaculatus* at the Eastern Marine Fisheries Development Center. *Thai Marine Fisheries Research Bulletin* 4:45–57.

SAS (Statistical Analysis System) Institute Inc. 1988. SAS/STAT User's Guide Release 6.03 Edition. SAS Institute Inc., Cary, North Carolina, USA.

SAS (Statistical Analysis System) Institute Inc. 1996. SAS Version 6.12 for Windows. SAS Institute, Inc., Cary, North Carolina, USA.

Sokal, R.R. and Rohlf, F.J. 1981. *Biometry*. New York, New York, USA: W.H. Freeman. 859 pp.

Støttrup, J.G. and Attramadal, Y. 1992. The influence of different rotifer and *Artemia* enrichment diets on growth, survival and pigmentation in turbot (*Scophthalmus maximus* L.) larvae. *Journal of the World Aquaculture Society* 23(4):307–316.

Støttrup, J.G. and Norsker, N.H. 1997. Production and use of copepods in marine fish larviculture. *Aquaculture* 155:231–247.

Støttrup, J.G., Richardson, K., Kirkegaard, E., and Pihl, N.J. 1986. The cultivation of *Acartia tonsa* Dana for use as live food source for marine fish larvae. *Aquaculture* 52:87–96.

Støttrup, J.G., Shields, R., Gillespie, M., Gara, M.B., Sargent, J.R., Bell, J.G., Henderson, R.J., Tocher, D.R., Sutherland, R., Naess, T., Jensen, A.M., Naas, K., van der Meeren, T., Harboe, T., Sanchez, F.J., Sorgeloos, P., Dhert, P., and Fitzgerald, R. 1998. The production and use of copepods in larval rearing of halibut, turbot and cod. *Bulletin of the Aquaculture Association of Canada* 4:41–45.

Takeuchi, T., Toyota, M., and Watanabe, T. 1992. Comparison of lipid and n-3 highly unsaturated fatty acid incorporations between *Artemia* enriched with various types of oil by direct method. *Nippon Suisan Gakkaishi/Bulletin of the Japanese Society of Scientific Fisheries* 58(2):277–281.

Tamaru, C.S., Murashige, R., Lee, C.-S., Ako, H., and Sato, V. 1993. Rotifers fed various diets of baker's yeast and/or *Nannochloropsis oculata* and their effect on the growth and survival of striped mullet (*Mugil cephalus*) and milkfish (*Chanos chanos*) larvae. *Aquaculture* 110:361–372.

Toledo, J.D., Golez, M.S., Doi, M., and Ohno, A. 1999. Use of copepod nauplii during early feeding stage of grouper *Epinephelus coioides. Fisheries Science* 65(3):390–397.

Torres, G., Gimenez, L., and Anger, K. 2002. Effects of reduced salinity on biochemical composition (lipid, protein) of zoea 1 decapod crustacean larvae. *Journal of Experimental Marine Biology and Ecology* 277:43–60.

van Ballaer, E., Amat, F., Hontoria, F., Leger, P., and Sorgeloos, P. 1985. Preliminary results on the nutritional evaluation of n-3 HUFA-enriched *Artemia* nauplii for larvae of the sea bass *Dicentrachus labrax. Aquaculture* 49:223–229.

van der Meeren, T. and Naess, T. 1993. How does cod (*Gadus morhua*) cope with variability in feeding conditions during early larval stages? *Marine Biology* 166:637–647.

Vilela, M.H. 1992. Mass culture and nutritional quality of the marine copepod *Tigriopus brevicornis Mueller. Boletim Instituto Nacional Investigaçäo das Pescas e do Mar, Lisboa* 17:49–63.

Watanabe, T. 1993. Importance of docosahexaenoic acid in marine larval fish. *Journal of the World Aquaculture Society* 24(2):152–161.

Watanabe, Y., Zenitani, H., Kimura, R., Sato, C., Okumura, Y., Sugisaki, H., and Oozeki, Y. 1998. Naupliar copepod concentrations in the spawning grounds of Japanese sardine, *Sardinops melanostictus,* along the Kuroshio Current. *Fisheries Oceanography* 7:101–109.

Webster, C.D. and Lovell, R.T. 1990. Response of striped bass larvae fed brine shrimp from different sources containing different fatty acid compositions. *Aquaculture* 90:49–61.

13
Studies on the Use of Copepods in the Semi-intensive Seed Production of Grouper *Epinephelus coioides*

Joebert D. Toledo, Ma. Salvacion Golez, and Atsushi Ohno

ABSTRACT

Previous studies by the authors have shown the feasibility of using copepods in the semi-intensive seed production of grouper *Epinephelus coioides*. Early-stage *E. coioides* larvae preferred to ingest copepod nauplii over rotifers, although their abundance is relatively low. Higher growth and survival were obtained in larvae provided with copepods than larvae fed with rotifers alone.

In this study, the authors tested various fertilization techniques for the mass production of zooplankton in ponds. The zooplankton population increased from an initial density of 86–148 ind/L to 1,524–3,186 ind/L 9–12 days after flooding. Major zooplankton were identified as rotifers, copepods, including cyclopoid and harpacticoid spp., and cladocerans. The calanoid copepod *Acartia tsuensis* can be propagated in 1-ton tanks in mixed species of microalgae alone or in combination with baker's yeast. Density of *A. tsuensis* increased from 60 ind/L at stocking to about 900 ind/L 3–5 days thereafter. A prototype collector for copepod eggs and nauplii was tested. The average daily numbers of eggs and nauplii collected varied (2,300–117,600), depending on the density of copepodids and adults in the holding container (1,000–8,000 ind/10-L container). Hatching rates of collected eggs ranged from 34 to 89%. Collected eggs may be stored at low temperature (4–10°C) for up to 7 days. Duration and viability of eggs stored at low temperature were highly affected by the presence of protozoans. *A. tsuensis* eggs survived freezing to −20°C (0.3–1.7%) only at the cleavage stage, suggesting the feasibility of cryopreservation. Use of pond-grown zooplankton, particularly the calanoid copepods, for the seed production of grouper is discussed.

INTRODUCTION

Various methods are used for growing larval fish in aquaculture. These include (a) intensive production in tanks with high larval densities and a supply of live prey or formulated feed; (b) semi-intensive production in large tanks with medium larval densities and feeding harvested natural zooplankton (mostly copepods), sometimes together with *Artemia* spp.; or (c) extensive production in ponds or lagoons with low larval densities and natural production of plankton (van der Meeren and Naas 1997). In many cases, the use of copepods as live prey increased survival and growth rates of the target species.

Groupers are among the most popular live reef fish traded in Hong Kong and other Asian cities. The increase in the demand for live grouper in the live reef fish trade has resulted in the widespread practice of destructive fishing methods, such as cyanide fishing. Aquaculture of grouper is perceived as an alternative to restricted fishing and the declining supply of these species from the

Corresponding author: Joebert D. Toledo, Aquaculture Department, Southeast Asian Fisheries Development Center, Tigbauan, Iloilo 5021, Philippines; Phone: +63-33-336-2965; Fax: +63-33-512-9070; E-mail: jdtoledo@aqd.seafdec.org.ph.

wild. Hatchery production is still considered to be one of the major bottlenecks in the growth of grouper aquaculture.

Many researchers have suggested that the small mouth size of first-feeding grouper larvae contributes to the difficulty of providing them with suitable food in aquaculture (Hussain and Higuchi 1980; Maneewong et al. 1986; Kayano 1988).

Successive spawning of grouper *Epinephelus suillus* (*E. coioides*) at the Aquaculture Department of the Southeast Asian Fisheries Development Center in the Philippines (Toledo et al. 1993) led to the development of intensive larval techniques in the hatchery (Duray et al. 1996). The protocol involved the use of screened (mesh size, 90 μm) rotifers *Brachionus plicatilis* at the early-feeding stage at a density of 20 ind/mL. Newly hatched *Artemia* nauplii were introduced at 21 days after hatching (DAH). Survival at 21 DAH ranged from 7 to 20%.

In a series of studies, Toledo et al. (1996, 1997, 1999) showed better growth and survival in early-stage larvae of the grouper *E. coioides* fed nauplii of copepods than for larvae provided with rotifers only. Early-stage *E. coioides* larvae have been shown to preferentially ingest copepod nauplii at 1–32/L among rotifers, which are much more abundant at about 5,000/L. The higher growth and survival of larvae provided with copepods were attributed to the superior nutrient quality of copepods over rotifers. Although rotifers and copepods contain closely similar amounts of 20:5*n*-3 eicosapentaenoic acid (EPA) and 22:6*n*-3 docosahexaenoic acid (DHA), levels in *Acartia* sp. and *Pseudodiaptomus annandalei* were several fold higher than those in rotifers. By providing copepod nauplii at the early feeding stages, a mean survival rate of 3.4% at harvest (36 DAH) was obtained in pilot scale production trials (Toledo et al. 1999).

Variability in the production of zooplankton in relation to the food demands of larval fish has been a major problem. For Asian sea bass (*Lates calcarifer*), copepod production in the pond is enhanced with a combination of lucerne pellets and diammonium phosphate added 1 week before stocking of newly hatched larvae. Sea bass larvae were initially fed a combination of rotifers and copepod nauplii and then were switched to copepodids and copepod adults as they grew (Rimmer and Russell 1998). This method, however, is greatly affected by abrupt weather changes such as rainfall. In a modified semi-intensive method, adult copepods and copepodids are introduced in the tanks to propagate copepod nauplii several days before stocking of mangrove red snapper (*Lutjanus argentimaculatus*) larvae (Doi and Singhagraiwan 1992) and grouper larvae (Toledo et al. 1997, 1999).

The purpose of the current study was to develop techniques for the propagation of zooplankton in earthen ponds that could serve as a source of live prey for late larval stages of grouper. Attempts to propagate *A. tsuensis* in 1-ton fiberglass tanks and methods to collect and store *A. tsuensis* eggs at low temperature are described. A measure to control the infestation of protozoans in stored eggs is proposed.

MATERIALS AND METHODS

Development of Culture Techniques of Zooplankton in Ponds

Nine 200-m^2 ponds at Punta Pulao, Dumangas, were used to develop mass culture techniques for zooplankton. Ponds were filled with water of salinity 32 ppt from a tidal creek. Incoming water was screened with a 0.8- to 1.0-mm mesh net to minimize entry of planktivores but allow entry of the natural population of copepods. The water contained the naturally occurring level of zooplankton. Fertilizer was added to ponds with three replicate ponds per treatment. Treatment I was a modification of the method of Ohno and Okamura (1988). Chicken manure was applied evenly as a basal fertilizer at 500 kg/ha. Chemical fertilizers containing urea, ammonium sulfate, and ammonium phosphate were added after filling the pond with water, and every 3 days thereafter, at a rate of 2.7, 4.0, and 6.0 kg/ha, respectively. Treatment II followed the recommendations of the Southeast Asian Fisheries Development Center (SEAFDEC) for brackish-water ponds by single application of fertilizers composed of 2 tons chicken manure, 25 kg urea, and 50 kg ammonium phosphate per hectare. Treatment III was a modification of the method of Geiger (1983) using highly soluble fertilizers. Rice bran and liquid inorganic fertilizer were added weekly at a rate of 300 kg and 50 L per hectare, respectively. Pond preparation followed the methods described by Baliao et al. (1999). Ponds were filled to an initial water depth of 1 m. The culture period was 45 days.

To monitor the quality and quantity of plankton, 10-L water samples were collected every Monday and Thursday from various points of each pond using a 150-cm length of 5-cm diameter polyvinyl chloride (PVC) pipe fitted with a swing valve at one end. One liter of the water sample was passed through a 60-μm mesh to retain plankton. Filtrates were fixed in 5% formalin for identification and quantification of zooplankton following Grice (1961) and Ohno (1996). Another 20 mL of the water sample was filtered on filter paper fitted in a plastic syringe for chlorophyll *a* determination, as described by Suzuki and Ishimaru (1990).

Temperature, salinity, and water transparency were monitored in each pond every Monday, Wednesday, and Friday between 0800–0900 hours.

Mass Culture of Copepods in Tanks

The experimental setup was constructed modifying the methods of Støttrup et al. (1986) and Schipp et al. (1999). Adult *A. tsuensis* copepods were collected from ponds by exploiting the strong phototactic response of these copepods (Toledo et al. 1999). Two 150-W light bulbs were set about 60 cm (2 feet) above the pond water starting at 1830 hours. Adults and copepodids of *A. tsuensis* are highly phototactic. A scoop net with a mesh size of 0.4–0.60 mm was used to scoop out the copepods aggregating near the lights. Collected copepods were held in 5 L of pond water in oxygenated plastic bags, brought to the laboratory, and washed through meshes of various sizes to remove detritus, mysids, fish larvae, and other unwanted material. Collected copepodids and adults comprised 92% *A. tsuensis*, 7% *P. annandalei*, and 1% was a mixture of cyclopoid and harpacticoid copepod species.

Adults and copepodids were transferred into six 1,000-L oval fiberglass production tanks at a starting density of 60 ind/L. Moderate aeration was provided in each tank. Three production tanks were fed daily with a mixture of *Nannochloropsis* sp., *Tetraselmis* sp., and *Chaetoceros* sp. at a final density of 300,000 cells/mL. The remaining production tanks were fed daily with half the amount of a mixture of the same algae (75,000 cells/mL) and baker's yeast (0.5 g/100,000 ind). A moderate airlift system kept the algae and baker's yeast suspended in the water column.

The numbers of copepod eggs, nauplii, copepodids, and adults were determined at 0800–0900 hours each day for 15 days by collecting water samples from each tank, as described previously. Temperature, salinity, and dissolved oxygen were monitored daily in each tank before the addition of food.

Development of Egg Collection Techniques

A prototype collector for *A. tsuensis* eggs was developed (Fig. 13.1). Copepodids and adult *A. tsuensis* were stocked in a 10-L capacity plastic container suspended in the upper portion of a 30-L conical tank. The bottom of the 10-L container was removed and replaced with a 100-μm net to retain *A. tsuensis* adults and copepodids, while allowing passage of *A. tsuensis* eggs and microalgae. *A. tsuensis* eggs and microalgae were directed upward to the 10-L container by a 1.25-cm diameter PVC pipe. A 5-cm diameter PVC pipe fitted with a 60-μm net positioned at the outflow of the airlift system retained the *A. tsuensis* eggs while microalgae re-entered the system.

Copepodids and adults of *A. tsuensis* were collected from nearby fishponds following the procedures described by Toledo et al. (1999). These were randomly distributed in 15 egg collectors at a stocking density of 100, 200, 400, 600, or 800

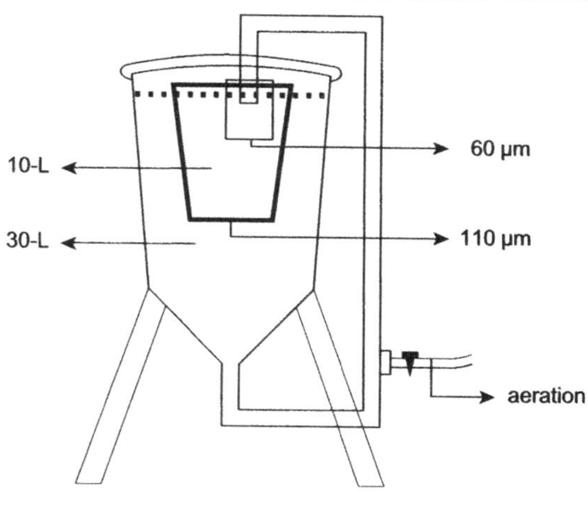

Figure 13.1. Prototype egg collector for *Acartia tsuensis*.

ind/L. Three replicates were established for each stocking density. *Tetraselmis* sp., *Chaetoceros* sp., and *Nannochloropsis* sp. in the collectors were maintained at 300,000 cells/mL every 0800 hours. Copepod eggs were collected and counted each day at 0800 hours and 1800 hours for 5 days after stocking. Aliquot samples of collected eggs were transferred in 3-mL cell well plates placed on top of a rotary shaker for incubation and hatching.

Low Temperature Storage Experiments

To obtain *A. tsuensis* embryos that were close to the same stage of development, fresh batches of collected *A. tsuensis* adults and copepodids were stocked in five egg collectors. Eggs were collected simultaneously 2 hours after stocking. Collected eggs were washed and pooled in a 250-mL beaker with seawater (salinity 34–36 ppt). The beaker was placed on top of a rotary shaker at 50 revolutions per minute (rpm) to maintain water movement until use.

To examine the low temperature tolerance of early developmental stages of *A. tsuensis*, samples were taken of embryos at the cleavage and invagination stages, early (about 10 hours after egg collection when nauplii are discernible in the eggs) and at late nauplii (about 12 hours after egg collection when nauplii are fully formed in the eggs). These were transferred from the 250-mL beaker to 50-mL beakers containing seawater alone or 10% dimethyl sulfoxide (DMSO) in seawater. Preliminary experiments showed that immersion of *A. tsuensis* eggs in 10% DMSO in seawater did not significantly affect subsequent hatching of eggs. From each 50-mL beaker, aliquot samples of eggs (n = 98–240) were transferred in 1-mL Eppendorf vials and immediately stored overnight in a refrigerator (4–10°C) or freezer (−20°C). The Eppendorf vials were taken out of storage the following day and placed in a gently stirred water bath at 22–25°C. Stored eggs were pipetted from the Eppendorf vials and transferred to 3-mL cell well plates. The eggs were washed with several changes of seawater to remove the DMSO. The cell well plates were then placed on top of a rotary shaker for incubation and hatching. The number of hatched nauplii and number of unhatched eggs were counted 24 hours after incubation. The hatching rate was expressed as the proportion of eggs that hatched into nauplii.

To examine the viability of *A. tsuensis* eggs after storage under refrigeration, eggs were collected, washed, and pooled in a 50-mL beaker with seawater. Aliquot samples were transferred in 1-mL Eppendorf vials and stored under refrigeration for 1, 3, 5, or 7 days. Replicate samples (not refrigerated) were incubated separately and served as the untreated control. After storage, eggs were handled for hatching as described previously.

A mini-bioassay was conducted to control the protozoans affecting the duration and viability of *A. tsuensis* eggs after cold storage. Collected eggs were pooled in 50-mL beakers. Samples of collected eggs were transferred to a 60-μm strainer and dipped for 3 minutes in a seawater bath containing 0, 5, 10, or 20 ppm formalin. Eggs were washed in several changes of seawater to remove the formalin. Aliquot samples were then transferred in Eppendorf vials and stored in a refrigerator for 1 week. Eggs were handled for hatching as described previously.

Semi-Intensive Seed Production of Grouper

Larval rearing runs were conducted using copepods in the seed production of *E. coioides* by modifying the method of Toledo et al. (1999). Feeding and water management procedures are summarized in Fig. 13.2. Two to six 5-m^3 tanks were stocked with newly hatched grouper larvae at a density of 2–5 ind/L. To propagate copepod nauplii in larval tanks, *A. tsuensis* and/or *P. annandalei* copepodids and adults were inoculated in larval rearing tanks at 60–80 ind/L 2 to 3 days before stocking of viable grouper eggs or newly hatched larvae. From day 2 to day 18, larvae were provided with rotifers (*Brachionus rotundiformis*) at increasing densities of 2–10/mL. The rotifers had been either enriched with commercial enrichment emulsions or fed with a mixture of *Isochrysis* sp. From days 15 until metamorphosis, larvae were fed to satiation with *Artemia* (nauplii and metanauplii) that had been treated with commercial emulsions for at least 6 hours before harvest. In addition to *Artemia*, from day 20 pond-grown zooplankton comprising a mix of species were provided daily at a density of 50–100 ind/L.

After hatching in salinity of 34–36 ppt, larvae were stocked in 5-ton tanks in 20–24 ppt. *Nannochloropsis* sp. was added into the tanks 2–3 days

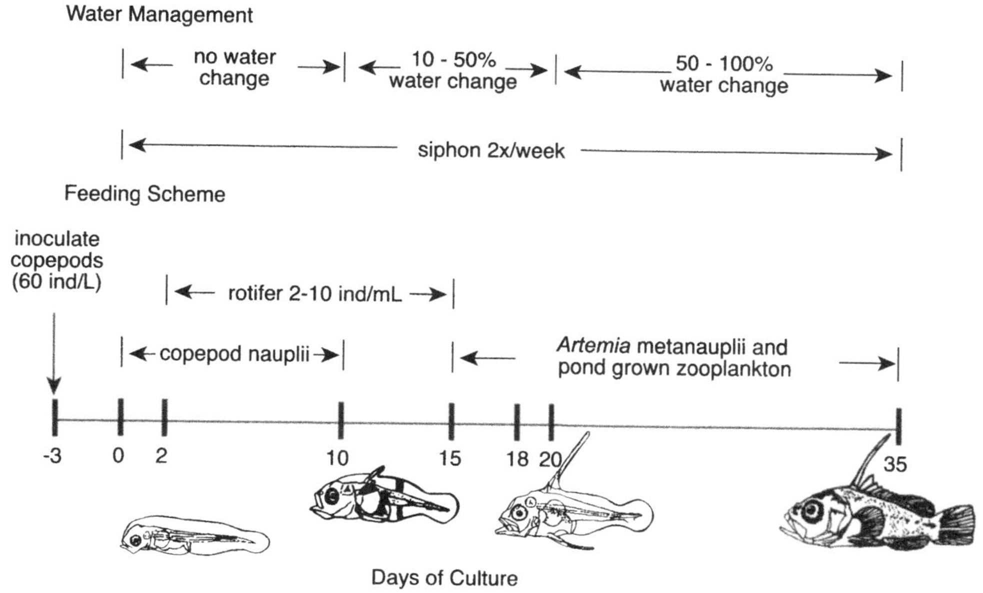

Figure 13.2. Water management and feeding schemes for semi-intensive rearing of grouper larvae.

before stocking and daily thereafter, if needed, to maintain density in the rearing water to at least 1×10^5 cells/mL. About 10% of the rearing water is changed, starting 11 DAH, and gradually increased to 50% at 25 DAH. At least 100% of the rearing water is changed starting at 26 DAH until harvest. Finely chopped fish meat or formulated feeds are introduced, starting at 30 DAH. Survival was determined 35–40 DAH by counting the number of larvae harvested. Larval rearing runs were replicated in time.

Statistical Analysis

Statistical analysis was performed using analysis of variance followed by Duncan's multiple range test to examine significant differences between means at $P < 0.05$.

RESULTS

Culture of Zooplankton in Ponds

The quantity and quality of zooplankton in earthen ponds depended on the combinations of organic and inorganic fertilizers used. In all treatments, the density of zooplankton rapidly increased from initial levels of 86–148 ind/L 1 day after flooding to 1,524–3,186 ind/L 9–12 days later (Fig. 13.3). The greatest increase in density

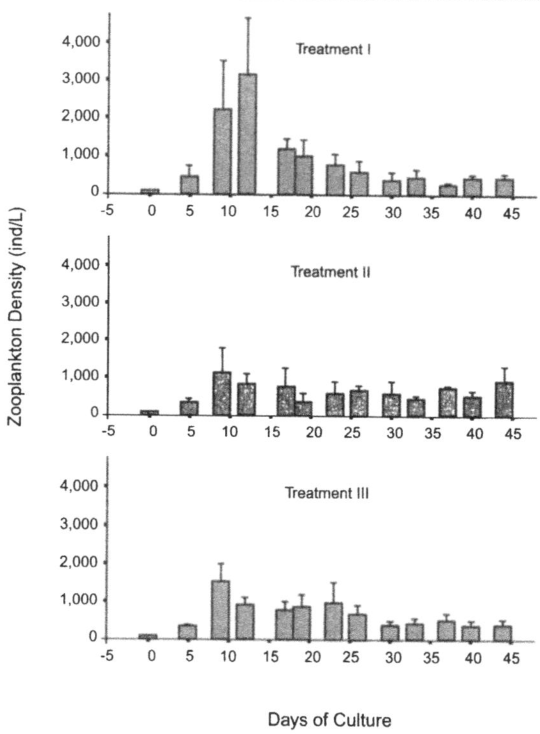

Figure 13.3. Total number of zooplankton in ponds with various combinations of fertilizers.

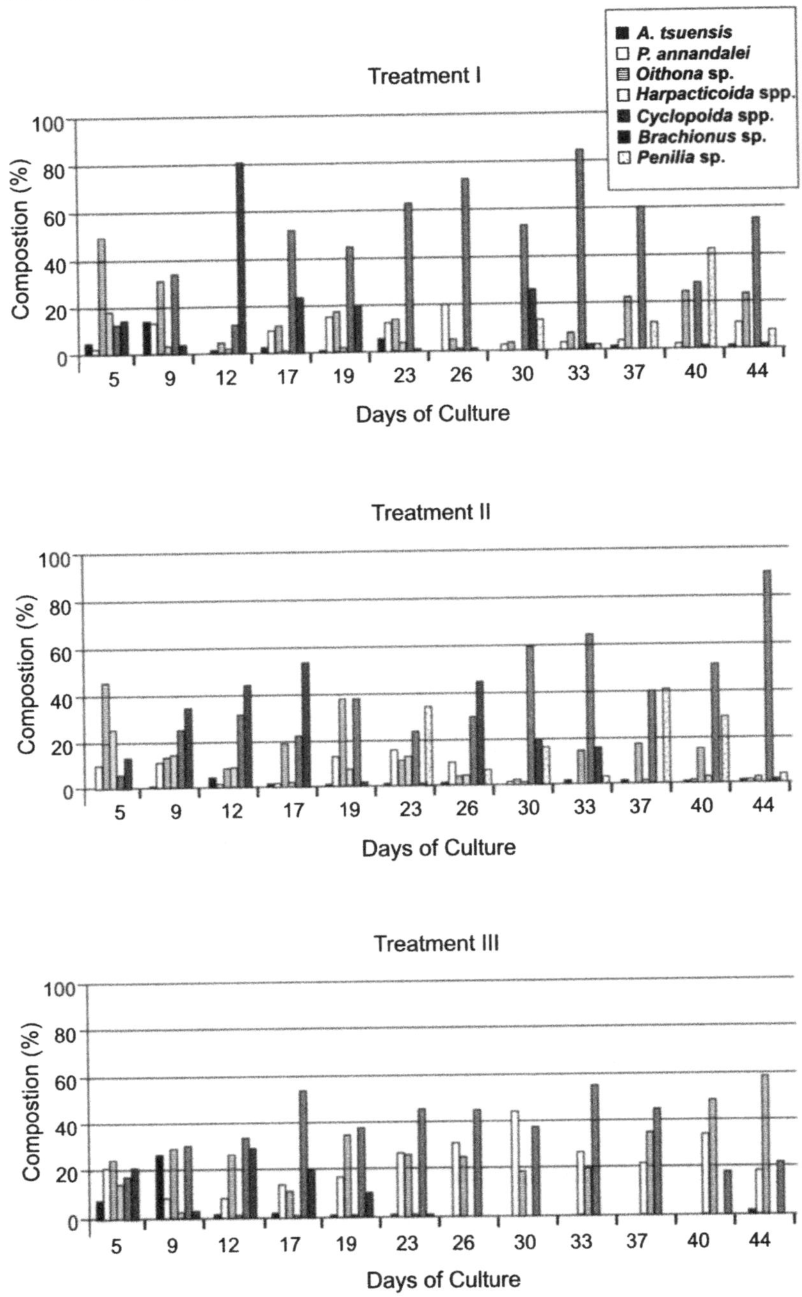

Figure 13.4. Zooplankton composition.

was observed in ponds fertilized weekly with a combination of urea, ammonium sulfate, and ammonium phosphate. In all treatments, zooplankton gradually decreased from 2 weeks after flooding to 405–939 ind/L at the end of the experiment.

Initial composition of zooplankton in the ponds is shown in Fig. 13.4. Peak levels of *A. tsuensis* (25%) and *P. annandalei* (44%) were observed in Treatment III 9 and 30 days after flooding, respectively. The highest density of *Brachionus* sp. was observed in Treatment II at the second and

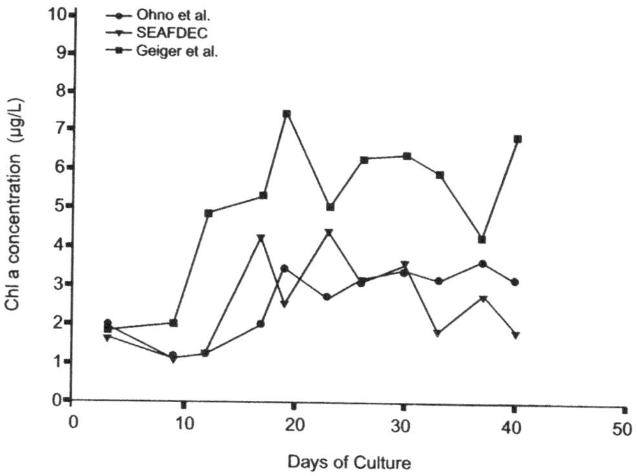

Figure 13.5. Chlorophyll *a* (Chl *a*) level in ponds fertilized with various combinations of fertilizers.

fourth week after flooding, and decreased to less than 2% at the end of the study. An increasing density of *Penilia* sp. was observed in ponds fertilized with chicken manure (Treatments I and II) starting 3 to 4 weeks after flooding.

Salinity dropped from 32 ppt after flooding to a minimum of 22 ppt during the first 2 weeks following rainfall and overcast skies and gradually increased to a maximum of 34 ppt at the end of the experiment. Temperature and water transparency during the experiment varied from 29 to 34°C and 32 to 78 cm, respectively. Chlorophyll *a* levels after flooding ranged from 1.6 to 2.0 μg/L and increased to maximum levels of 3–7 μg/L 2–3 weeks thereafter (Fig. 13.5).

Culture of Copepods in Tanks

Copepodids and adults of *A. tsuensis* fed a mixture of algae alone or in combination with baker's yeast appeared to propagate well under tank conditions (Fig. 13.6). The density of *A. tsuensis*, including various naupliar stages, rapidly increased 3–5 days after stocking from 60 ind/L (copepodids and adults) to 930 ind/L in tanks fed with mixed microalgae alone and to 1,800 ind/L in tanks fed a combination of algae and baker's yeast. A decrease in density was observed thereafter until the end of the experiment. Average daily temperature, salinity, and dissolved oxygen varied from 27.6–29.8°C, 34–36 ppt, and 4.2–6.5 ppm, respectively.

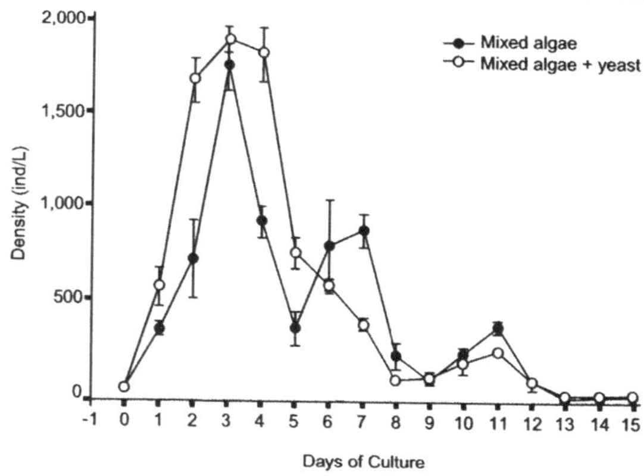

Figure 13.6. Density of *Acartia tsuensis* in 1-ton tanks fed with algae alone or in combination with bread yeast.

Development of Egg Collection Technique

The prototype collector was shown to be efficient for collecting *A. tsuensis* eggs. The daily total number of eggs collected varied from 2,259 to 117,000 eggs per collector, depending on the day of collection or the density of *A. tsuensis* adults and copepodids in the tanks (Fig. 13.7). The number of eggs collected increased until the third day and gradually decreased thereafter. The highest average number of eggs collected (117,000 eggs) was noted on the third day after stocking in tanks stocked initially with 600 ind/L. Hatching rates of

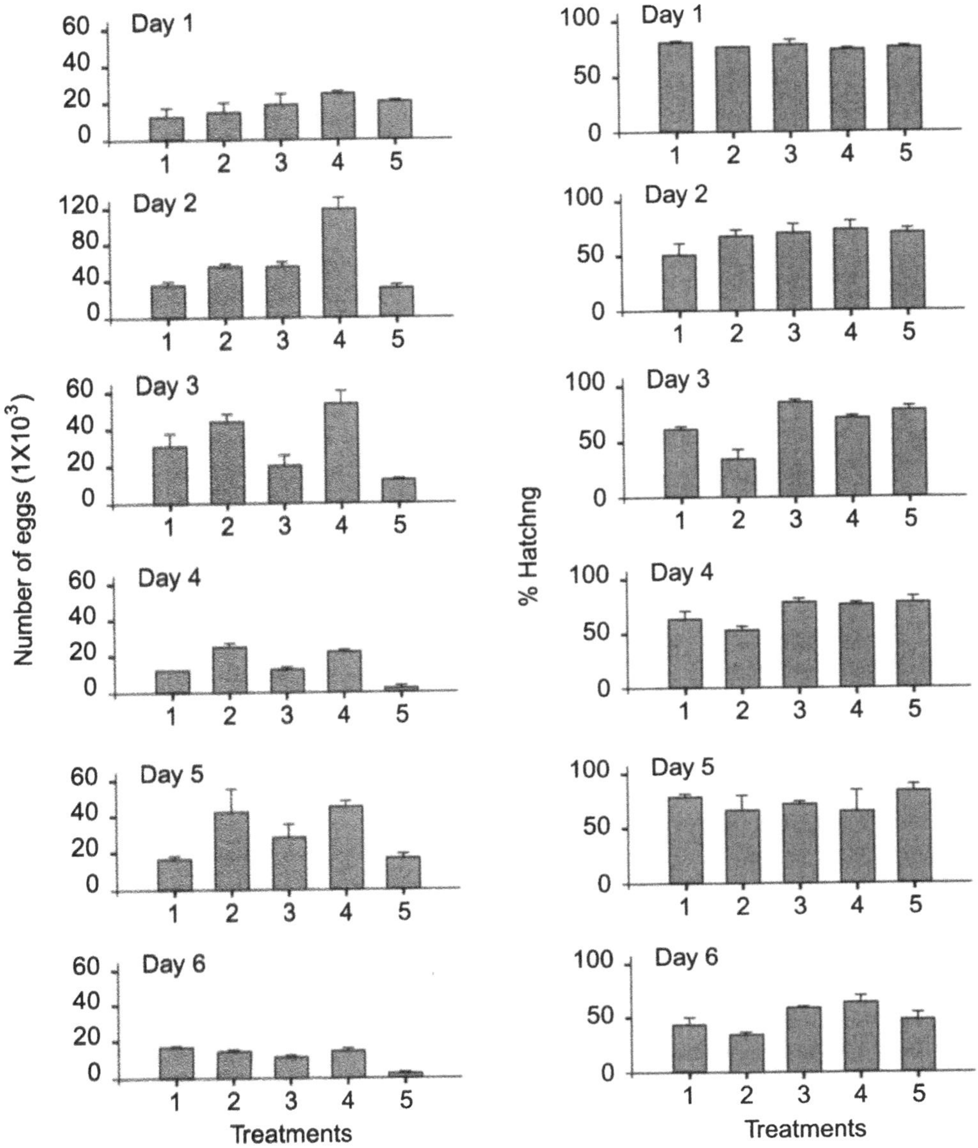

Figure 13.7. Egg production and hatching rate, using prototype egg collector.

collected eggs ranged from 34 to 89% and appeared to gradually decrease several days after stocking.

Low Temperature Storage

Collected *A. tsuensis* eggs developed to cleavage, invagination, early nauplii, and late nauplii stages at 2, 6, 10, and 12 hours, respectively, after incubation at an air temperature of 22°C. About 50% of the eggs hatched 15 hours after collection. More than 90% of the eggs hatched 18 hours after collection.

Viability of *A. tsuensis* eggs after low temperature storage varied according to the stage of development and the storage temperature (Fig. 13.8). Very low hatching was observed for embryos at the cleavage stage in seawater alone (0.28%) or in 10% DMSO in seawater (1.67%), stored below freezing (−20°C). No invaginating embryos or embryos with formed nauplii survived freezing even with the addition of DMSO. Regardless of the stage, survival rates of embryos stored under refrigeration (4–10°C) in seawater (22–31 ppt) were significantly higher than for

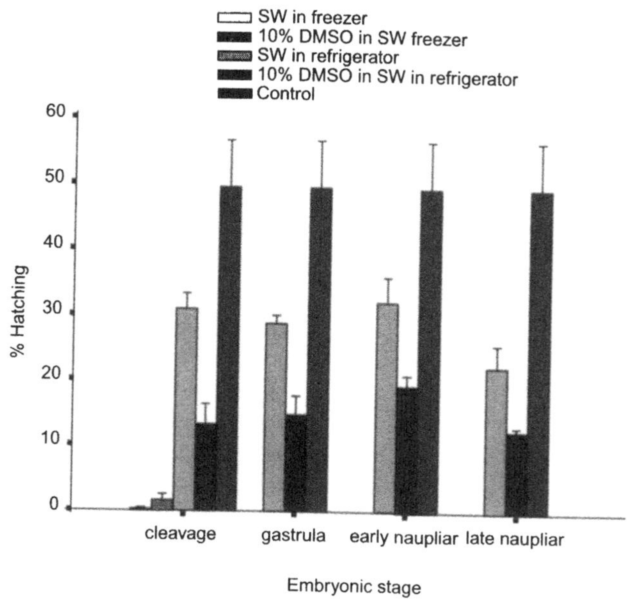

Figure 13.8. Low-temperature tolerance of *Acartia tsuensis* embryos at various developmental stages. SW, seawater.

other treatments. Embryos under refrigeration in 10% DMSO in seawater had significantly lower hatching rates than those in seawater alone. This indicates toxic effects of DMSO. Mean survival of *A. tsuensis* eggs in the untreated control group was significantly higher, regardless of stage or storage system tested.

Variable hatching was observed in *A. tsuensis* eggs after storage under refrigeration (Table 13.1). In the first run, there was no significant difference in the hatching rates of eggs stored under refrigeration for 1, 3, 5, or 7 days (68.5–76.7%) compared with the untreated control (77.9%). In the second run, a significant decrease in hatching rate was observed 1 day after storage under refrigeration (53.8%) compared with the untreated control (88.2%). The hatching rate further dropped to 4.6% after 3 days of storage, and no eggs hatched after 5 and 7 days of storage. High numbers of protozoans were observed aggregating on the surface of eggs used in the second run.

A formalin bath appears to be effective in controlling the infestation of protozoans in refrigerated *A. tsuensis* eggs (Table 13.2). The average hatching rate of eggs dipped in 0 ppm formalin for 3 minutes before storage was significantly lower (70.5%) than those bathed in formalin solutions of 5 (89.6%), 10 (95.2%), or 20 ppm (91.6%) and of

Table 13.1. Hatching rates (%) of *Acartia tsuensis* stored under refrigeration for various durations

	Trial	
Days	1st	2nd
Control	77.93 ± 4.84^a	88.25 ± 2.25^a
1	76.72 ± 4.13^a	53.75 ± 5.45^b
3	72.32 ± 4.57^a	4.56 ± 0.69^c
5	71.50 ± 4.13^a	0.00 ± 0.00^c
7	68.50 ± 4.65^a	0.00 ± 0.00^c

Note: For each column, figures with different superscripts are significantly different ($P < 0.05$). Figures are for the mean ± S.E.M. of four replicates.

Experimental treatments in the second trial were contaminated with protozoans.

eggs in the untreated (not refrigerated) control (92.6%). The number of protozoans in the eggs after storage progressively decreased as the concentration of formalin increased (personal observation). Very few protozoans were observed on embryos treated with 10 and 20 ppm formalin.

Larval Rearing Runs

A total of 20 production batches were conducted in 2002. Six batches were aborted or discarded after high cumulative mortalities. Mean survival

Table 13.2. Hatching rates of *Acartia* eggs dipped in various concentrations of formalin and stored under refrigeration

Treatment	Hatching rate (%)
Control	92.57 ± 2.03[a]
0 ppm	70.52 ± 3.13[b]
5 ppm	89.55 ± 1.16[a]
10 ppm	95.17 ± 1.28[a]
20 ppm	91.55 ± 2.46[a]

Note: Figures are for the mean ± S.E.M. of three replicates. Figures with different superscripts are significantly different ($P < 0.05$).

at harvest was 9.07% (1.10–49.42%) in 2002. Water temperature and salinity varied from 26.8 to 29.2°C and 22 to 34 ppt, respectively.

DISCUSSION

Zooplankton have been widely used in the seed production of a variety of freshwater fishes such as striped bass *Morone saxatalis* (Geiger and Parker 1985; Geieger et al. 1985), walleye *Stizostedion vitreum* (Tice et al. 1996; Myers et al. 1996), and paddlefish *Polyodon spathula* (Mims et al. 1995a, 1995b). To enhance zooplankton blooms, ponds are fertilized with organic fertilizers such as alfalfa, cottonseed meal, rice bran, or animal manure alone or in combination with inorganic fertilizers. Larvae are introduced about a week after fertilization, when zooplankton of various sizes are available as food to sustain growth. Larvae are harvested 3–4 weeks after stocking once the supply of zooplankton in the pond is almost exhausted. Larvae then begin to take formulated food. A similar technique was developed and has been modified in the seed production of marine fish species such as Atlantic cod *Gadus morhua*, Atlantic halibut *Hippoglossus hippoglossus* L., sea bass *Dicentrarchus labrax*, and sea bream *Sparus aurata* (see van der Meeren and Naas 1997). Zooplankton in the latter system was collected from extensive lagoons or fertilized ponds before being fed to the larvae. In the present study, the density of various species of zooplankton, mostly cyclopoid copepods, calanoid copepods, harpacticoid copepods, and rotifers, rapidly increased in all types of fertilization regimes tested 1–2 weeks after flooding. The density of zooplankton decreased thereafter, even with moderate to high levels of chlorophyll *a* (1.7–7.2 μg/L) in the pond water. Increasing numbers and sizes of predators such as gobies (Gobiidae) and other fish species were observed in all ponds starting 2–3 weeks after flooding. It will be interesting to examine the effect of organic compounds (e.g., saponin) used to eliminate fish predators on the reproduction of zooplankton.

In the present study, the percentage composition of zooplankton species varied among treatments and between ponds within the same treatment, although the same source of water was used during flooding. The cyclopoid copepods *Oithona* sp. and *Apocyclops* sp. dominated the species composition after flooding and almost throughout the duration of the study, regardless of the fertilization scheme used. Rotifer density gradually decreased in all treatments to almost nil at end of the experiment, with a corresponding increase in the percentage composition of other species. It is interesting to note that *Penilia* sp. started to propagate at the middle of this study only in ponds fertilized with chicken manure. Whether the decreasing trend in the total number of zooplankton starting 2–3 weeks after flooding is due to predation or cannibalism is not known. The results indicate the complex ecology of small ponds and the difficulty of predicting the outcome of ecological processes, especially when the starting conditions are not completely known.

Ohno and Okamura (1988) propagated *A. tsuensis* in a large tank (24 m^3) by pumping pond water with 61–74% *A. tsuensis* into the tank. *Acartia tsuensis* and *P. annandalei* are collected from plankton blooms in brackish water fish and shrimp ponds in Taiwan (Liao et al. 2001) and used as feed during the late stage of larval rearing. Collection and sale of copepods has become a profitable sub-business in Taiwan. In Europe, zooplankton (mainly copepods) harvested from naturally enclosed or semienclosed lagoons by mechanized plankton filters are used for the semi-intensive seed production of marine fish (van der Mereen and Naas 1997). Zooplankton propagated as described in the present study could be partially harvested during peak density for larval feed for the latter stages of *E. coioides* and other marine fish species. Studies to determine the percentage of zooplankton population to be harvested to sustain productivity of the remaining population should be elucidated.

Støttrup et al. (1986) reported the cultivation of a temperate species of the calanoid copepod, *A. tonsa*, for use as food for late larval stages of marine fish. Schipp et al. (1999) described a hatchery method for the culture of tropical species of a calanoid copepod *Acartia* sp. Payne and Rippingale (2000) examined several algal species for the culture of the egg-bearing temperate calanoid copepod *Gladioferens imparipes*. Støttrup and Norsker (1997) developed a continuous production unit for the harpacticoid copepod, *Tisbe holothuriae*. In all cases, copepods were fed various species of microalgae. The authors have shown in their present work that half the ration of microalgae for *A. tsuensis* could be replaced with baker's yeast without a significant decrease in reproduction. Algae and yeast are known as good sources of the types of fatty acids and amino acids thought to be important in crustacean nutrition (Ahlgreen et al. 1990; Harisson 1990). Whether the yeast was used by protists that were then eaten by the copepods was not clear in the present study.

The prototype egg collector described in the present work appears to be efficient in collecting *A. tsuensis* eggs. The highest number of eggs collected (117,000) was noted in 10-L tanks stocked with 60,000 ind. At a hatching rate of 70%, about 9,000 nauplii/L can be harvested from our present setup. Støttrup et al. (1986) reported daily productivity of 95,000 eggs for *A. tonsa* in 200-L culture, while Schipp et al. (1999) obtained a peak production of *Acartia* spp. nauplii of 5,150 ind/L. Differences in figures obtained could be attributed to species differences and the density of copepodids and adults used.

Low-temperature preservation of marine invertebrate embryos such as a sea urchin Echinoidea (Asahina and Takahashi 1978), blue mussel *Mytilus edulis* (Toledo et al. 1989), and the euryhaline rotifer *Brachionus plicatilis* (Toledo and Kurokura 1990) has been reported. Exposure of copepods to critical environmental factors such as low temperature and dissolved oxygen triggers production of quiescent eggs. These eggs resume normal embryonic development after incubation to optimal conditions (Marcus 1996). The authors have shown in the present study that embryonic development and hatching of *A. tsuensis* eggs could be delayed at low temperature (4–10°C). No significant change in hatching rates was observed in eggs incubated in seawater alone and stored under refrigeration, regardless of the embryonic stage at the time of storage. A significant decrease in eggs stored with 10% DMSO suggests toxicity of DMSO after prolonged exposure of eggs at low temperature. Extended incubation of eggs with DMSO could have prolonged the exposure of the eggs to high osmotic pressure, thereby causing death, as shown by Toledo (1990) in rotifer embryos.

In the present study, very few embryos (0.22–1.52%) survived freezing to −20°C. These were at the cleavage stage. The slight increase in survival of cleavage stage eggs frozen with 10% DMSO indicates some degree of protection by DMSO against freezing damage to eggs. Freezing of mouse embryos at the cleavage stage allows a reasonable chance for blastomeres damaged by freezing to develop into live offspring (Polge and Willadsen 1978). In contrast blue mussel (Toledo et al. 1989) and rotifer embryos (Toledo and Kurokura 1990) at late embryonic stages were reported to have a higher freezing tolerance than those at the earlier stages.

Less than 2% of the *A. tsuensis* eggs in the present work survived freezing to −20°C, while more than 80% of rotifer embryos at the symmetrical stage (similar to the early embryo stage in *A. tsuensis*) survived at the same temperature (Toledo and Kurokura 1990). Aside from differences in embryonic stage, high variation in survival after freezing could be also attributed to the cooling protocol used. The slow and controlled cooling rates used probably allowed the rotifer embryos to dehydrate to minimize intracellular ice crystal formation, which could damage the embryos upon thawing (Mazur 1970). In contrast *A. tsuensis* eggs in Eppendorf vials were immediately stored in a freezer. The cooling rate to dehydrate a cell before freezing is inversely proportional to the cell size (Mazur 1970). It is interesting to note that *Brachionus* sp. and *A. tsuensis* eggs have a similar egg size. Further studies to improve the survival of *A. tsuensis* eggs after freezing should be continued.

A. tsuensis eggs in the present study stored under refrigeration significantly retained their viability for up to 1 week of storage. Viability was only affected by the contamination of ciliated protozoans. The use of microfiltered seawater as a storage medium is ineffective for controlling the adherence of protozoans to collected eggs. Contamination of protozoans in the present study was controlled by dipping *A. tsuensis* eggs in 20 ppm formalin in seawater. Payne et al. (personal communication) stored the nauplii of an embryo-bearing temperate species, *G. imparipes*, in a refrigerator for 4 weeks without reduction in viability after storage. Payne and Rippingale (2001) stored nauplii for 42 days at 4°C and 8°C with low mortality rates for the first 14 days. After 42 days, surviving nauplii grew to become healthy adults when put at higher temperatures. The results of the present study clearly suggest that low-temperature preservation of *A. tsuensis* eggs is possible. The duration of successful storage at low temperatures should be investigated. A successful protocol for the low temperature storage and hatching of *A. tsuensis* eggs will help to improve the availability of copepod nauplii for first-feeding marine fish larvae.

The results of larval rearing runs of the present study confirm earlier reports (Toledo et al. 1997, 1999) of the feasibility of using copepod nauplii for the early stages of grouper larvae. The average survival of the 16 production runs of 9% (range, 1–49%) at 35–40 DAH is higher than the 3.4% previously reported by Toledo et al. (1999). Increase in larval survival at harvest could be attributed to the addition of copepod nauplii, mostly *P. annandalei* and *A. tsuensis*, during the early stage, the "enrichment" of rotifers and *Artemia*, and the supplementation of mixed zooplankton grown in earthen ponds to *Artemia* during the late larval stage. Near the end of harvest (> 30 DAH), metamorphosing grouper larvae in the present work began to swim near the bottom of the tank. Because of their benthic habitat, harpacticoid copepods (*Tisbe* spp.) collected from the pond appear to be an appropriate live prey for grouper larvae at this stage.

Larvae that were discarded from the present work had high cumulative mortalities with clinical signs of viral nervous necrosis (VNN). From day 12 onward, moribund larvae swimming listlessly near the water surface with abrupt swirling movements were commonly observed. Mortalities were associated with VNN, as described by Maeno et al. (2002). Fertilized eggs as well as larvae at various ages from the discarded batches were positive for VNN by cell culture and reverse transcription-polymerase chain reaction. Histopathological observations revealed vacuoles in the retina and brain of moribund larvae.

ACKNOWLEDGMENTS

This study was partially funded by the Australian Center for Agricultural Research (FIS/97/73), Japan International Cooperation Agency, and Southeast Asian Fisheries Development Center, Aquaculture Department (Nr-03-F98T and Nr-01- F2000T). The authors are grateful to Ms. Sharon Pedrajas-Gomez and Mr. Jesus Rodriguez, Jr. for technical assistance.

REFERENCES

Ahlgreen, G., Lundstedt, L., Brett, M., and Forsberg, C. 1990. Lipid and food quality of some phytoplankton for cladoceran zooplankter. *Journal of Plankton Research* 12:809–918.

Asahina, E. and Takahashi, T. 1978. Freezing tolerance in embryos and spermatozoa of sea urchin. *Cryobiology* 15:122–127.

Baliao, D.D., de los Santos, M.A., Rodriquez, E.M., and Ticar, R.B. 1999. Grouper culture in brackish water ponds. Aquaculture Extension Manual No. 24. Aquaculture Department, Southeast Asian Fisheries Development Center, Tigbauan, Iloilo, Philippines.

Doi, M. and Singhagraiwan, T. 1992. Biology and culture of the red snapper, *Lutjanus argentimaculatus*. Eastern Marine Fisheries Development Center, Bangkok, Thailand and Japan International Cooperation Agency, Tokyo, Japan. 51 pp.

Duray, M.N., Estudillo, C.B., and Alpasan, L.G. 1996. Larval rearing of the grouper *Epinephelus suillus* under laboratory conditions. *Aquaculture* 150:63–76.

Geieger, J.C., Turner, C.J., Fritzmayer, K., and Nichols, W.C. 1985. Feeding habits of larval and fingerling striped bass and zooplankton dynamics in fertilized rearing ponds. *Progressive Fish-Culturist* 47:213–223.

Geiger, J.G. 1983. A review of pond zooplankton production and fertilization in striped bass rearing ponds. *Aquaculture* 35:353–369.

Geiger, J.G. and Parker, N.C. 1985. Survey of striped bass hatchery management in the Southeastern United States. *Progressive Fish-Culturist* 47:1–13.

Grice, G.D. 1961. Calanoid copepods from equatorial waters of the Pacific Ocean. Fishing Bulletin No. 186. *Fishery Bulletin* 61:171–264.

Harisson, K.E. 1990. The role of nutrition in maturation, reproduction and embryonic development of decapod crustaceans: a review. *Journal of Shellfish Research* 9:1–28.

Hussain, N.A. and Higuchi, M. 1980. Larval rearing and development of the brown-spotted grouper, *Epinephelus tauvina* (Forskal). *Aquaculture* 19:339–350.

Kayano, Y. 1988. Development of mouth parts and feeding in the larval and juvenile stages of red spotted grouper *Epinephelus akaara*. *Bulletin of the Okayama Prefecture Fisheries Experiment Station* 2:52–55.

Liao, I.C., Su, H.M., and Chang, E.Y. 2001. Techniques in finfish larviculture in Taiwan. *Aquaculture* 200:1–31.

Maeno, Y., de la Pena, L.D., and Cruz-Lacierda, E.R. 2002. Nodavirus infection in hatchery-reared orange-spotted grouper *Epinephelus coioides*: first record of viral nervous necrosis in the Philippines. *Fish Pathology* 37:87–89.

Maneewong, S., Akkayanot, P., Pongmaneerat, J., and Iizawa, J. 1986. Larval rearing and development of grouper, *Epinephelus malabaricus* (Bloch and Schneider). Report of Thailand and Japan Joint Coastal Aquaculture Research Project No. 2. Japan International Cooperation Agency, Tokyo, Japan. pp 39–42.

Marcus, N.H. 1996. Ecological and evolutionary significance of resting eggs in marine copepods; past, present, and future studies. *Hydrobiologia* 320:141–152.

Mazur, P. 1970. Cryobiology: the freezing of biological systems. *Science* 168:939–949.

Mims, S.D., Clark, J.A., Williams, J.C., and Bayne, D.R. 1995a. Factors influencing zooplankton production in organically fertilized ponds for culture of paddlefish, *Polyodon spathula*. *Journal of Applied Aquaculture* 5:29–44.

Mims, S.D., Clark, J.A., Williams, J.C., and Lovishin, L.L. 1995b. Food selection by larval paddlefish *Polyodon spathula* supplied with rice bran to promote production of live foods, with prepared diets, or with their contribution in earthen ponds. *Journal of the World Aquaculture Society* 26:438–446.

Myers, J.J., Soderberg, R.W., Kirby, J.M., and Marcinko, M.T. 1996. Production of walleye, *Stizostedion vitreum*, in earthen organic ponds fertilized with organic and inorganic fertilizers and stocked at three rates. *Journal of Applied Aquaculture* 6:11–19.

Ohno, A. 1996. Guide for the identification of copepods and their life cycle. Tokyo University of Fisheries, Tokyo, Japan. 28 pp.

Ohno, A. and Okamura, Y. 1988. Propagation of the calanoid copepod, *Acartia tsuensis*, in outdoor tanks. *Aquaculture* 70:39–51.

Payne, M.F. and Rippingale, R.J. 2000. Evaluation of diets for culture of the calanoid copepod *Gladioferens imparipes*. *Aquaculture* 187:85–86.

Payne, M.F. and Rippingale, R.J. 2001. Effects of salinity, cold storage and enrichment on the calanoid copepod *Gladioferens imparipes*. *Aquaculture* 201: 251–262.

Polge, C. and Willadsen, S.M. 1978. Freezing eggs and embryos of farm animals. *Cryobiology* 15:370–373.

Rimmer, M.A. and Russell, D.J. 1998. Aspects of the biology and culture of *Lates calcarifer*. In *Tropical Mariculture*, edited by S.S. De Silva, pp. 449–476. San Diego, California, USA: Academic Press.

Schipp, G.R., Bosmans, J.M.P, and Marshall, A.J. 1999. A method for hatchery culture of calanoid copepods, *Acartia* spp. *Aquaculture* 174:81–88.

Støttrup, J.G. and Norsker, N.H. 1997. Production and use of copepods in marine fish larviculture. *Aquaculture* 155:231–247.

Støttrup, J.G., Richardson, K., Kirkegaard, E., and Pihl, N.J. 1986. The cultivation of *Acartia tonsa* Dana for use as a live food source for marine fish larvae. *Aquaculture* 52:87–96.

Suzuki, R. and Ishimaru, T. 1990. An improved method for the determination of phytoplankton chlorophyll using N, N-Dimethylformamide. *Journal of the Oceanographic Society of Japan* 46:190–194.

Tice, B.J., Soderberg, R.W., and Kirby, J.M. 1996. Growth and survival of walleyes reared in ponds fertilized with organic and inorganic fertilizers. *Progressive Fish-Culturist* 58:135–139.

Toledo, J.D. 1990. Cryopreservation of marine invertebrate embryos. M. Sc. Thesis, Hiroshima University, Japan. 88 pp.

Toledo, J.D. 2002. Studies on the seed production of grouper *Epinephelus coioides*. Ph.D. Thesis. Hiroshima University, Japan. 130 pp.

Toledo, J.D., Golez, S.N., Doi, M., Bravo, R.S., and Hara, S. 1996. Preliminary studies on the rearing of the red-spotted grouper, *Epinephelus coioides* larvae using copepod nauplii as initial food. *UPV Journal of Natural Science* 5:119–129.

Toledo, J.D., Golez, S.N., Doi, M., and Ohno, A. 1997. Food selection of early grouper, *Epinephelus coioides*, larvae reared by the semi-intensive method. *Suisanzoshoku* 45:327–337.

Toledo, J.D., Golez, M.S.N., Doi, M., and Ohno, A. 1999. Use of copepod nauplii during the early feeding stage of grouper *Epinephelus coioides*. *Fisheries Science* 65:390–397.

Toledo, J.D. and Kurokura, H. 1990. Cryopreservation of the euryhaline rotifer *Brachionus plicatilis* embryos. *Aquaculture* 91:385–394.

Toledo, J.D., Kurokura, H., and Kasahara, S. 1989. Preliminary studies on the cryopreservation of the blue mussel embryos. *Nippon Suisan Gakkaishi* 55:1161.

Toledo, J.D., Nagai, A., and Javellana, D. 1993. Successive spawning of grouper, *Epinephelus suillus* (Valenciennes), in a tank and a floating net cage. *Aquaculture* 115:361–367.

van der Meeren, T. and Naas, K.E. 1997. Development of rearing techniques using large enclosed ecosystems in the mass production of marine fish fry. *Reviews in Fisheries Science* 5:367–390.

14
Culture of Copepods and Applications to Marine Finfish Larval Rearing in Taiwan

Huei-Meei Su, Shin-Hong Cheng, Tzyy-Ing Chen, and Mao-Sen Su

ABSTRACT

Aquaculture in Taiwan annually produces billions of finfish fry. This is achieved mostly in outdoor pond systems where larvae feed on bloomed zooplankton. The zooplankton is dominated by copepods, mainly *Apocyclops royi* and *Pseudodiaptomus annandalei* (formally *Schmackeria dubia*). Zooplankton is produced in dedicated earthen ponds or as a by-product of finfish or shrimp culture. Such zooplankton, however, can contain pathogens such as viral nervous necrosis (VNN), parasitic *Amyloodinium* sp., and gill flukes such as *Pseudorhabdosynochus* sp. These pathogens have caused mass mortality of fish larvae, especially among larval groupers. A pathogen-free system to mass produce copepods is urgently needed. The copepods *A. royi* and *P. annandalei* have potential to be mass cultured in pathogen-free systems. The effects of food type, temperature, salinity, and density on the development and reproduction of these copepods were investigated. This chapter briefly describes the mass production of copepods and their application to finfish larval rearing on farms in Taiwan and reviews the status of copepod use for grouper larval culture.

INTRODUCTION

The first success in marine finfish propagation in Taiwan was achieved with grey mullet, *Mugil cephalus* (Liao et al. 1969). In 1983, the first spontaneous spawning and successful larval rearing of milkfish (*Chanos chanos*) was achieved in Pingtung County (Lin 1984). Fry production of malabar grouper (*Epinephelus malabaricus*) soon followed, using an indoor larval rearing system (Lin et al. 1986). Using an outdoor pond larval-rearing technique, mass seed production of black sea bream (*Acanthopagrus schlegeli*) was accomplished in Tainan County (Lee et al. 1989). Building on techniques developed for these species, larviculture for over 60 marine finfish species is now possible in Taiwan (Liao et al. 2001).

Taiwan's mild climate, its close proximity to markets, the use of clustered aquatic farms, and its investment in research and extension of knowledge to farmers has enabled it to become a seed production center for the Asia Pacific Economic Cooperation (APEC) area. About 1,000 family farms produce most fry in outdoor systems with ponds of 0.1–0.3 ha, while about 150 farms use more intensive and expensive indoor systems to produce higher-value species such as groupers.

It is estimated that 200 million fingerlings of milkfish (3 cm), 45 million red-spotted grouper *Epinephelus coioides* (3 cm), 25 million short dorsal fin pompano *Trachinotus ovatus* (4 cm), 25 million pink snapper *Lutjanus erythropterus*, 2 million emperor red snapper *Lutjanus sebae* (5 cm), and 600,000 king grouper *Epinephelus lanceolatus* (9 cm) were exported to Southeast Asia and The People's Republic of China in 2002. In total, nearly a half billion fry of over 10 species are produced annually for both export and domes-

Corresponding author: Huei-Meei Su, Tungkang Marine Laboratory, Fisheries Research Institute, Council of Agriculture R.O.C., Tungkang, Pingtung, 92845 Taiwan; Phone: 886-8-832-4121; Fax: 886-8-832-0234; E-mail: hmsu@mail.tfrin.gov.tw

Table 14.1. Marine finfish larval-rearing season and fry production in 2002 for export and domestic use in Taiwan

Species	J	F	M	A	M	J	J	A	S	O	N	D	Export	Domestic use
Acanthopagrus latus (Yellow fin seabream)	●	●	●	ø	ø	ø	ø	ø	ø	ø	●	●	ø	1×10^6
Chanos chanos (Milkfish)	ø	ø	ø	●	●	●	●	●	●	●	ø	ø	2×10^8	1×10^8
Epinephelus coioides (Red-spotted grouper)	●	●	●	●	●	●	●	●	●	●	●	●	4.5×10^7	2×10^7
E. lanceolatus (King grouper)	ø	ø	ø	ø	●	●	●	●	●	●	ø	ø	6×10^5	2×10^5
E. tukula (Potato grouper)	ø	ø	ø	ø	●	●	●	ø	ø	ø	ø	ø	ø	4×10^4
Lutjanus erythropterus (Pink snapper)	ø	●	●	●	●	●	●	●	●	●	●	ø	2.5×10^7	1.5×10^7
L. sebae (Emperor red snapper)	ø	ø	●	●	●	●	●	●	●	●	●	ø	2×10^6	ø
L. stellatus (White-spotted snapper)	ø	ø	ø	●	●	●	●	●	●	●	●	ø	ø	5×10^6
Polynemus plebeius (Fivefinger threadfin)	ø	ø	●	●	●	ø	ø	●	●	●	ø	ø	ø	1×10^7
Pomadasys kaakan (Head grunt)	ø	●	●	●	ø	ø	ø	ø	ø	ø	ø	ø	ø	3×10^6
Rachycentron canadum (Cobia)	ø	ø	●	●	●	●	●	●	●	●	ø	ø	ø	1×10^6
Sparus sarba (Goldlined seabream)	●	●	●	ø	ø	ø	ø	ø	ø	ø	ø	●	ø	1×10^6
Trachinotus ovatus (Short dorsal fin pompano)	ø	ø	●	●	●	●	ø	ø	ø	ø	ø	ø	2.5×10^7	2×10^5

Note: Filled circle, activity; slashed circle, no activity.

tic use (Table 14.1). Capacity exists within the industry to increase supply should demand increase.

Production of such large quantities of fry requires significant, integrated infrastructure. This includes a production and retail supply chain for chemicals and aquaculture equipment as well as all aspects of fish production from brood stock and fertilized-egg providers to suppliers of larvae and providers of live food. Sufficient quantities of high-quality live food, of the correct size and provided at the right time, determine the success of commercial-scale larviculture.

For marine finfish larviculture, live food includes microalgae, oyster trocophores, rotifers, copepods, and subadult *Artemia*. Both copepods and *Artemia* have a wide range of sizes as they develop from egg to adult. Only copepods, however, are small enough for the critical initial and early stages of larval feeding. Therefore, the size range of developmental stages of copepods enables them to be used as live food for a longer period of larval development. Productive and reliable copepod production techniques are currently a key factor limiting the expansion and diversification of finfish larviculture. This paper reviews culture techniques for copepods and the application of copepods for marine finfish larviculture practiced in Taiwan.

CULTURE OF COPEPODS IN OUTDOOR PONDS AND APPLICATION FOR COMMERCIAL FRY PRODUCTION

Copepod Mass Culture Techniques in Outdoor Ponds

Copepod blooms regularly occur in brackish-water ponds used for fish grow-out, shrimp grow-out, or both. In such systems the copepods may be harvested as the principal product or as a by-product of the fish or shrimp. The collection and sale of copepods has become a profitable aquaculture sub-business in Taiwan. There are about 10 main dealers, with ponds located in southern Taiwan: Tainan, Kaohsiung, and Pingtung Counties. The daily demand for copepods has decreased gradually in recent years due to a decreasing demand for fry, while the number of copepod producers has increased, due to the greater profitability of raising copepods than fish or shrimp.

Two factors control the yield of copepods. One is eliminating the organisms that prey on copepods; the other is supplying enough food for the copepods. To control competing species and those that prey on copepods, tea seed cake with 10% saponin and chemicals (such as bleaching powder) are used to clean pond water before starting to culture the copepods. Carnivorous fish are stocked with the copepod to eat the small fish and shrimp, such as 3- to 5-cm Gobiidae and *Palaemon* (*Exopalamon*) *carincauda*, which ingest large numbers of copepods. When copepods are to be the main product, 1,000 fish are usually reared per 0.6 ha. Common species are red drum (*Sciaenops ocellatus*), gold-lined sea bream (*Sparus sarba*), mangrove red snapper (*Lutjanus argentimaculatus*), long dorsal fin pompano (*Trachinotus blochii*), head grunt (*Pomadasys kaakan*), and tiger shrimp (*Penaeus monodon*) or kuruma shrimp (*Penaeus japonicus*). When copepods are to be a by-product, fish are stocked at a much higher density. For example, 35,000 spotted butterfish (*Scatophagus argus*) and 300,000 white shrimp (*Litopenaeus vannamei*) were cultured in one 0.7-ha pond. To provide enough food for the copepods, an algal bloom must be maintained. Fish feed, which includes trash fish; "fish soluble," liquid feed made from fish; powder diets for fish, shrimp, or eel; and fermentative liquid from dead fish, can be directly input as food for the copepods, but also input indirectly by providing nutrients for maintaining an algal bloom. The color of the pond water is used as an indication of algal concentration, and additional nutrients are added to boost algal growth when the pond color is too light.

Pond sizes range from 0.1 to 5 ha, although they are more commonly 0.3–0.6 ha. They are unlined, with a bottom of clay or fine sand or a mixture; the color of the water is easily maintained with clay, due to microorganisms and minerals associated with clay as fertilizers for algae bloom and to hold the water, while the odor is controlled by waste being leached through the sand. Wheel paddles are used to aerate the water and to drive the water current for collection of copepods in plankton nets. Salinity of the pond water varies from 10 to 30 ppt and is optimal at 15–20 ppt. The difference in salinity between the larval-rearing water and the copepod-culture water should not be larger than 10 ppt.

The copepod culture ponds are dried in the sun for 1 to 2 weeks. When water is pumped from the ocean surface, it is first filtered in 100 mesh (300-µm) nets. Chemicals and tea seed cake are then applied to clean the water. Water from underground wells is used directly. In general, it takes about 15 days and not more than 30 days for the nauplii to appear. In the most rapid cases, nauplii occurred in 3 days and copepodids in 15 days. Usually, copepods are harvested at 0300–0400 hours using a 100–200 mesh (300-µm) plankton net fixed to the wheel paddle for 20–30 minutes. When the amount of harvest decreases, the harvest shifts to another pond. Generally, daily harvest continues for 7–15 days. Copepods are transported to farmers in aerating tanks by car in the morning. After waiting for another 7–15 days, harvesting can resume. The yield is higher from April to November in Taiwan, when solar radiation is high. Typhoons and heavy rain cause a decrease of production. Daily yield of copepods per hectare range from 20 to 200 kg, with an average of 40–80 kg. In recent years, about 30,000 kg of frozen copepods per year has been exported and more can be provided if demand increases.

Application of Pond-Cultured Copepods for Commercial Fry Production

Fry production in Taiwan is conducted in indoor and outdoor systems (Liao et al. 2001). The indoor system was adapted from technology used in shrimp postlarvae production hatcheries. Since indoor systems are more stable and controlled, they are better for rearing fragile fish such as groupers or during seasons when the weather can adversely affect outdoor systems. Using indoor systems and tank volumes of 30–50 m^3, grouper fry survival rates as high as 30% are possible. Alternatively, the survival rates in outdoor systems are inconsistent, but occasionally crops of 500,000 grouper fry have been produced from a 3,000-m^2 outdoor pond. Although the outdoor system is less productive, the lower stocking density, lower capital costs, and a simpler management regime makes the outdoor system more popular among the family-style aqua farmers in Taiwan. In addition, with the current low price of fingerlings, producing them in indoor systems would be too expensive. Outdoor systems dominate in southern Taiwan, with its favorable weather conditions and availability of highly productive land. Mass production of seed for many marine finfish is achieved in this region using such systems.

Copepod nauplii (about 110 μm) naturally occur in the outdoor larval-rearing ponds and are smaller than the gape of fish larvae, like giant grouper. Their presence in these ponds has enabled the successful culture of a number of grouper species. During the late stage of larval rearing, there are not enough naturally occurring copepods, so additional copepods are provided from other ponds. To avoid possible transfer of diseases from fish that have been cultured with these copepods, the supplemental copepods are only collected from ponds used to culture shrimp. As a result, the live-food production units are very large. In one case, an annual production of 3 million grouper fingerings (2–3 cm) was achieved from five to six crops in Tainan County between April and October using three 0.3-ha larval-rearing ponds, nine rotifer culture ponds (450 m^2), and nine copepod culture ponds (7 × 1 ha plus 2 × 4 ha).

In the indoor system, copepods are fed to groupers from 7 mm to 2–3 cm, i.e., as an *Artemia* replacement, not for first feeding. The quantity needed for the fish can be checked at night, using nets to collect the copepods in the larval rearing ponds. For 10,000 grouper fry at 1–2 cm, about 1–1.5 kg wet weight of copepods is needed and fed to the fry two to four times daily. In one case, in which 50,000 fry at 2.5 cm were harvested in a 30-m^3 pond, a total of 180 kg wet weight of copepods was provided for 10 days. Such a system, however, requires intensive monitoring of water quality if mass mortality of larvae is to be avoided because of the accumulation of wastes, including dead copepods and feces, which lead to disease problems. In one severe case, more than 90% of the fry died in a few days.

Disease Problems Associated with Pond-Cultured Copepods in Grouper Larviculture

Techniques for rearing larval grouper in the laboratory have been established in Taiwan since 1985 (Lin et al. 1986). In 1995 commercial seed production was over 10 million fry. The survival rate, however, was less than 1%, and production fluctuated due to viral nervous necrosis disease (VNN) and iridovirus-like infection (Chi et al. 1997; Chao and Pang 1997). In general, oyster fertilized eggs, rotifers, and copepods (in that order) were applied to feed the larvae. Many studies have identified live food as a critical source of microbial loading in the marine fish hatchery and a potential vector for disease transmission to larvae. Lo et al. (1996) reported white spot syndrome baculovirus detected in copepods. To clarify potential problems associated with using copepods as live food, the authors conducted grouper larval rearing in the laboratory. Although their results showed the superiority (higher survival and growth) of copepods as live feed, there were certain disease problems involved in using them (Su et al. 2002, 2003). In 31 batches of grouper larval reared in 1998–2002, no disease problems occurred in 2-cm fry, which, using enriched *Artemia* as live food, grew to 9 cm. Two cases, however, (described below) occurred in four batches using pond-cultured copepods as food.

In 1999, grouper (*Epinephelus coioides*) larvae reared in 3,000-L tanks were fed with the same feed regime from the day of hatching (day 0) to day 22, and then copepods or enriched *Artemia* were provided separately in different tanks until the fry were harvested. The copepods were collected from fish grow-out ponds and were mainly either *Apocyclops royi* or *P. annandalei*, or a mixture of both genera. *Artemia* nauplii were enriched with commercial emulsions (DC DHA; Selco) for 24 hours before use. Fry fed with copepods grew faster and reached harvest size (2 cm) at day 30, while those fed with *Artemia* were suitable for harvest at day 37, 1 week later. The harvested fry were transferred to and reared in a closed recirculating system and fed with copepods from day 42 to day 47 with commercially formulated diets. At day 77, disease problems occurred, and some fry died every day thereafter. At day 98, gill fluke *Pseudorhabdosynochus* sp. appeared. From day 77 to day 112, about half of the fry died. Using polymerase chain reaction (PCR) techniques, rotifers were shown to be nervous necrosis virus (NNV) negative, copepods were nested PCR positive, fry at day 75 were nested PCR positive, and living fry at day 83 were also nested PCR positive, but the succumbing fry were first-run PCR positive. Later, the most severe case occurred in 2001, when the fry were fed with copepods for only 2 days. Due to the combined infection of parasitic *Amyloodinium* sp. and NNV, all fish died. In addition, among three batches of fry brought from commercial hatcheries and reared in the laboratory, in two cases gill fluke disease appeared after 1-month culture, and in the third case, NNV occurred in the first week with more than 50% dying after 1 month.

The high nutritional value and low cost of wild-caught copepods makes them a preferred live-food species for farmers using outdoor pond systems. Nevertheless, a high risk of disease from pathogen transfer exists, which is the main cause of the low and erratic survival of groupers in larviculture. Further study is needed to determine how to disinfect copepods to be used as live or frozen feed.

EXPERIMENTAL STUDIES ON CULTURE OF COPEPODS

In 1988, studies on copepods began in Taiwan (Shiao 1988) on the feeding selectivity of the calanoid copepod, *P. annandalei* (former called *Schmackeria dubia*). Chang (1992) studied the physioecology of the cyclopoid copepod, *Apocyclops royi*, with the purpose of elucidating its life cycle relative to temperature, salinity, and food density, in addition to increased understanding of culture factors on ingestion and metabolism. Both studies were originally designed to provide a database for estimating the production of copepods in natural water sources. The results indicated that both copepod species had potential for mass production as live food for finfish larvae. In fact, copepods collected from outdoor ponds and used as live food were mainly *A. royi* and *P. annandalei* (Fig. 14.1).

To develop an efficient mass culture system for *A. royi*, the effects of food type, temperature, salinity, and stocking density on the production and life cycle were studied (Chang 1992; Cheng et al. 1999, 2001; Hsu 2000; Hsu et al. 2001). In addition, the effects of adding shrimp diet or fermentative liquid on the production of *P. annandalei* were tested (Cheng 2002).

CULTURE OF *APOCYCLOPS ROYI*

Apocyclops royi is a cyclopoid copepod species found in the Tan-Shui River estuary and brackish-water shrimp ponds in I-Lan and Tainan, Taiwan (Chang et al. 1991). It undergoes six nauplii and six copepodid stages in its development, with sizes and dry weights ranging from 110 to 265 μm and 0.233 to 0.772 μg, respectively, for nauplii stages, and 425 to 1034 μm and 0.805 to 2.408 μg, respectively, for copepodid stages (Chang and Lei 1993).

Effects of Temperature

Effects of temperature on reproduction tested by Chang (1992), Cheng et al. (1999), and Hsu (2000) showed the optimal temperature is 30°C and ranged from 15 to 35°C. Hsu (2000) reported that the time to maturation, the active spawning period, and longevity decreased significantly with increasing temperature (25°C, 30°C, 35°C), and ranged from 7 to 11 days, 12 to 24 days, and 22 to 41 day in females and 21 to 23 day in males, respectively (Table 14.2). The total number of offspring produced in a lifetime (166–222 nauplii/female) and their sex ratio (48–54% female) did not significantly differ at the three temperatures, although both were the highest at 30°C (Table 14.2). The average number of offspring produced by one female per day, however, was significantly higher at 30°C and 35°C than at 25°C (i.e., 12.4, 14.2, and 8.0 nauplii/female/day), respectively. Using egg sac–carrying females instead of newly hatched nauplii, and adding algal feed daily without renewing the culture medium, Cheng et al. (1999) found that at 30 ppt the nauplii ceased to grow at 15°C and grew to maturation in a shorter period than that reported in Hsu (2000), that is, 9 days (11 days in Hsu) at 25°C, and 5 days (8 days in Hsu) at 30°C.

Effects of Salinities

Chang (1992) studied the effect of six salinities on reproduction, using 20-mL test tubes contain-

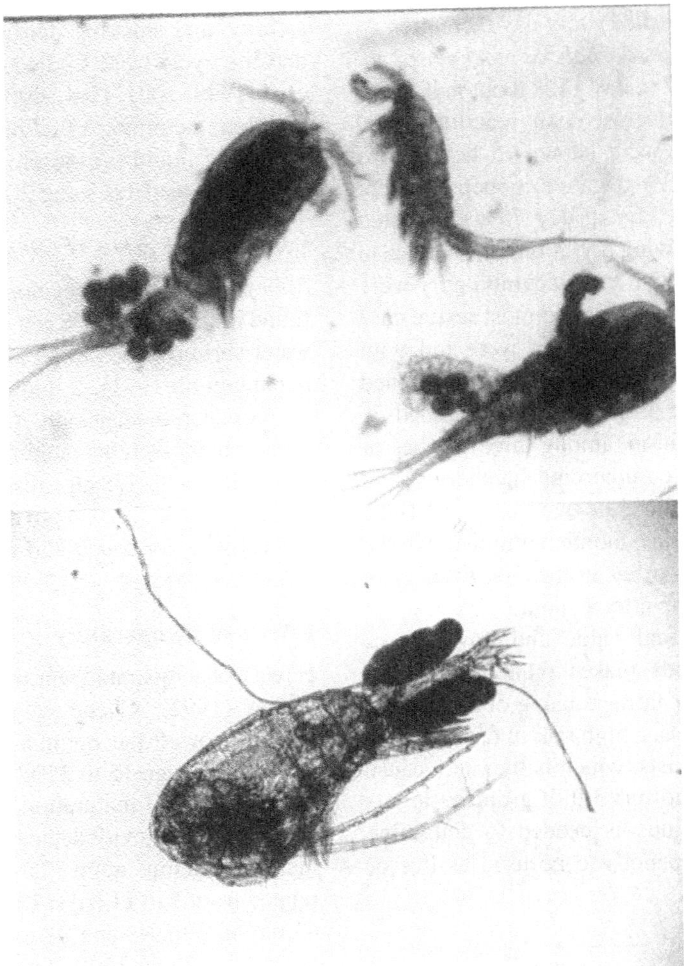

Figure 14.1. *Apocyclops royi* (upper) and *Pseudodiaptomus annandalei* (lower) collected from outdoor culture ponds.

ing 10 mL of medium and 10 newly hatching nauplii, and indicated that development time, longevity, and brood size were not significantly different from 10 to 35 ppt. In contrast Cheng et al. (1999) found that the development of *A. royi* varied with salinity. They cultured one female (carrying egg sacs) at six salinities (10–60 ppt), using 50-mL beakers containing 30 mL of culture medium with *Isochrysis galbana,* added daily (7.5×10^5 cells/mL), without aeration or water exchange. Cultured at 30°C, the nauplii grew to maturity in 4 days at 10 and 20 ppt, in 5 days at 30 ppt, in 6 days at 40 ppt, and in 8 days at 50 ppt, while growth ceased at 60 ppt.

At the optimal salinity (10 ppt) and temperature (30°C), and provided with enough food and space, the female was shown to produce one pair of egg sacs daily and yield at least six pairs of egg sacs after one mating (Cheng et al. 1999). Upon the cessation of egg production, the female can be reactivated to produce eggs if a male is introduced into the culture.

Effects of Food Types

Hsu (2000) and Hsu et al. (2001) examined the effects of food types on the growth and reproduction of *A. royi*. Four species of microalgae, *Tetraselmis chui*, *Chaetoceros muelleri*, *I. galbana*, and *Nannochloropsis oculata*, and seven types of commercial artificial feeds (applied at a rate of 100 ppm) were tested: BP-1000E and GSP-500 (both products from Far East Microalgae Ind. Co. Ltd.,

Table 14.2. Reproductive characteristics of *Apocyclops royi* fed with *Chaetoceros muelleri* at three temperatures

Reproductive characteristics		Temperature (°C) 25	30	35
Development time (d)	N I–C I	5.3 ± 0.6[b]	4.0 ± 0.2[a]	4.0 ± 0.2[a]
	N I–A	10.7 ± 0.8[c]	7.8 ± 0.4[b]	6.7 ± 0.6[a]
Survival rate (%)	N I–C I	89.6 ± 8.6	95.8 ± 4.8	97.9 ± 4.2
	N I–A	89.6 ± 8.6	84.7 ± 8.2	94.4 ± 9.6
Sex ratio (% female)	—	50.7 ± 8.8	53.6 ± 14.5	48.1 ± 8.9
Life span (d)	Male	23.2 ± 8.9[a]	21.1 ± 3.6[a]	20.6 ± 7.0[b]
	Female	40.6 ± 9.2[a]	32.4 ± 9.4[b]	22.1 ± 7.4[c]
Spawning period (d)	—	23.5 ± 3.9[a]	17.7 ± 2.7[b]	11.7 ± 0.6[c]
Offspring per female	Total	197 ± 17	222 ± 43	166 ± 35
	Nauplii/d	8.0 ± 0.9[b]	12.4 ± 3.2[a]	14.2 ± 2.5[a]

Source: Modified from Hsu (2000).

Note: Values are mean ± S.E., with different superscript letters in each row indicating significant differences. N I, nauplius stage I; C I, copepodid stage I; A, female bearing egg sac; d, day.

Taipei, Taiwan), Micro-Mac ω30 and Algamac-2000 (both products from Aquafauna Bio-Marine Inc., Hawthorne, CA, USA), DHA-Protein Selco, Culture Selco, and Baker's yeast (Yeast Ind. Co., Ltd., Zhanghua, Taiwan). Testing was at 25°C, under a 16-hour light (40 μE/m^2/sec)/8-hour dark regimen and 25 ppt. Newly hatched nauplii, less than 8 hour after hatching, were cultured in a 24-well culture vessel, each well containing 2.5 mL medium of alga or diet and each food duplicated in 36 wells. The culture medium was renewed and *A. royi* were transferred to a new well daily until they reached the copepodid stage. The *A. royi* in 12 wells were pooled into one 100-mL beaker containing 30 mL culture medium. Growth and reproduction were checked under a stereomicroscope, and when the population matured, all the nauplii produced were collected, fixed, and counted daily. The results showed that reproduction of copepods fed with algae, with the exception of *N. oculata*, which has an indigestible thick cell wall, was better than with artificial diets. When fed *T. chui*, *A. royi* matured earlier (8.7 day after hatching) and produced more offspring, averaging 250 nauplii (13.5 nauplii/female/day), than copepods fed with the other 10 diets (Fig. 14.2). Among the copepods fed artificial diets, those fed with Culture Selco had the highest survival (70.8%) at maturation, comparable to those fed with algae (77.8–89.3%), but it took the Culture Selco–fed copepods 17.5 days to mature. Those fed with Micro-Mac ω30 had the highest number of offspring (59.3 nauplii/female; 9.1 nauplii/female/day). Although only one female survived, *Apocyclops royi* fed with GSP-500 produced a higher number of offspring (160 nauplii). The copepods that fed on *T. chui*, which contains low amounts of eicosapentaenoic acid (EPA) and no docosahexaenoic acid (DHA) (Su et al. 1988), grew faster to maturation and produced more offspring, indicating that the highly unsaturated fatty acid content may be not critical for reproduction of *A. royi*. In addition, reproductive characteristics of *A. royi* fed with *I. galbana* (rich in DHA) or *C. muelleri* (rich in EPA) appeared to be almost the same, except that the latter resulted in more offspring (197 nauplii/female). Some special nutritional constituent may be the key factor (Jónasdóttir 1994). Further investigation is needed to clarify this.

Effects of Stocking Densities

To find the optimal copepod inoculation density, 1, 5, 10, and 20 ovisac-bearing females were reared at 10 ppt and 30°C, with *I. galbana* added daily. After 9 days, the average density of offspring produced in a 50-mL beaker was 24, 33, 24, and 12 individuals (ind)/mL, respectively (Cheng et al. 2001). In the group with five females inoculated on the first day, the number of ovisac-bearing females increased to 13 on the sixth day, and to 74 ind on the ninth day (Table 14.3). In addition, 490 stage-I to -III nauplii, 490 stage-IV to -VI nauplii, and 747 copepodids and adults (in

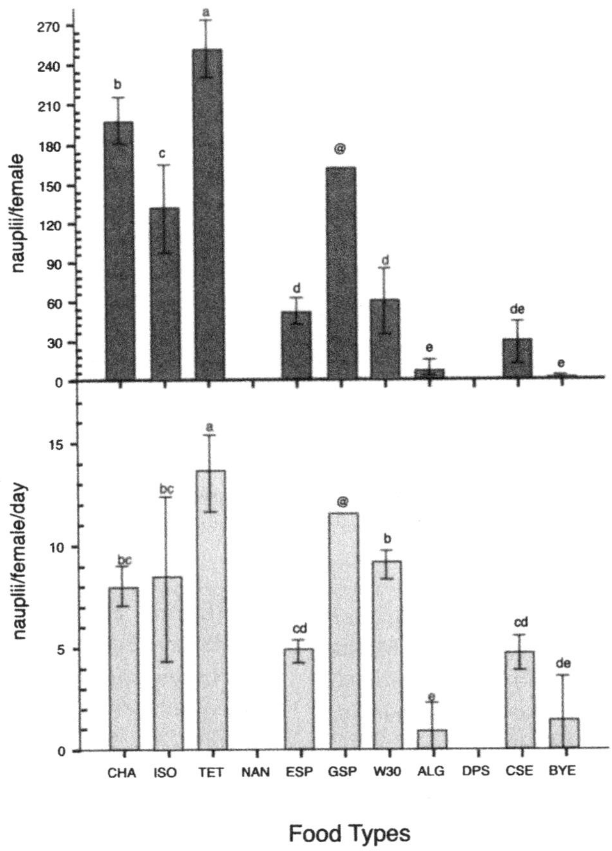

Figure 14.2. Average number (mean ± SD) of nauplii per day (below) and in life time (above) produced by one female *Apocyclops royi* fed with microalgae or diets (different superscript letters indicate significant differences; $P < 0.05$). @, not included in statistical analysis due to only one surviving female; CHA, *Chaetoceros muelleri*; ISO, *Isochrysis galbana*; TET, *Tetraselmis chui*; NAN, *Nannochloropsis oculata*; ESP, BP-1000E; GSP, GSP-500; W30, Micro-Macω30; ALG, Algamac-2000; DPS, DHA-Protein Selco; CSE, Culture Selco; BYE, baker's yeast. Cited from Hsu et al. (2001).

total 1801 ind) survived in 49-mL water volume at a density of 37 ind/mL on the ninth day.

From the experiments mentioned above, the authors conclude the optimal culture conditions for *A. royi* are at 10–20 ppt and 30°C, with a starting density of five females per 30 mL, providing an algal culture of *Tetraselmis* sp., *Isochrysis* sp., or *Chaetoceros* sp. daily and renewing the culture medium every 10 days. These results agree well with the observation that the optimal salinity of outdoor pond culture is 15–20 ppt, and daily harvest continued for 7–15 days.

Culture of *Pseudodiaptomus annandalei*

A female *P. annandalei* bearing two egg sacs is about 1,300 μm long. The size of the egg sac ranges from 350 to 425 μm. Each egg sac contains about 17 eggs with a diameter of 100 μm at maturation. At 25°C and 15 ppt, newly hatched nauplii are 125 μm, at copepodid stage CI they are 400 μm, and mature at day 6–7. Compared with *A. royi*, *P. annandalei* matured at a lower temperature, were larger with a higher biomass at harvest, and stayed near the bottom of the culture tank.

Table 14.3. Offspring produced by five egg-bearing female *Apocyclops royi* cultured in 30 mL of medium at 10 ppt, 30°C, and 12L/12D cycle with *Isochrysis galbana* added daily for 10 days

Elapsed day	Volume (mL)	Females and offspring					Density (ind./mL)
		EF	NEF	N	C + A	T	
1	30	5	—	—	—	5	< 1
2	32	1	4	60	—	65	2
3	34	2	3	88	—	93	3
4	36	3	2	148	47	200	5
5	38	5	—	172	30	207	5
6	40	13	—	177	140	330	8
7	43	18	—	655	247	920	21
8	46	7	—	530	375	912	20
9	49	74	—	980	747	1,801	37
10	52	85	—	673	792	1,550	30
11	55	35	—	900	288	1,223	22

Note: 12L/12D, 12-hour light/12-hour dark cycle; EF, egg sac–bearing female; NEF, female with no egg sac; N, nauplius stage; C, copepodid stage; A, adult, T, total individuals.

The first study on *P. annandalei* culture in Taiwan was conducted by Shiao (1988), who fed the copepod with three microalgae, *Cryptomonas* sp., *Tetraselmis* sp., and *Isochrysis* sp., to find its feeding selectivity in terms of particle size. *Cryptomonas* sp., *Tetraselmis* sp., and *Isochrysis* sp. range in size from 11.9 to 27.9, 6.5 to 11.9, and 3.0 to 6.5 μm, respectively. The active selection potential was significantly higher at copepodid stages than at naupliar stages, and the ingestion rate was higher for particles larger than 8 μm. When provided with a mixture of the three types of algae, *P. annandalei* rarely ingested *Isochrysis* sp., while female *P. annandalei* preferred *Cryptomonas* sp. This demonstrated that *P. annandalei* is a suspension feeder and feeds both passively and actively, depending on the concentration and size of particle present in the medium.

To reduce the demand for microalgae for the mass culture of *P. annandalei*, Cheng (2002) tried using artificial food. In 500-mL beakers at 25°C, under a 12-hour light/12-hour dark regimen and 15 ppt, five females bearing egg sacs were reared with microalgae (*I. galbana*) supplied daily. Powder diet for kuruma prawns at a concentration of 0.0125, 0.0250, or 0.0500 g/L was provided once at the start of culture. Copepods matured 1 day earlier and produced more nauplii in the culture with that diet, with the highest production at 0.05 g/L. The water quality, however, was difficult to maintain, and the results varied widely. As a result, a fermentative liquid was prepared instead of the powdered diets and added daily. The fermentative liquid is prepared as follows: 20 g of powder diet for eels or 10 g diet for prawns, 20 g zeolite (commercial product), 150 mL water are mixed in a 1,000-mL flask and autoclaved. After cooling, 20 g of enzymes from plants (commercial product) and 20 g of probiotic A (commercial product) are added and put at 32°C under a fluorescent lamp for 2–3 days, and then stored at −18°C. Ten grams of the frozen mixture is mixed with 1,000 mL sterilized water and aerated for 1 day. Each 2 mL of the clear suspending liquid is added four times daily to a 400-mL copepod culture medium. In experiment I, the total number of individuals produced in 10 days was nearly two times higher in the culture fed with a combination of *I. galbana* and fermentative liquid than in that fed *I. galbana* alone (Table 14.4). In contrast, providing the fermentative liquid without *I. galbana* resulted in very poor production. In experiment II, by adding *I. galbana* for the first 3 days, followed by the fermentative liquid for 7 days, more nauplii and copepodids of *P. annandalei* were produced than by adding *I. galbana* alone daily (Table 14.4). A total of 2,029 ind were produced in the group receiving the fermentative liquid, which was 2.42 times higher than that in the group receiving algae (837 ind). The number of nauplii and copepods produced were 1.98 and 10.28 times higher, respectively. The number of

Table 14.4. Offspring produced by five female *P. annandalei* cultured for 10 days in 400 mL of medium at 15 ppt and 25°C with *Isochrysis galbana*, fermentative liquid, or both, added

Experiment	I		II	
Isochrysis galbana	Day 1–10	Day 1–10	Day 1–10	Day 1–3
Fermentative liquid	—	D1–D10	—	Day 4–10
Nauplii	1,049 ± 249	1,320 ± 551	701 ± 251	1,387 ± 489
Copepodids	153 ± 92	1,087 ± 635	54 ± 46	555 ± 220
Adults	40 ± 9	35 ± 17	33 ± 9	41 ± 4
Adults with egg sacs	51 ± 12	99 ± 38	49 ± 28	46 ± 7
Average total number	1,287 ± 282	2,542 ± 1230	837 ± 273	2,029 ± 273

Note: Mean ± SD; $n = 5$.
Source: Modified from Cheng (2002).

Table 14.5. Some reproductive characteristics of *Apocyclops royi* and *Pseudodiaptomus annandalei* at optimum conditions

	Apocyclops royi	*Pseudodiaptomus annandalei*
Nauplii (μm)	110–265	125–320
Copepodid (μm)	425–1,034	400–1,280
Female (μm)	1,010	1,300
Egg sac (μm)	ND	350–425
Mature	4–5 days	6–7 days
Temperature	30°C	25°C
Salinity	10–20 ppt	15 ppt
Nauplii produced in lifetime	270	ND
Brood size	18.7	17
Longevity (female)	32.4 days	ND
Longevity (male)	21.1 days	ND
Maximum density in culture(/mL)	37 ind	5 ind

Note: ND, not determined.

females bearing egg sacs, however, was 94% of that in the group fed algae alone. Thus, adding fermentative liquid promoted development and maturation, but had a slightly negative effect on egg bearing. The fermenting liquid may be input directly as feed or indirectly, by allowing the algae to bloom. Studies to find more nutritional artificial diets to reduce the dependence on algae are still ongoing.

CONCLUSION

The shorter time to maturation for *A. royi* (4–5 d) and *P. annandalei* (6–7 days) compared with *Acartia* sp. (10 days; Ohno and Okamura 1988), *Tigriopus japonicus* (10–14.5 days; Hagiwara et al. 1995), and *Tisbe furcata* (12–22 days; Abu-Rezq et al. 1997) indicates that these two copepods have a high potential for mass culture. In fact, more than 350 metric tons of copepods, mainly *A. royi* and *P. annandalei*, were produced in 2000 in Taiwan. Some reproductive characteristics of *A. royi* and *P. annandalei* are listed in Table 14.5.

Fish larvae that are fed copepods cultured in outdoor ponds with fish are at risk from introduced disease problems such as VNN and parasites. Further studies are needed to determine how to mass culture copepods more hygienically, yet in an economically efficient manner. Major research subjects include selection of the right species of copepods, suitable food for copepods, facilities for rearing copepods, and management of copepod production systems. In this paper, the former two issues are addressed, but the critical and difficult missions of developing suitable equipment and controllable operations still need to be established. Studies using intensively cultured *A. royi* and *P. annandalei* as larval feeds under controlled conditions are ongoing. Alternatively, one solution is to provide only the naupliar stage of copepods for the rearing of fish larvae and substitute the copepodid stage with inert diets or *Artemia*.

ACKNOWLEDGMENTS

The authors thank the organizers of the workshop on Culture of Copepods and Applications to Marine Finfish Larval Rearing, particularly Dr. Cheng-Sheng Lee and Ms. Pat O'Bryen; for the warm hospitality to H.M. Su; and to The Oceanic Institute and for its generous support of travel and accommodations. The authors acknowledge Mr. Chia Hsing Hsu, Mr. R.L. Chou, and aquatic farmers in Taiwan for providing the information. Thanks are also owed to Dr. Mike Rimmer and Dr. Atsushi Hagiwara for reviewing the manuscript.

REFERENCES

Abu-Rezq, T.S., Yule, A.B., and Teng, S.K. 1997. Ingestion, fecundity, growth rates and culture of the harpacticoid copepod, *Tisbe furcata*, in the laboratory. *Hydrobiologia* 347:109–118.

Chang, W.B. 1992. Studies on the physioecology of a copepod, *Apocyclops royi* (family: Cyclopoidae). Ph.D. thesis. National Taiwan University, Taiwan, Republic of China. (In Chinese).

Chang, W.B. and Lei, C.H. 1993. Development and energy content of a brackish-water copepod, *Apocyclops royi* (Lindberg) reared in a laboratory. *Bulletin of the Institute of Zoology, Academia Sinica* 32:62–81.

Chang, W.B., Wu, F.L., Wu, K.S., and Lei, C.H. 1991. Study on the zooplankton of shrimp ponds in Taiwan. *COA Fisheries Series* 28:53–76. (In Chinese).

Chao, C.B. and Pang, V.F. 1997. An outbreak of an iridovirus-like infection in cultured grouper (*Epinephelus* spp.) in Taiwan. *Journal of Chinese Society of Veterinary Science* 23:411–422. (In Chinese).

Cheng, S.H. 2002. The effect of adding artificial fermentative liquids on the reproduction of *Pseudodiaptomus annandalei*. Annual Report of Taiwan Fisheries Research Institute, Keelung, Taiwan. (In Chinese).

Cheng, S.H., Chen, H.C., Chang, S.L., Chen, T.I., and Liao, I C. 2001. Study on the optimal density of mass culture in copepod *Apocyclops royi*. In *Book of Abstracts. 6th Asian Fisheries Forum 2001*, p.58. Kaoshiung, Taiwan, 25–30 November 2001. Quezon, Philippines: Asian Fisheries Society.

Cheng, S.H., Chen, H.C., Su, M.S. and Ho, J.S. 1999. Effects of temperature and salinity on the maturation in *Apocyclops royi* (Cyclopidae, Cyclopoida). In *Book of Abstracts. The 7th International Conference on Copepoda*, p. 80. Curitiba, Brazil, 25–31 July 1999. Chetumal, Mexico: World Association of Copepodologists.

Chi, S.C., Lo, C.F., Kou, G.H., Chang, P.S., Peng, S.E., and Chen, S.N. 1997. Mass mortalities associated with viral nervous necrosis (VNN) disease in two species of hatchery-reared grouper, *Epinephelus fuscogutatus* and *Epinephelus akaara* (Temminck & Schlegel). *Journal of Fish Diseases* 20:85–193.

Hagiwara, A., Lee, C.S., and Shiraishi, D.J. 1995. Some reproductive characteristics of the broods of the harpacticoid copepod *Tigriopus japonicus* cultured in different salinities. *Fisheries Science* 61:618–622.

Hsu, C.H. 2000. Effects of food types and temperature on the development and reproduction of *Apocyclops royi* (Copepoda, Cyclopoida). Master's dissertation. National Sun Yat-Sen University, Taiwan, Republic of China. (In Chinese).

Hsu, C.H., Su, H.M., and Chen, I.M. 2001. Effects of food types on the development and reproduction of *Apocyclops royi*. In *Larvi '01—Fish & Shellfish Larviculture Symposium*, edited by C.I. Hendry, G. Van Stappen, M. Wille, and P. Sorgeloos, pp. 250–253. Ghent, Belgium, 3–6 September 2001. Oostende, Belgium, European Aquaculture Society, Special Publication No. 30.

Jónasdóttir, S.H. 1994. Effects of food quality on the reproductive success of *Acartia tonsa* and *Acartia hudsonica*: laboratory observations. *Marine Biology* 121:67–81.

Lee, J.L., Chang, C.S., and Ting, Y.Y. 1989. Promotion and extension of aquaculture-studies on mass seed producing of black sea bream. *Bulletin of Taiwan Fisheries Research Institute* 47:253–258. (In Chinese).

Liao, I C., Huang, C.P., Lin, M.C., Hou, Y.W., Huang, T.L., and Tung, I.H. 1969. Artificial propagation of grey mullet, *Mugil cephalus* Linnaeus. *Joint Community Rural Reconstruction, Fishery Service* 8:10–20. (In Chinese).

Liao, I C., Su, H.M., and Chang, E.M. 2001. Techniques in finfish larviculture in Taiwan. *Aquaculture* 200:1–31.

Lin, K.J., Yen, C.L., Huang, T.S., Liu, C.Y., and Chen, C.L. 1986. Experiment of fry nursing of *Epinephelus salmonoides* (Lacepede) and its morphological study. *Bulletin of Taiwan Fisheries Research Institute* 40:219–240. (In Chinese).

Lin, L.T. 1984. Studies on the induced breeding of milkfish, *Chanos chanos*, reared in ponds. *China Fisheries Monthly* 378:3–29. (In Chinese).

Lo, C.F., Ho, C.H., Peng, S.E., Chen, C.H., Hsu, H.C., Chiu, Y.L., Chang, C.F., Liu, K.F., Su, M.S., Wang, C.H., Kou, G.H. 1996. White spot syndrome baculovirus (WSBV) detected in cultured and captured shrimp, crabs and other arthropods. *Diseases of Aquatic Organisms* 27:215–225.

Ohno, A. and Okamura, Y. 1988. Propagation of calanoid copepod, *Acartia suensis*, in outdoor tanks. *Aquaculture* 70:39–51.

Shiao, C. M. 1988. Feeding selectivity of a marine copepod *Schmackeria dubia* (Kiefer, 1936). Master's dissertation, National Taiwan University, Taiwan, Republic of China. (In Chinese).

Su, H.M., Lei, C.H., and Liao, I C. 1988. The effect of environmental factors on the fatty acid composition of *Skeletonema costatum*, *Chaetoceros gracilis* and *Tetraselmis chuii*. *Journal of the Fisheries Society, Taiwan* 15:21–34. (In Chinese).

Su, H.M., Wang, S.S., Hsu, C.H., Chen, T.I., and Su, M.S. 2002. Why we use and don't use copepods for grouper larviculture? In *Book of Abstracts. The 8th International Conference on Copepoda*, p.167. Keelung, Taiwan, 21–26 July 2002. Keelung, Taiwan: World Association of Copepodologists.

Su, H.M., Wang, S.S., Hsu, C.H., Chen, T.I., and Su, M.S. 2003. A comparison of copepods and *Artemia* nauplii for grouper larviculture. *Journal of Taiwan Fisheries Research* 11:39–46. (In Chinese).

15
Copepods as a Live Feed for Striped Trumpeter *Latris lineata* Larvae

David T. Morehead, Stephen C. Battaglene, Ephrime B. Metillo, Matthew P. Bransden, and Graeme A. Dunstan

ABSTRACT

Copepods were harvested from estuaries and coastal areas of southeast Tasmania for examination as an alternative or supplementary larval feed to rotifers *Brachionus plicatilis* and brine shrimp *Artemia* sp. One dominant calanoid copepod, *Acartia tranteri,* was isolated from wild harvests and cultured on microalgae. Egg production from *A. tranteri* was greatest when fed *Rhodomonas salina,* and population growth was highest when provided a mixed diet of *R. salina* and *Isochrysis* sp., suggesting that docosahexaenoic acid–rich flagellates that cover a wide size range may be the preferred microalgae for successful copepod production. Intensive cultivation of *A. tranteri* and other copepods at a scale sufficient for use as the primary live feed for larvae proved difficult. Harpacticoid (mainly *Tisbe* sp.) and calanoid (mainly *Acartia* sp.) copepods were separately cultured and their fatty acid composition analyzed. Copepods had distinct lipid profiles compared with *Artemia* and rotifers. Most notably, copepod total lipid consisted mainly of polar lipid, while rotifers and *Artemia* enriched on commercial preparations were rich in triacylglycerides. Trials with striped trumpeter *Latris lineata* larvae suggest the use of copepods improved survival, but it was unclear if this was a result of the copepods or the microbial flora that was introduced with them.

INTRODUCTION

Problems in the production of juvenile marine fish occur in a number of species (Støttrup et al. 1986; van der Meeren and Naas 1997; Payne and Rippingale 2001), and malformation of cultured fish is a common affliction (Divanach et al. 1996; Støttrup 2000). Mortality and malformations of larval fish have often been attributed to dietary deficiencies or imbalances in three essential polyunsaturated fatty acids (PUFA): docosahexaenoic acid (DHA); eicosapentaenoic acid (EPA); and arachidonic acid (ARA) (Bell 1998; Sargent et al. 1999; Koven et al. 2001). Rotifers (*Brachionus* spp.) and *Artemia,* which are often used as live feeds for marine fish larvae from first feeding, generally lack these essential fatty acids and need to be enriched (Sargent et al. 1999). Despite the use of commercially available enrichments, resulting in vastly improved essential fatty acid profiles of these live feeds, species such as Atlantic halibut *Hippoglossus hippoglossus* (Pittman 1996), groupers *Epinephelus coioides* (Toledo et al. 1999), tropical snappers (Tucker 1998), and striped trumpeter *Latris lineata* (Battaglene et al. 2002) continue to prove difficult to rear through the larval stage.

Marine copepods are a natural source of food for marine fish larvae (Hunter 1981; Bell 1998; Tucker 1998), and many fish species that have

Corresponding author: David T. Morehead, Marine Research Laboratories, Tasmanian Aquaculture and Fisheries Institute and Aquafin Cooperative Research Centre, University of Tasmania, Hobart, Tasmania, 7001, Australia; Phone: +61-3-6227-7297; Fax: +61-3-6227-8035; E-mail: David.Morehead@utas.edu.au.

proven difficult to rear through their larval stage with traditional live feeds have performed better when fed diets composed of, or including, copepods (Støttrup et al. 1986; Kraul 1991; Bell 1998; Tucker 1998; McKinnon et al. 2003). The nutritional properties of copepods, particularly the predominance of phospholipids over triacylglycerols, combined with their smaller size and ease of digestion by finfish larvae, have made their culture the subject of intense research (Zurlini et al. 1978; Kuhlmann et al. 1981; Støttrup et al. 1986; Ohno et al. 1990; Kraul 1991; Bell 1998). Reliable production, however, of large numbers of copepods remains difficult (Payne and Rippingale 2001).

Progress toward propagation of striped trumpeter *Latris lineata* (Family Latrididae) has been constrained by early larval mortality (Battaglene et al. 2002; Trotter et al. 2003), and most of the juveniles produced have exhibited jaw malformation (Cobcroft et al. 2001), swim bladder malformation, and skeletal deformities including kyphosis and vertebral fusion (Trotter et al. 2001). Little is known of the complex larval cycle of striped trumpeter in the wild, but the larvae are neustonic and they have an extended postlarval or "paperfish" stage lasting for 7–9 months (Furlani and Ruwald 1999; Jordan 2000). The present study describes efforts to collect and intensively culture copepods and investigates their potential suitability as a supplementary live feed for striped trumpeter larvae.

COPEPOD CULTURE

To increase the likelihood of collecting species of copepods that were amenable to intensive culture, collections were made from a diverse number of sites and habitats.

SOURCE OF COPEPODS

Copepods were collected on 10 occasions from three field locations in southeast Tasmania, Australia (Fig. 15.1), and on many occasions from rotifer culture tanks and from an outdoor cement raceway (120,000 L) at the Marine Research Laboratories (MRL).

Collection of Copepods: Field

Copepods were obtained from Orielton Lagoon by pumping water through a container fitted with 63-μm mesh screen. Assessment of the collections revealed native rotifers were the dominant species (73.9%), followed by ciliates (22.2%), other organisms (2.3%), and copepods (1.6%). A plankton net (250-μm mesh; 100 cm long with a 40-cm mouth) was used to collect samples from Mercury Passage, while a smaller collector (85-μm mesh; 15 cm diameter) was used to collect samples from Norfolk Bay. Species composition and abundance varied at different times for both sites, but always included calanoid copepods (mainly *Acartia* sp.).

To remove larger organisms and debris, plankton was rinsed through a 1,000-μm mesh screen upon arrival at the MRL. Collection, transport, and rinsing of the copepods were estimated to reduce survival by up to half.

Collection of Copepods: MRL

Harpacticoid copepods (mainly *Tisbe* sp.) resident in an aged rotifer *Brachionus plicatilis* culture reached densities of up to 0.5/mL and were harvested daily by passing a screen (250 μm) through the 1,800-L tank. Harvests of harpacticoid copepods (mainly *Tisbe* sp.) also took place from an outdoor raceway by pumping water from a depth of 1 m into a suspended 70-L container with 63-μm mesh screen walls housed within a 100-L container, with the number collected at night (~50,000) much higher than during the day (~1,000).

Evaluation of Copepod Collection

Generally, calanoid copepods were obtained in collections at field locations, and harpacticoid copepods were obtained in collections on site. The authors identified three main problems in using field-harvested copepods directly as a larval feed for marine fish.

First, it is difficult to ensure the reliable capture of sufficient quantities of copepods. Improved collection methods, such as larger nets or longer collection periods, would increase wild catches, but collection would still depend on the presence of natural plankton blooms and suitable weather conditions (Sargent et al. 1999).

The second problem is the difficulty in quickly sorting large numbers of copepods from other potentially pathogenic or nuisance organisms on a daily basis. On one occasion, for example, the introduction of toxic blue-green algae caused com-

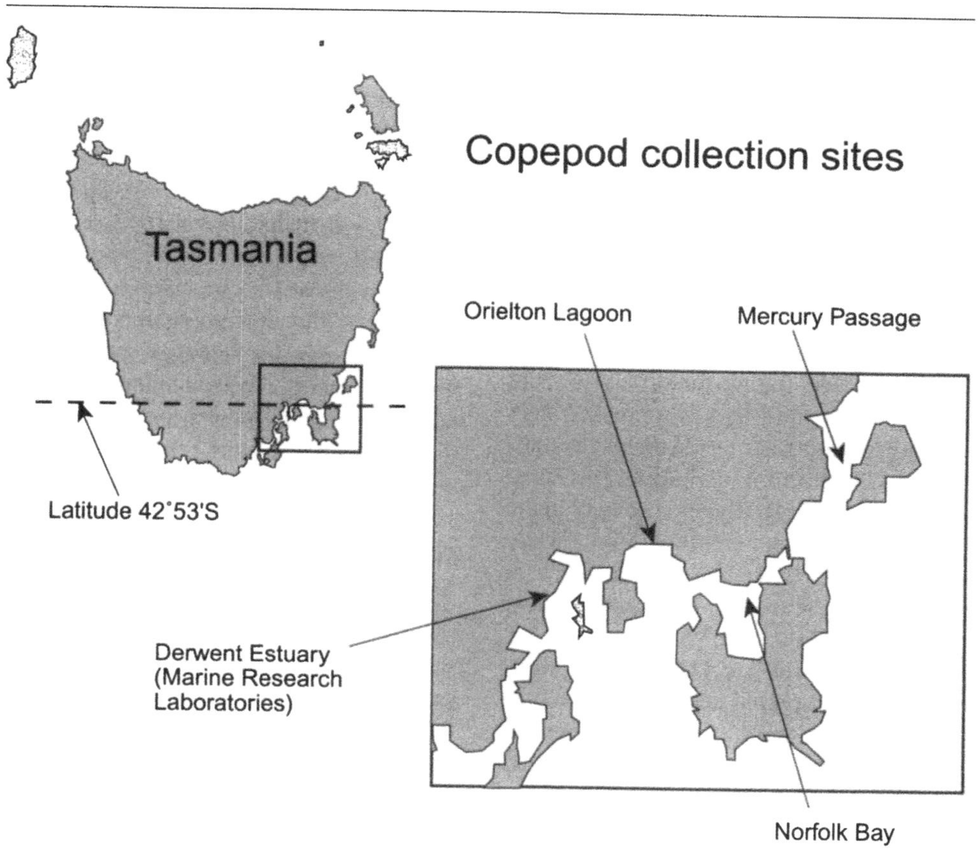

Figure 15.1. Copepod collection sites from southeast Tasmania.

plete mortality of striped trumpeter larvae (Thomas et al. 1992). A more routine problem, however, was the introduction of gelatinous zooplankton (medusae, siphonophores, ctenophores, salps, and larvaceans) and fish species that either directly prey on the striped trumpeter larvae or compete for food and space. Similar difficulties using wild zooplankton have been highlighted in other studies (Tucker 1998; Sargent et al. 2002).

The third problem is changes in the nutritional quality of the copepods with season, due largely to changes in the dominant phytoplankton species present (Sargent et al. 2002).

Culture of Copepods

Mixed copepod species were intensively reared on a number of occasions in tanks of various sizes (250, 1,000, 3,000, and 25,000 L), and a single species, *Acartia tranteri*, was targeted for fecundity and population-growth trials using different microalgae as feed.

Microalgae were obtained from the CSIRO Marine Laboratories Micro-algae Culture Collection (Tasmania, Australia). Species included the diatom *Chaetoceros muelleri* (CS176), the cryptomonad *Rhodomonas salina* (CS24/4), the prymnesiophyte *Isochrysis* sp. (Tahitian strain; CS177), and the eustigmatophyte *Nannochloropsis oculata* (CS-179/4). For intensive copepod culture, algae were produced in 400-L bags or 1,000-L tubs. For fecundity and population-growth trials, algae were produced in 2-L Schott bottles. All algal cultures were maintained at 22°C with continuous light and aeration that included 0.5% CO_2.

Intensive Calanoid Copepod Production

A 1,000-L tank was maintained at 17°C with an 18-hour light/6-hour dark photoperiod and stocked with 75,000 calanoid copepods. A 63-µm screen over the outlet allowed addition of algae (50 L of *N. oculata* daily) while retaining the copepods. After 2 weeks, and coinciding with the production of copepod nauplii, the algal feed was changed to a 100-L mix of *N. oculata*, *R. salina*, and *C. muelleri* (1:6:1 by volume of culture, respectively), together with 400 L of 1 μm filtered seawater. At the same time, the 63-μm mesh screen was changed to a 390-μm mesh to allow nauplii to be flushed to a second 1,000-L tank fitted with a 63-μm mesh screen on the outlet. After a further 2 weeks, a third 1,000-L tank was put in place to passively receive nauplii from both 1,000-L culture tanks during times of water and algae addition. This third tank was stocked with 3-day-old striped trumpeter larvae that were additionally fed 1–2 rotifers/mL daily from first feeding (day 6 after hatching). Copepod density in the 1,000-L tank culture tanks reached 1/mL, while nauplii density in the fish tank reached 0.2/mL at the time of larval first feeding and, over the next 10 days, varied from 0.1 to 0.3/mL.

A 3,000-L tank was stocked with approximately 1,000 adult calanoid copepods and fed *R. salina*, *Isochrysis* sp., and *C. muelleri* at a ratio of 2:1:1 (v/v/v) to maintain cell densities around 20,000–30,000 cells/mL. Water flow was maintained at 1,500 L/day and a 63-μm mesh screen over the outlet retained copepods. The tank was lightly aerated and maintained at 16–17°C with a 14-hour light/10-hour dark photoperiod. The copepod population underwent slow growth for a number of weeks and then began to increase. Because no larval fish were available for feeding trials, the copepod population was harvested for lipid analysis.

On a number of occasions indoor 25,000-L tanks were allowed to bloom with *N. oculata* prior to stocking with rotifers (1/mL) and mixed wild-caught plankton that included calanoid copepods. Copepod populations did not appear to increase within the tanks, but they remained present. On one occasion, a population of unidentified cnidarians dominated the tank fauna, and on all occasions other zooplankton species became established, including at least eight different teleost species (hatched from eggs), crustaceans, mollusks, and algae (including blooms of the dinoflagellate *Noctiluca scintillans*). Because these tanks were stocked with striped trumpeter larvae and contained other copepod predators, it is likely that copepod numbers were being continuously cropped.

Intensive Harpacticopid Copepod Production

Harpacticoid copepods (*Tisbe* sp.), harvested daily from an aged rotifer culture, were stocked into a 250-L cylindro-conical static tank that received an addition of approximately 20 L of algae daily (*N. oculata*, *R. salina*, and *Isochrysis* sp.; 2:1:1 v/v/v). The tank was maintained at 16°C, received no aeration and was kept in darkness. Harpacticoid copepods massed in the column near the surface and appeared to undergo rapid population growth. Daily collection for feeding to postflexion striped trumpeter larvae involved submersing a 5-L container into the copepod swarm, just under the surface. Samples of these harpacticoid copepods were also collected for lipid analysis.

Fecundity and Population Growth Trials

The authors chose the calanoid copepod *A. tranteri* for intensive culture experiments because it was the most common species among samples collected from wild harvests. *Acartia tranteri* isolated from plankton collected from Mercury Passage were transferred to a vessel of sterilized seawater and fed a mix of microalgae (*R. salina* and *Isochrysis* sp.). The stock culture was aerated, maintained at 16–17°C with a 14-hour light/10-hour dark photoperiod, and received fresh algae and a 50% water exchange daily. To determine the effect of different algal species on copepod fecundity, four dietary treatments were investigated: *R. salina* and *Isochrysis* sp. were fed separately or in combination (1:1 v/v), or the copepods were unfed (control). Fifteen to 17 mature females were placed in 200-mL vessels (n = 3–5 replicates) containing uniform microalgal concentrations (~0.015 mg/mL). After a 24-hour egg-laying period, all eggs were collected and counted as a measure of egg production per female.

Egg production was significantly higher when *A. tranteri* were maintained on *R. salina* or a combination of *R. salina* and *Isochrysis* sp. (14.7 ± 4.7 and 12.1 ± 2.0 eggs per female, respectively) than *Isochrysis* sp. alone (5.6 ± 2.2) or in the absence of algae (2.8 ± 1.2)(F = 13.98, df = 17, $P < 0.05$)(Fig. 15.2). The experiment highlights the importance of selecting the right algal species for egg production, with the larger-celled *R. salina* (5–12 μm in diameter) providing greater fecundity than the smaller *Isochrysis* sp. (3–5 μm

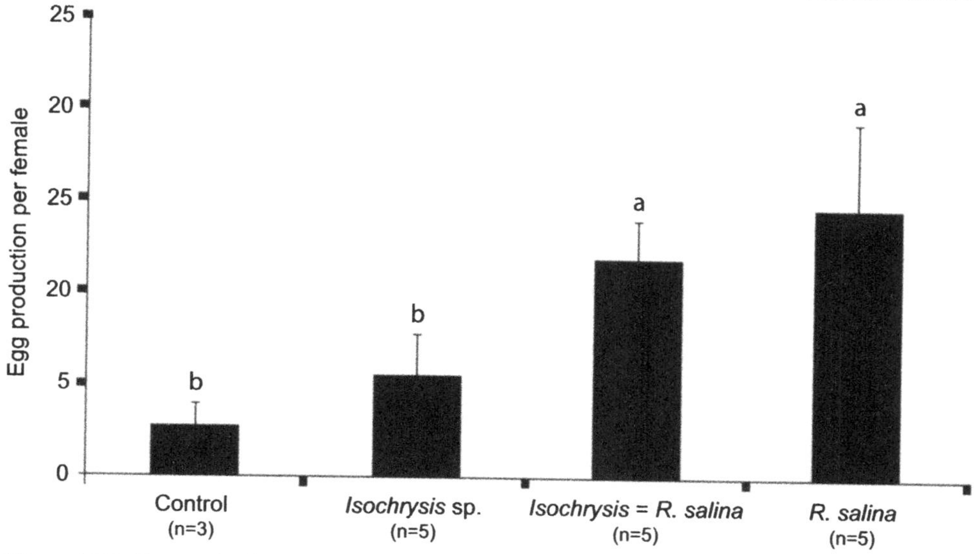

Figure 15.2. Egg production of *Acartia tranteri* provided with different dietary treatments (mean + SD). Different superscripts represent significantly different mean values ($P < 0.05$).

in diameter). The fecundity observed at 16–17°C was higher (14.7 eggs per female) than field estimates for *A. tranteri* at 20°C (12 eggs per female; Kimmerer and McKinnon 1987), although fecundity in other *Acartia* species has been recorded between 10 to 50 eggs per female at 15°C (Mauchline 1999). The observed egg production of *A. tranteri* in seawater alone probably reflects their feeding history and ability to convert stored substances into producing eggs (Donaghay 1985).

To examine population growth of *A. tranteri* on various algal diets, equal numbers of adults were added to cylindro-conical 30-L rearing tanks (n = 3 replicates) and allowed to lay eggs for a 24-hour period. One of five algal diets was used as the food source for each rearing tank, namely (a) *Isochrysis* sp., (b) *R. salina*, (c) *N. oculata*, (d) *R. salina* and *Isochrysis* sp. (1:1, v/v; mix I), and (e) *R. salina, Isochrysis* sp., and *N. oculata* (1:1:1, v/v/v; mix II). Algae were harvested daily from 4-day-old cultures, and algal concentrations in the copepod tanks were kept uniform and ranged from 0.014 to 0.089 mg/mL. Each day a 1-L sample was removed from each tank, and the number of copepods and their stage of development recorded.

The number of *A. tranteri* nauplii produced and their ongoing survival varied according to treatment (Fig. 15.3A). *Nannochloropsis oculata* resulted in only limited numbers of nauplii hatching, and none progressed to the copepodid stage. *Rhodomonas salina* resulted in higher numbers of nauplii being produced, but this was followed by a sudden decrease in numbers from day 3 after hatching, with almost complete mortality of nauplii by day 9; and again, none progressed to the copepodid stage. Mortality also coincided with microbial contamination, evident as a biofilm on the walls of the culture vessel and filamentous strands in the water column. Adults stocked at the start continued to supply nauplii to the remaining three treatments throughout the experiment. The two mixtures of microalgae, mix I (*R. salina* and *Isochrysis* sp., 1:1, v/v) and mix II (*R. salina, Isochrysis* sp., and *N. oculata,* 1:1:1, v/v/v), not only produced high numbers of nauplii but supported production for an extended period (Fig. 15.3A). Using *Isochrysis* sp. alone as a diet did not result in as high a number of nauplii being produced as *R. salina* and the mixed diets, but nauplii were produced over an extended period.

Copepodids were first detected on day 11 in the treatments containing mix I and mix II and on day 13 in the treatment containing only *Isochrysis* sp. (Fig. 15.3B). Second generation adults were first detected on day 25 in all three remaining treatments (Fig. 15.3C).

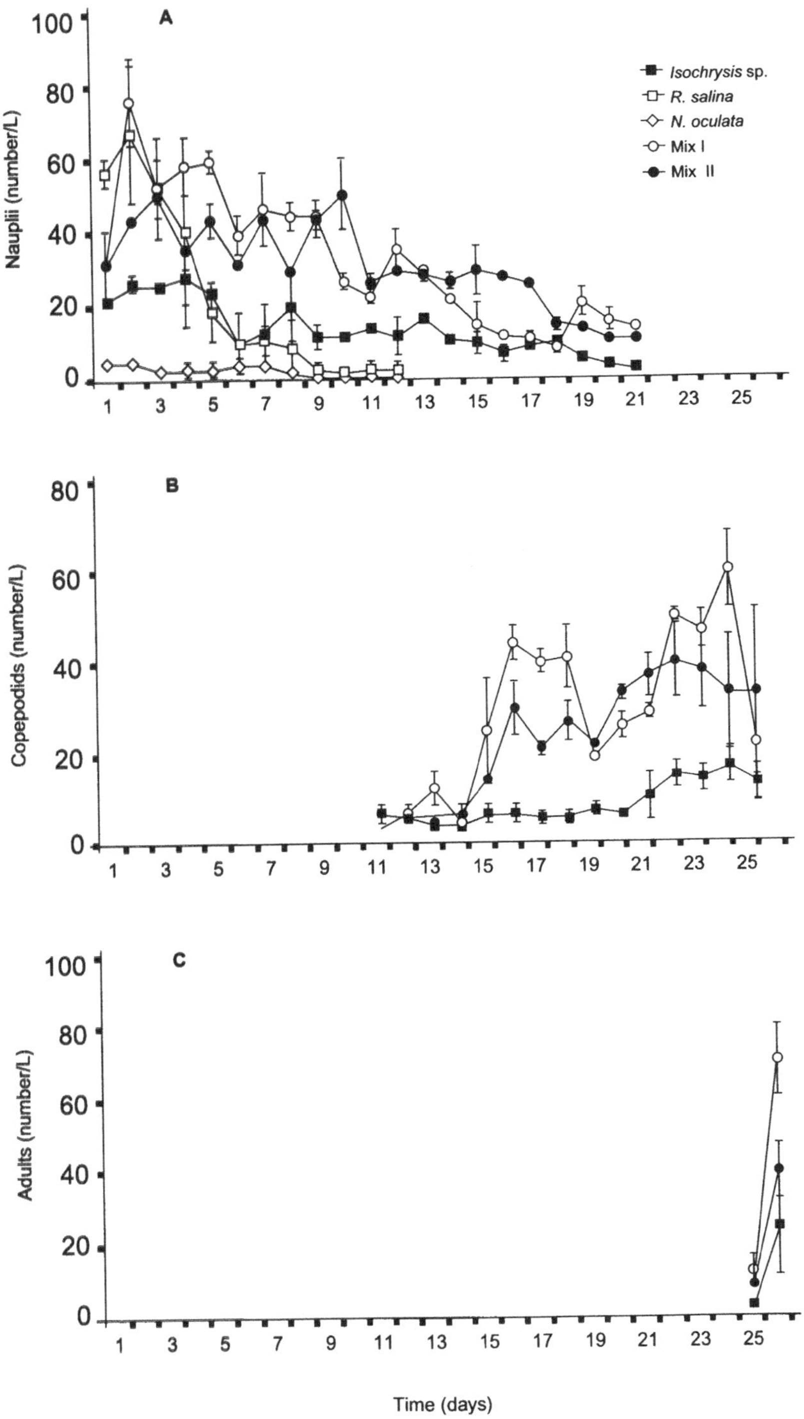

Figure 15.3. Population density of *Acartia tranteri* fed different species of microalgae. (A) nauplii (NI to NVI); (B) copepodids (CI to CVI); and (C) adults. Data are mean ± S.E.

Evaluation of Culture Methods

The mass culture systems used by the authors only produced sufficient numbers of copepods for supplementary feeding to larval striped trumpeter. Calanoid copepods reached densities of 1/mL in 1,000-L culture vessels, with all stages present in the water column. In contrast, adult harpacticoid copepods tended to inhabit tank surfaces rather than the water column and only entered the water column when disturbed. Harpacticoid copepods, however, did mass near the surface and within the column when cultured temporarily under static conditions and darkness, and vigorous feeding on harpacticoids by 4-week-old striped trumpeter larvae did occur immediately after they were introduced to the centre of the fish tanks. Culture techniques for copepods are similar to those for rotifers, but copepods can be inhibited by overcrowding (requiring more tank space) and take longer to double in number (Tucker 1998). Certain species are also cannibalistic (Ohno et al. 1990).

The provision of the right algal species, or a mix of algal species, was important for the culture of *A. tranteri*. The authors found a diet of the small-celled (3 µm in diameter) *N. oculata* alone did not support population growth of *A. tranteri*, a finding similar to that of Payne and Rippingale (2000) in the culture of *Gladioferens imparipes* and Rey et al. (2001) for *Calanus helgolandicus*. These authors noted that the indigestible cellulose walls of *N. oculata* render this microalga less suitable as food for calanoid copepods. Also, while *N. oculata* is a good source of EPA, it is deficient in DHA (Dunstan et al. 1993). In contrast *R. salina* is a good source of both EPA and DHA (G.A. Dunstan, unpublished data), and when fed to *A. tranteri*, it supported nauplii production. The high mortality of copepod nauplii in this treatment coincided with a bacterial bloom, but may have been related to the large size of *R. salina* (5–12 µm in diameter), reducing its availability to the nauplii. Our stock culture of *R. salina* was not axenic, which may account for the bacterial bloom. *Isochrysis* sp. also supported population growth of *A. tranteri*, with results similar to those obtained by Støttrup et al. (1999) for *A. tonsa* and by Payne and Rippingale (2000) for *G. imparipes*, who used *Isochrysis galbana* as the sole food for these copepods. The flagellates *R. salina* and *Isochrysis* sp. (mix I) are both good sources of DHA and cover a broad size range, making them not only nutritious, but accessible by nauplii, copepodids, and adults, which may explain why this combination supported good production and survival in *A. tranteri*. The addition of *N. oculata* as a feed (mix II) did not improve population growth, further suggesting that this nonflagellated alga is not appropriate for *A. tranteri* culture. Kleppel (1993), however, emphasized that dietary diversity is often a key to the procurement of a nutritionally complete ration among the calanoid copepods. This is supported by Schipp et al. (1999), who used *Rhodomonas* sp., *Tetraselmis* sp., and *Isochrysis* sp. to produce sufficient nauplii and other stages of the tropical copepod *Acartia* sp. for the feeding of golden snapper larvae, *Lutjanus johnii*.

Chemical Composition of Live Feed

Lipid class and fatty acid analyses of cultured calanoid and harpacticoid copepods and nonenriched and enriched rotifers and *Artemia* were performed by initially extracting total lipid from freeze-dried samples using the method of Bligh and Dyer (1959). An aliquot of total lipid was analyzed for lipid class composition (wax ester/hydrocarbon [WE/HC]; triacylglyceride, [TAG]; free fatty acid [FFA]; sterol [ST]; and polar lipid [PL]) using an Iatroscan MK V (Volkman and Nichols 1991). Fatty acid methyl esters (FAME) were formed from an aliquot of total lipid with methanolic HCl and then analyzed with a Hewlett Packard 5890A gas chromatograph fitted with a HP-5 cross-linked methyl silicone-fused silica capillary column (50 m, 0.32 mm internal diameter). Rotifers and *Artemia* were enriched with Algamac 2000 and Algamac 3050, respectively, according to manufacturer's specifications (Aquafauna Bio-Marine, Hawthorne, CA, USA).

Polar lipid was the dominant lipid found in both harpacticoid and calanoid species, representing about 70% of the total lipid (Table 15.1). Nonenriched rotifers had a similar lipid class profile to harpacticoid and calanoid copepods, but with a lower proportion of free fatty acids and a greater proportion of wax ester/hydrocarbon. Nonenriched *Artemia* nauplii were also low in free fatty acids, but high in triacylglycerides.

Table 15.1. Lipid class fractions (%) and total lipid content (mg/g) for different live feed types ($n = 2$)

Live feed	Harpacticoid copepods	Calanoid copepods	Rotifer	Rotifer	*Artemia* nauplii	*Artemia* nauplii
Enrichment	Algae	Algae	None	AM2000	None	AM3050
Wax esters/hydrocarbons	3.19	0.61	9.28	8.96	0.56	0.52
Triacylglycerides	18.51	7.57	5.33	35.39	37.49	52.02
Free fatty acids	8.29	9.56	1.43	6.45	0.68	1.70
Sterols	3.90	3.96	4.00	2.59	3.07	2.44
Polar lipid	66.11	78.29	79.58	46.40	58.20	43.33
Total (mg/g)	63.38	80.69	63.80	151.56	69.25	102.81

Enrichment of rotifers and *Artemia* increased the total lipid concentration and resulted in a relative increase in triacylglycerides and free fatty acids and a relative decrease in polar lipid and sterols.

The highest proportion and total amount of DHA was found in enriched rotifers, followed by calanoid and then harpacticoid copepods, and the highest DHA/EPA ratio (7.23:1) was found in enriched rotifers (Table 15.2). The *n*3/*n*6 ratio was highest for copepods, with calanoids (6.18:1) having a ratio twice that of harpacticoids (3.08:1). As a percentage, EPA was relatively high in copepods compared with rotifers and *Artemia*. There was a large difference in the percentage of ARA between the harpacticoid (2.27%) and the calanoid (0.42%) copepods.

Sargent and coworkers (Sargent et al. 1989, 1999) have long argued that the biochemical composition of live feed should reflect that of the endogenous reserves of the larva. Striped trumpeter eggs contain a high polar lipid content (60%; Morehead et al. 2001), suggesting a diet high in polar lipid may have particular benefits to the developing larvae. The benefits of feeding diets high in phospholipid have been shown in several larval fish species (see review by Coutteau et al. 1997). While the exact mechanisms for improved growth and survival with dietary phospholipid remain unclear (e.g., increased digestibility, higher retention of fatty acids, improved lipid transport), it has been proposed that fish larvae probably have an absolute dietary requirement for phospholipid, because they have limited ability to biosynthesize phospholipid de novo (Coutteau et al. 1997; Sargent et al. 1999). Our data show copepods to have a high proportion of phospholipid (~70%) compared with enriched rotifers and *Artemia* (~45%), with the latter containing much higher amounts of triacylglycerides (~40%). The authors also found the calanoid copepods (mainly *Acartia* sp.) reared on cultured microalgae to have substantially higher total lipid content (81 mg/g) than *Acartia tsuensis* (42 mg/g; Toledo et al. 1999) and *Acartia sinjiensis* (18 mg/g; McKinnon et al. 2003). This supports the assertion by McKinnon et al. (2003) that temperate copepod genera appear to have substantially higher lipid content than their tropical counterparts.

While unenriched rotifers also contain a high proportion of phospholipid (~80%), their absolute amounts of *n*-3 PUFA and particularly DHA are generally lower than the dietary requirement for *n*-3 PUFA of most marine fish larvae so far examined (Izquierdo 1996), thus limiting their suitability for striped trumpeter. Striped trumpeter eggs also have a relatively high DHA/EPA ratio (4:1), a high *n*3/*n*6 ratio (9:1), and comprise 2% ARA (Morehead et al. 2001). Our work found enriched rotifers provide the highest DHA/EPA ratio (> 7:1), while the calanoid copepods provided the highest *n*-3/*n*-6 ratio (6:1). Enriched rotifers had the highest amounts of ARA (2.6 mg/g dry matter [DM]) and DHA (25.2 mg/g DM) of all the live feeds examined. Calanoid copepods provided higher amounts of EPA (7.4 mg/g DM) and DHA (9.6 mg/g DM) than enriched *Artemia* nauplii (4.2 and 8.3 mg/g DM, respectively), while the harpacticoid copepods provided more EPA and DHA than unenriched rotifers and *Artemia* nauplii.

Should maximizing the amount of DHA be identified as a key goal in finfish nutrition, initial feeding with rotifers could be followed by

Table 15.2. Proportion of fatty acids for different live-feed types ($n = 2$)

Live feed	Harpacticoid copepods	Calanoid copepods	Rotifer	Rotifer	*Artemia* nauplii	*Artemia* nauplii
Enrichment	Algae	Algae	None	AM2000	None	AM3050
Fatty acid (% of total)						
14:0	0.00	2.29	2.02	7.44	0.34	1.40
16:0	10.39	18.00	15.70	22.73	6.29	11.78
18:0	12.08	4.41	4.68	2.36	2.86	3.38
Total SFA[a]	29.48	28.15	26.37	34.37	10.46	17.62
16:1*n*-7	3.59	1.25	8.46	6.14	2.37	2.90
18:1*n*-9[b]	9.68	10.58	17.06	4.88	26.13	30.38
18:1*n*-7	4.53	1.74	4.26	3.69	4.41	4.87
20:1*n*-9	2.18	0.40	2.28	0.27	0.63	0.70
22:1*n*-11	0.85	0.08	0.61	0.12	0.00	0.24
Total MUFA[c]	25.97	18.08	38.19	16.71	34.90	40.74
18:2*n*-6	4.35	3.30	6.31	2.53	2.48	2.86
20:4*n*-6 (ARA)	2.27	0.42	3.63	2.59	0.83	2.22
20:3*n*-6	0.00	0.05	1.48	0.60	0.09	0.22
20:2*n*-6	0.00	0.04	2.55	1.49	0.07	0.00
22:5*n*-6	2.95	2.39	1.46	8.76	0.00	3.95
Total *n*-6[d]	10.40	7.03	13.96	15.76	3.47	9.25
20:5*n*-3 (EPA)	12.19	16.54	4.41	3.46	2.31	4.82
20:4*n*-3	1.41	0.43	1.27	0.89	0.18	0.49
22:6*n*-3 (DHA)	17.70	21.61	5.23	25.03	0.05	9.53
22:5*n*-3	0.67	0.94	2.12	1.11	0.00	0.26
Total *n*-3	32.12	43.41	13.61	30.78	2.53	15.10
n-3/*n*-6	3.08	6.18	1.07	1.75	1.07	1.70
Total PUFA[e]	44.54	53.77	35.44	48.92	8.32	27.32
TFA (mg/g)	25.04	44.58	24.37	100.74	55.16	87.37
20:4n-6 (ARA) (mg/g)	0.57	0.19	0.88	2.61	0.46	1.94
20:5n-3 (EPA) (mg/g)	3.05	7.37	1.07	3.49	1.27	4.21
22:6n-3 (DHA) (mg/g)	4.43	9.63	1.27	25.22	0.03	8.33
DHA/EPA	1.45	1.31	1.15	7.23	0.02	1.98
DHA/EPA/ARA	8:5:1	51:39:1	1:1:1	14:2:1	0:3:1	2:2:1

Note: SFA, saturated fatty acid; MUFA, monounsaturated fatty acid; PUFA, polyunsaturated fatty acid; TFA, total fatty acids; ARA, arachidonic acid; EPA, eicosapentaenoic acid; DHA, docohexaenoic acid.

[a]Total SFA also includes 15:0, 17:0, 20:0, 22:0, and 24:0.

[b]Includes 18:3*n*-3.

[c]Total MUFA also includes 16:1*n*-9, 16:1*n*-7t, 16:1*n*-5, 17:1, 18:1*n*-7t, 18:1*n*-5, and 20:1*n*-7.

[d]Total *n*-6 also includes 18:3*n*-6 and 22:4*n*-6.

[e]Total PUFA includes C16 PUFA, 18:3*n*-6, C21 PUFA, 22:4*n*-6 and C24 PUFA.

calanoid copepod spp., rather than *Artemia*. Recent work, however, by Bransden et al. (2004) suggests that the DHA requirement for larval striped trumpeter during the rotifer feeding stage is approximately 13 mg/g DM rotifer. Obtaining this level of DHA may require enrichment of calanoid copepods (cf. Table 15.2). This could be done by feeding copepods microalgal species selected and cultured to maximize their DHA content (e.g., Dunstan et al. 1993) or enriching copepods with high DHA microparticulate diets, as are used for rotifers and *Artemia*.

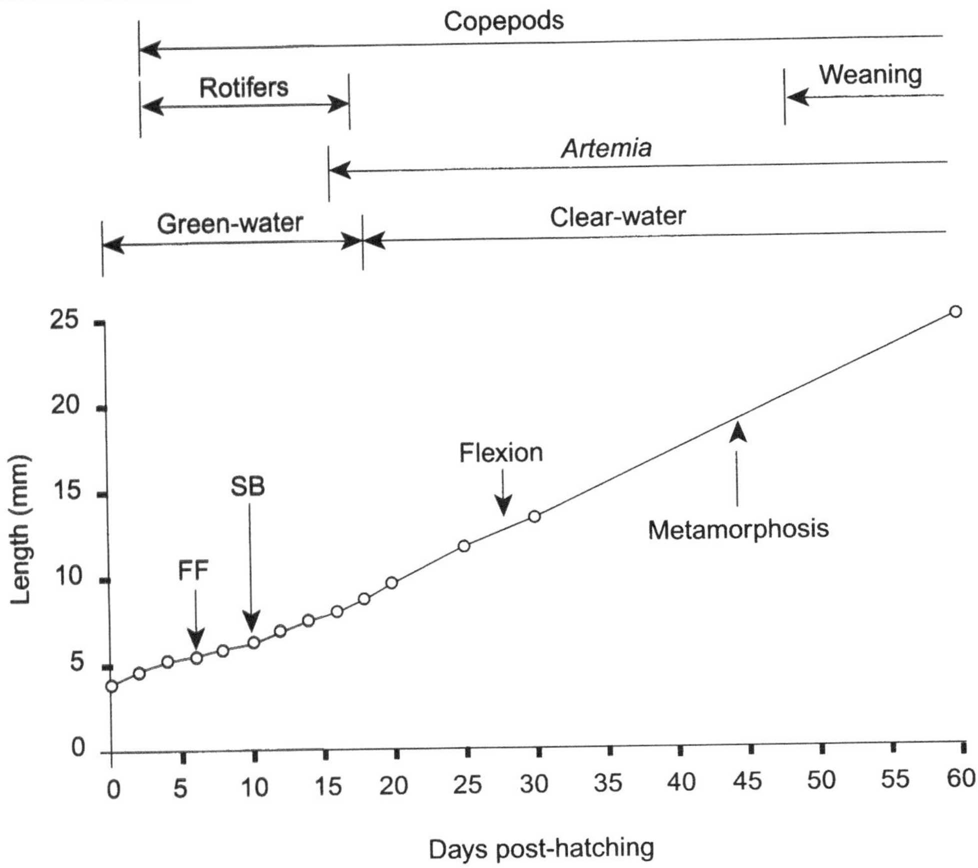

Figure 15.4. Generalized culture technique and development of striped trumpeter larvae using initial green water and supplementary copepod feeding. FF, first feeding; SB, swim bladder inflation.

LARVAL STRIPED TRUMPETER REARING

Attempts to rear striped trumpeter larvae began in 1989, and, to date, only small numbers of larvae have been successfully reared through metamorphosis. Mortality of larvae is often associated with first feeding, swim bladder inflation, and flexion, with continuing mortality occurring up to and through metamorphosis until around 70–80 days after hatching. Mortality of postlarval-stage fish thereafter is low, but rates of malformation in these postlarval-stage fish (especially jaw deformity) can be high.

Rearing Conditions

Striped trumpeter are usually fed rotifers from first feeding (day 6) to around day 20 after hatching (Fig. 15.4). Different rearing techniques have been employed that involve either clear water with flow-through (rotifers added twice daily at 5/mL), pseudo-green water with flow-through at night only (rotifers fed daily at 10/mL), or static green water (rotifers maintained at 10–20/mL). Freshly harvested wild plankton (comprising mostly copepods) and cultured copepods (calanoid and harpacticoid species) have been used as a supplementary feed to rotifers and are usually stocked prior to first feeding, and then added as availability allows. Larval rearing has been conducted in tanks of various sizes (1,000, 2,000, 3,000, and 25,000 L), and *N. oculata* has been used to create pseudo-green-water (daily addition) or static green-water cultures. *Artemia* nauplii (Great Salt Lake, Utah, cysts, Grade A; packaged by INVE Aquaculture) are fed from around day 18 onward, at 0.2–2.0/mL twice daily. Green-water technique, if used, usually ceases around the time of *Artemia* feeding. Weaning to a particulate diet begins soon after larvae have metamorphosed into postlarval-stage fish (day 45–60; Fig 15.4).

Rearing Success

Striped trumpeter larvae have been reared through metamorphosis to day 100 after hatching on 12 occasions at the MRL. The first two successful trials in 1994 and 1997 used static green water and resulted in 67 and 20 postlarval-stage fish being produced, respectively (Table 15.3). In 1999, three cohorts were produced (*n* = 46, 120, and 155), and all initially used a green-water technique and included the use of wild-harvested copepods as a supplementary feed. Up until 2000, clear-water trials had been unsuccessful at producing postlarval-stage fish. In 2001, however, two cohorts of postlarval-stage fish were produced for the first time using clear water and antibiotics (daily 50 ppm oxytetracycline water bath). In 2002, five cohorts of 100-day-old postlarval-stage fish were produced using a variety of culture conditions, including the addition of copepods as a live feed on one occasion.

The use of clear-water culture techniques for rearing striped trumpeter postlarval-stage fish through to day 100 were unsuccessful (0 of 34 trials). Rearing postlarval-stage fish to day 100 was achieved on a number of occasions when initial green-water culture techniques were applied (4 of 19 trials; 21%). The supplementary use of copepods as a live feed source (4 of 10 trials; 40%) or the use of antibiotics (4 of 8 trials; 50%) considerably increased the likelihood of successfully rearing postlarval-stage fish.

Evaluation of Copepods as a Live Feed

Even though there was a lower abundance of copepods in the water, compared with rotifers or *Artemia*, examination of larval striped trumpeter gut contents revealed copepod carapaces on many occasions, suggesting copepods were being eaten and possibly preferentially targeted. The authors found the supplementary use of copepods increased the likelihood of successfully rearing a cohort of striped trumpeter through metamorphosis and through to day 100. Their use, however,

Table 15.3. Number and duration (day of termination after hatching) of striped trumpeter trials conducted from 1994 to 2002 and culture technique employed

	Trial duration (days)									
Trial	1994	1995	1996	1997	1998	1999	2000	2001	2002	Total
1	36	—	45	35	88	29	29(GW)	26(GW)	100(AB)	—
2	100(GW)	—	26	19(GW)	44	28	32(C)	34(GW,C)	100(GW)	—
3	17	—	35	31(GW)	30	28	96	25(GW)	100(GW)	—
4	35	—	7	33(GW)	40	19(GW)	37(GW)	13(GW)	100(GW,C)	—
5	35	—	25	100(GW)	14	100(GW,C)	36(GW,C)	100(AB)	100(GW,AB)	—
6	25	—	28	77(GW)	26	19	23(GW)	37(AB)	25(GW)	—
7	33	—	27	—	20	34(GW,C)	17(GW,C)	21(AB)	29(GW,AB)	—
8	30	—	—	—	20	100(GW,C)	36(GW)	24(AB)	—	—
9	—	—	—	—	14	25	—	100(AB)	—	—
10	—	—	—	—	13	31	—	—	—	—
11	—	—	—	—	48(GW)	100(GW,C)	—	—	—	—
12	—	—	—	—	30(GW)	48(GW,C)	—	—	—	—
13	—	—	—	—	30	—	—	—	—	—
14	—	—	—	—	47	—	—	—	—	—
Total CW	7	0	7	1	12	6	1	0	0	34
Total GW	1	0	0	5	2	1	4	3	3	19
Total C	0	0	0	0	0	5	3	1	1	10
Total AB	0	0	0	0	0	0	0	5	3	8
Total trials	8	0	7	6	14	12	8	9	7	71
Total fish*	67	0	0	20	0	321	0	48	1,101	1,557

Note: Trials without coding are clear water only. CW, clean water; GW, green water only; C, copepods added to tanks; AB, antibiotic treatment.

*Total number of fish produced to day 100.

did not appear to improve growth or reduce the level of malformation in the resulting fish.

CONCLUSION

Calanoid and harpacticoid copepods can be sourced from coastal waters and the hatchery environment and either used immediately or intensively cultured (up to 1/mL) for use as a supplementary feed to larval fish. Problems associated with wild harvesting include reliability of supply, introduction of undesirable organisms, and fluctuating nutritional value of the copepods.

Successful intensive culture of copepods requires provision of the right algal species. It was important for successful culture of *A. tranteri* to use *R. salina*, which increased fecundity, and a mix of *R. salina* and *Isochrysis* sp. to support high population growth.

Many factors impact upon striped trumpeter larval survival. Some environmental factors, such as temperature and photoperiod, can be adequately controlled. Other factors, however, such as the quality of the live feed and the introduction of pathogens, are more difficult to control. As a result, the benefit of providing copepods to striped trumpeter larvae was difficult to assess. The use of copepods, however, did appear to improve larval survival, possibly through improved larval nutrition and the introduction of a better microbial flora.

REFERENCES

Battaglene, S.C., Morehead, D.M., Nichols, P.D., Brown, M., and Carson, J. 2002. Why are striped trumpeter (*Latris Lineata*) larvae so difficult to culture—is it lipid nutrition? In *26th Annual Larval Fish Conference American Fisheries Society*, edited by H. Browman, p. 33. Bergen, Norway, 22–26 July 2002. Bergen, Norway: Early Life History Section of the American Fisheries Society.

Bell, J.G. 1998. Current aspects of lipid nutrition in fish farming. In *Biology of Farmed Fish*, edited by K. Black and A.D. Pickering, pp. 114–145. Sheffield, U.K.: Sheffield Academic Press.

Bligh, E.G. and Dyer, W.G. 1959. A rapid method of total lipid extraction and purification. *Canadian Journal of Biochemistry and Physiology* 37:911–917.

Bransden, M.P., Dunstan, G.A., Battaglene, S.C., Cobcroft, J.M., Morehead, D.T., Kolkovski, S., and Nichols, P.D. 2004. Influences of dietary n-3 long chain PUFA on body concentrations of 20:5n-3, 22:5n-3 and 22:6n-3 in larvae of a marine teleost fish from Australian waters, the striped trumpeter (*Latris lineata*). *Lipids* 39:215–222.

Cobcroft, J.M., Pankhurst, P.M., Sadler, J., and Hart, P.R. 2001. Jaw development and malformation in cultured striped trumpeter *Latris lineata*. *Aquaculture* 199(3-4):267–1197.

Coutteau, P., Geurden, I., Camara, M.R., Bergot, P., and Sorgeloos, P. 1997. Review on the dietary effects of phospholipids in fish and crustacean larviculture. *Aquaculture* 155:149–164.

Divanach, P., Boglione, C., Menu, B., Koumoundouros, G., Kentouri, M., and Cataudella, S. 1996. Abnormalities in finfish mariculture: an overview of the problem, causes and solutions. In *Handbook of Contributions Presented at the International Workshop on Seabass and Seabream Culture: Problems and Prospects*, pp. 45–66. Verona, Italy, October 16–18. Oostende, Belgium: European Aquaculture Society.

Donaghay, P.L. 1985. An experimental test of the relative significance of food quality and past feeding history to limitation of egg production of the estuarine copepod *Acartia tonsa*. *Archiv fur Hydrobiologie Beihtrage Ergebnisse Limnologie* 21:235–245.

Dunstan G.A., Volkman J.K., Barrett S.M., and Garland C.D. 1993. Changes in the lipid composition and maximisation of the polyunsaturated fatty acid content of three microalgae grown in mass culture. *Journal of Applied Phycology* 5:71–83.

Furlani, D.M. and Ruwald, F.P. 1999. Egg and larval development of laboratory reared striped trumpeter *Latris lineata* (Schneider 1801) (Percoidei: Latrididae) from Tasmanian waters. *New Zealand Journal of Marine and Freshwater Research* 33:153–162.

Hunter, J.R. 1981. Feeding ecology and predation of marine fish larvae. In *Marine Fish Larvae*, edited by R. Lasker, pp. 33–77. Seattle, Washington, USA: University of Washington Sea Grant.

Izquierdo, M.S. 1996. Essential fatty acid requirements of cultured marine fish larvae. *Aquaculture Nutrition* 2:183–191.

Jordon, A. 2000. Fisheries biology and assessment of striped trumpeter. In *Fishing Today: Tasmanian Fishing Industry News* 13(3):19–20.

Kimmerer, W.J. and McKinnon, A.D. 1987. Growth, mortality, and secondary production of the copepod *Acartia tranteri* in Westernport Bay, Australia. *Limnology and Oceanography* 32(1):14–28.

Kleppel, G.S. 1993. On the diets of calanoid copepods. *Marine Ecology Progress Series* 99:183–195.

Koven, W., Barr, Y., Lutzky, S., Ben-Atia, I., Weiss, R., Harel, M., Behrens, P., and Tandler, A. 2001. The effect of dietary arachidonic acid (20:4n-6) on growth, survival and resistance to handling stress in gilthead seabream (*Sparus aurata*) larvae. *Aquaculture* 193:107–122.

Kraul, S. 1991. Hatchery methods for Mahimahi, *Coryphaena hippurus*, at Waikiki Aquarium. In *CRC Handbook of Mariculture: Volume II, Finfish Aquaculture*, edited by J.P. McVey, pp. 241–250. Boca Raton, Florida, USA: CRC Press.

Kuhlmann, D., Quantz, G., and Witt, U. 1981. Rearing of turbot larvae (*Scophthalmus maximus* L.) on cultured food organisms and postmetamorphosis growth on natural and artificial food. *Aquaculture* 23:183–196.

Mauchline, J. 1999. The biology of calanoid copepods. *Advances in Marine Biology* 33:1–720.

McKinnon, A.D., Duggan, S., Nichols, P.D., Rimmer, M.A., Semmens, G., and Robino, B. 2003. The potential of tropical paracalanid copepods as live feeds in aquaculture. *Aquaculture* 223:89–106.

Morehead, D.T., Hart, P.R., Dunstan, G.A., Brown, M., and Pankhurst, N.W. 2001. Differences in egg quality between wild striped trumpeter (*Latris lineata*) and captive striped trumpeter that were fed different diets. *Aquaculture* 192:39–53.

Ohno, A., Takahashi, T., and Taki, Y. 1990. Dynamics of exploited populations of the calanoid copepod, *Acartia tsuensis*. *Aquaculture* 84:27–39.

Payne, M.F. and Rippingale, R.J. 2000. Evaluation of diets for culture of the calanoid copepod *Galdioferens imparipes*. *Aquaculture* 187:85–96.

Payne, M.F. and Rippingale, R.J. 2001. Intensive cultivation of the calanoid copepod *Gladioferens imparipes*. *Aquaculture* 201(3-4):329–342.

Pittman, K. 1996. Rearing halibut in Norway: present practices and challenges. *Marine and Freshwater Research* 47:243–249.

Rey, C., Harris, R., Irigoien, X., Head, R., and Carlotti, F. 2001. Influence of algal diet on growth and ingestion of *Calanus helgolandicus* nauplii. *Marine Ecology Progress Series* 215:151–165.

Sargent, J., Henderson, R.J., and Tocher, D.R. 1989. The lipids. In *Fish Nutrition*, edited by J.E. Halver, pp. 153–218. San Diego, California, USA: Academic Press.

Sargent, J., McEvoy, L., Estevez, A., Bell, G., Bell, M., Henderson, J., and Tocher, D. 1999. Lipid nutrition of marine fish during early development: current status and future directions. *Aquaculture* 179:217–229.

Sargent, J.R., Tocher, D.R., and Bell, J.G. 2002. The lipids. In *Fish Nutrition*, edited by J.E. Halver and R.W. Hardy, pp. 181–257. San Diego, California, USA: Academic Press.

Schipp, G.R., Bosmans, J.M.P., and Marshall, A.J. 1999. A method for hatchery culture of tropical calanoid copepods, *Acartia* spp. *Aquaculture* 174:81–88.

Støttrup, J.G. 2000. The elusive copepods: their production and suitability in marine aquaculture. *Aquaculture Research* 31:703–711.

Støttrup, J.G., Bell, J.G., and Sargent, J.R. 1999. The fate of lipids during development and cold storage of eggs in the laboratory-reared calanoid copepod *Acartia tonsa* Dana, and in response to different algal diets. *Aquaculture* 176:257–269.

Støttrup, J.G., Richardson, K., Kirkegaard, E., and Pihl, N.J. 1986. The cultivation of *Acartia tonsa* Dana for use as a live food source for marine fish larvae. *Aquaculture* 52:87–96.

Thomas, C.W., Wikely, D.M., and Searle, L.D. 1992. Harvesting and monitoring zooplankton from Orielton Lagoon. Investigating the potential use of wild harvested zooplankton for the culture of native marine finfish. In *Marine Farming Information Series*. Publication No. 7, Department of Primary Industry Tasmania, Australia. 21 pp.

Toledo, J.D., Golez, M.S., Doi, M., and Ohno, A. 1999. Use of copepod nauplii during early feeding stage of grouper *Epinephelus coioides*. *Fisheries Science* 65:390–397.

Trotter, A.J., Pankhurst, P.M., and Hart, P.R. 2001. Swim bladder malformation in hatchery-reared striped trumpeter *Latris lineata* (Latridae). *Aquaculture* 198:41–54.

Trotter, A.J., Pankhurst, P.M., Morehead, D.T., and Battaglene, S.C. 2003. Effects of temperature on initial swim bladder inflation and related development in cultured striped trumpeter (*Latris lineata*) larvae. *Aquaculture* 221:141–156.

Tucker, J.W. 1998. *Marine Fish Culture*. Norwell, Massachusetts, USA: Kluwer Academic Publishers. 750 pp.

van der Meeren, T. and Naas, K.E. 1997. Development of rearing techniques using large enclosed ecosystems in the mass production of marine fish fry. *Reviews in Fisheries Science* 5:367–390.

Volkman, J.K. and Nichols, P.D. 1991. Application of thin layer chromatography-flame ionisation detection to the analysis of lipids and pollutants in marine and environmental samples. *Journal of Planar Chromatography* 4:19–26.

Zurlini, G., Ferrara, I., and Nassogne, A. 1978. Reproduction and growth of *Euterpina acutifrons* (Copepoda: Harpacticoida) under experimental conditions. *Marine Biology* 46:59–64.

16
Intensive Cultivation of a Subtropical Paracalanid Copepod, *Parvocalanus* sp., as Prey for Small Marine Fish Larvae

Robin J. Shields, Tomonari Kotani, Augustin Molnar, Kimo Marion, Jon Kobashigawa, and Larren Tang

ABSTRACT

A small paracalanid copepod, *Parvocalanus* sp., was isolated from Hawaiian coastal waters and subjected to a series of laboratory experiments to ascertain the effects of different combinations of microalgae (*Chaetoceros* sp. and *Isochrysis* sp.) on copepod survival, growth, and fecundity. Adult copepods exhibited the highest survival when fed *Chaetoceros* sp., whereas fecundity was highest (up to 21 offspring per adult per day) when *Chaetoceros* sp. was offered in combination with *Isochrysis* sp. The viability of nauplii produced by adult copepods fed *Chaetoceros* sp. was lower than from adults fed *Isochrysis* sp., in terms of survival and growth. *Chaetoceros* sp. was also an inferior diet for nauplii. Copepodid stages of *Parvocalanus* sp. exhibited similar survival on all diet combinations but had the greatest growth when fed *Isochrysis* sp. Large quantities of *Parvocalanus* sp. were also cultured routinely in 400-L tanks using a mixed diet of *Chaetoceros* sp. and *Isochrysis* sp. Copepod population densities fluctuated cyclically in these cultures, with nauplius densities ranging from approximately 1 to 30/mL. Up to 49 million *Parvocalanus* sp. nauplii were harvested per 400-L tank during preliminary 30-day trials to assess nauplius harvesting techniques. Red snapper, *Lutjanus campechanus,* larvae offered *Parvocalanus* sp. nauplii exhibited significantly greater survival to day 7 after hatching (50.3%) compared with larvae fed ss-type rotifers, *Brachionus plicatilis* (2.6%), and were significantly larger in size, confirming the efficacy of *Parvocalanus* sp. nauplii as a first feed for small subtropical marine fish larvae.

INTRODUCTION

The high nutritional value of copepods for marine fish larvae has been verified both in terms of copepod biochemical composition (Evjemo and Olsen 1997; McKinnon et al. 2003) and efficacy as larval feeds. Several harpacticoid copepod species have been evaluated experimentally for rearing larvae of European turbot, *Scopthalmus maximus* (Støttrup and Norsker 1997); Dover sole, *Solea solea* (Heath and Moore 1997); and greenback flounder, *Rhombosolea tapirina* (Marshall and Purser 1999). Although able to be cultured at higher densities than the calanoid or cyclopoid copepods (Støttrup and Norsker 1997), the harpacticoid copepods have yet to be adopted for mass larviculture, perhaps because their epibenthic grazing habit restricts prey availability to pelagic marine fish larvae.

Calanoid copepods of the genus *Acartia* have been found to be effective diets for larvae of *S. maximus* (Støttrup et al. 1986; Urup 1994) and several lutjanid snapper species (Doi et al. 1997a;

Schipp et al. 2001), while extensively cultured copepods (mixed calanoid species) have been successfully used to rear larvae of red snapper, *Lutjanus campechanus* (Ogle et al. 2000) and red-spotted grouper, *Epinephelus coiodes* (Doi et al. 1997b; Toledo et al. 1999).

Having identified the high dietary value of marine calanoid copepods, the challenge for researchers and aquaculturists has been to identify suitable species and methods for mass culture. Commercial production of juvenile tropical marine fish currently relies mainly on extensive rather than intensive production of copepods, with large-scale zooplankton production in ponds being widespread in Southeast Asia (for example, Liao et al. 2001). The extensive culture approach has the advantage of requiring only simple rearing facilities and of offering species and size diversity to meet the changing developmental needs of the fish larvae. These advantages are especially useful in locations lacking technical resources or for new fish species for which exact dietary requirements are unknown (Dhert et al. 1997). Copepod productivity in extensive systems, however, tends to be highly variable, due to lack of control over plankton abundance/species composition or physical environmental conditions. The open nature of the rearing system also offers a pathway for introduction of fish pathogens, for example, viral nervous necrosis (Liao et al. 2001).

Indoor intensive culture systems offer an alternative means of producing large quantities of marine calanoid copepods. Intensive culture systems use less water volume than extensive mesocosms and can operate independently of outdoor environmental conditions, although they require close control over dietary input and water quality (Støttrup 2000). Støttrup et al. (1986) successfully reared the temperate calanoid, *Acartia tonsa,* for multiple generations in 250- to 400-L volume rearing tanks. Adult *A. tonsa* densities in this system, however, were limited to just 50–100/L.

Based on these and similar findings, it was initially questioned whether calanoid copepods can attain sufficient population densities to allow intensive culture. This assumption has been revised following the discovery that much higher culture densities (up to 5,000 copepods/L) and nauplius yields can be obtained for warm-water calanoid copepods, including *Acartia* sp. (Schipp et al. 1999), *Gladioferens imparipes* (Payne and Rippingale 2001), and *Oithona* sp. (Lipman et al. 2001). McKinnon et al. (2003) have more recently recommended the use of tropical paracalanids (*Bestiolina similis, Parvocalanus crassirostris*) as larval feeds, based on their essential fatty acid composition, small size, susceptibility of young stages to predation, and absence of cannibalistic behavior by adult stages.

The current study on the culture of *Parvocalanus* sp. stems from this research group's interest in rearing subtropical marine fish larvae that require very small prey during first feeding. Findings are presented on the dietary requirements and culture attributes of *Parvocalanus* sp. and the efficacy as *Parvocalanus* sp. nauplii as prey for small marine fish larvae.

MATERIALS AND METHODS

All studies were carried out at The Oceanic Institute (Oahu, HI, USA). *Parvocalanus* sp. stocks were isolated during April 2001 from zooplankton samples collected in Kaneohe Bay using a 50-cm diameter plankton net with 120-µm mesh aperture, by pipetting individual adult copepods into 250-mL Erlenmeyer flasks containing clean seawater. The seawater used for all studies was supplied from a lava rock well at a salinity of 34 ppt. Microalgae, sourced from the Provasoli-Guillard National Center for Culture of Marine Plankton, was cultured indoors in 500-mL flasks, 15-L polyethylene carboys, and 150-L plastic cylinders using the batch-culture technique described by Sato (1991), with F/2 nutrient media (Kent Pro Culture Professional F/2 Algal Culture Formula; Kent Marine Inc., Acworth GA, USA; or Cellpharm Cell-hi F2P; Cellpharm Ltd., Worcestershire, UK). Microalgae cell densities were calculated by direct observation of samples in a Neubauer hemocytometer, using a Carl Zeiss ZE155 compound microscope. Routine copepod counts were performed on Lugol's iodine-fixed specimens at ×40 magnification, using an Olympus SZX-12 stereo-zoom fitted with a ×1 objective lens and ×10 eyepieces. Water temperature, pH, and oxygen concentration were measured routinely using a WTW 340i multiparameter handheld meter (WTW Measurement Systems Inc., Fort Myers, FL, USA).

Maintenance of Stock and Intermediate Copepod Cultures

The original *Parvocalanus* sp. populations in 250-mL flasks were expanded and transferred into 30-L and 100-L transparent acrylic tanks. These tanks were located in separate, continuously lit rooms at a temperature of 25–27°C. At start-up, each container was filled to 60% volume with 5-µm filtered, ultraviolet light-sterilized seawater. Gentle aeration was provided to the 30-L containers via an open glass capillary tube, while each 100-L container was aerated using a single 3.8 cm^2 air stone. Continuous lighting was provided by a combination of fluorescent and tungsten lamps, providing 500-lux light intensity at the water surface for the 30-L containers and 1,000 lux for the 100-L containers. Copepods (mixed size classes) were stocked into each container at a density of at least 1/mL. *Chaetoceros* sp. and *Isochrysis* sp. (Tahitian strain, CCMP 1324) microalgae were then added to provide a cell density of 150,000/mL for each species. The same quantity of microalgae was subsequently added once per day until the culture container was full (5–7 days), at which time all of the copepods were collected on a 28-µm nylon screen, rinsed, and stocked into a clean container.

Stage-Specific Effects of Different Microalgae Combinations

Initial trials were carried out to evaluate different microalgae species (*Isochrysis* sp., *Chaetoceros* sp., *Nannochloropsis* sp., and *Tetraselmis* sp.) for culturing *Parvocalanus* sp. (data not shown). Based on comparative measurements of copepod survival, growth and fecundity, *Chaetoceros* sp. and *Isochrysis* sp. were selected for optimization of feeding regimen. A series of replicated experiments was conducted to define the copepod's stage-specific microalgae requirements. These experiments were carried out using 1-L glass beakers located in a continuously lit, air-conditioned room at a temperature of 26.0 ± 1.0°C. Copepod dimensions (prosome width and length) were calculated from images recorded with an Olympus Camedia C-2000Z digital camera attached to an Olympus BX51 compound microscope (for nauplii), or an Olympus SZX-12 stereo-zoom microscope (for copepodids and adults). The images were analyzed using Sigmascan Pro software (SPSS Inc., Chicago, IL, USA).

Diet Experiment I

Adult *Parvocalanus* sp. were offered *Chaetoceros* sp. and *Isochrysis* sp. alone or in combination, as follows:

Group 1: *Chaetoceros* sp. (300,000 cells/mL)
Group 2: *Isochrysis* sp. (500,000 cells/mL)
Group 3: *Chaetoceros* sp. (150,000 cells/mL) and *Isochrysis* sp. (250,000 cells/mL)

Fifty adult copepods were individually stocked into each of nine beakers containing 500 mL of seawater. Three beakers were assigned at random to each diet group and the appropriate quantity of microalgae added. Total numbers of adult copepods, eggs, and nauplii were quantified daily over the 4-day experimental period by pouring the entire contents of each beaker onto two nylon screens with 70-µm and 20-µm aperture sizes, to retain adult copepods and eggs/nauplii, respectively. The adult copepods from each beaker were rinsed into a 4.5-cm diameter Petri dish, counted, returned live to their beaker (which had been cleaned and refilled), and then fed. Copepod eggs and nauplii were fixed using Lugol's iodine, counted, and discarded.

Diet Experiment II

Diet experiment II investigated the effect of adult copepod feeding history on the viability of *Parvocalanus* sp. nauplii receiving different diets. Adult *Parvocalanus* sp. were fed either *Chaetoceros* sp. (at a density of 300,000 cells/mL), or *Chaetoceros* sp. and *Isochrysis* sp. (100,000 cells/mL and 200,000 cells/mL, respectively). From each adult diet group, 1,000 nauplii were stocked into each of six glass beakers containing 1 L of seawater. The beakers were then selected at random to receive either *Chaetoceros* sp. or *Isochrysis* sp. (three beakers per group), and the required quantity of microalgae was added to each to make the following diet groups:

Group 1: Adults fed *Chaetoceros* sp.
Group 1a: Nauplii fed *Chaetoceros* sp. (200,000 cells/mL)
Group 1b: Nauplii fed *Isochrysis* sp. (400,000 cells/mL)

Group 2: Adults fed *Chaetoceros* sp. and *Isochrysis* sp.
Group 2a: Nauplii fed *Chaetoceros* sp. (200,000 cells/mL)
Group 2b: Nauplii fed *Isochrysis* sp. (400,000 cells/mL)

Copepod population density was checked daily in all beakers over the 6-day experimental period by volumetric sampling. Ten nauplii per beaker were also sampled each day for length and width measurement and to assess developmental stage.

Diet Experiment III

Diet experiment III tested the effects of different microalgae combinations on copepodid survival and growth rates. Copepodids for this experiment (mixed developmental stages) were obtained from a *Parvocalanus* sp. stock fed a mixture of *Chaetoceros* sp. and *Isochrysis* sp. (100,000 cells/mL and 200,000 cells/mL, respectively). One thousand copepodids were stocked into each of nine 1-L glass beakers and assigned at random to the following diet groups (three beakers per group):

Group 1: *Chaetoceros* sp. (200,000 cells/mL)
Group 2: *Isochrysis* sp. (400,000 cells/mL)
Group 3: *Chaetoceros* sp. (100,000 cells/mL) and *Isochrysis* sp. (200,000 cells/mL).

Copepodid density was checked daily in all beakers over the 6-day experimental period by volumetric sampling. Ten copepodids per beaker were also measured for prosome length and width each day.

Pilot-Scale Copepod Culture

Pilot scale culture of *Parvocalanus* sp. was carried out using a system of eight semisquare polyethylene tanks (106 × 106 × 70 cm; Polytank Inc., Litchfield, MN, USA) operated at a volume of 400 L. Preliminary trials were conducted with mixed copepod size classes to ascertain long-term culture stability, after which a two-phase production cycle was tested to determine harvestable quantities of copepod nauplii.

In all cases, the culture tanks were filled with 5 µm of filtered seawater and gently aerated using four evenly spaced 3.8 cm^2 air stones. Continuous illumination was provided by ceiling-mounted fluorescent fixtures, supplemented by a 60-W tungsten lamp clamped to the side of each tank. Mean light intensity at the water surface was 490 lux. Water temperature was maintained in the range of 26–28°C using a single 300-W immersion heater per tank. Microalgae was added to the culture tanks once per day. Copepod abundance was calculated daily by collecting five 100-mL water samples per tank, pooling the water samples, concentrating them on a 28-µm nylon screen, fixing the collected specimens with Lugol's iodine, then staging and enumerating the copepods using a stereozoom microscope (minimum two counts per tank).

During preliminary trials, mixed size classes of *Parvocalanus* sp. were stocked into the tanks at a density of 3–8/mL. A 5-day culture cycle was applied, at the end of which the entire copepod population was collected on a 28-µm nylon screen, rinsed, and restocked into a clean container. Trials included testing the effects of different microalgae combinations on copepod population density:

Chaetoceros sp.: 300,000 cells/mL (2 tanks)
Chaetoceros sp.: 200,000 cells/mL (2 tanks)
Isochrysis sp.: 100,000 cells/mL (2 tanks)

Following this initial experimentation, a standard feeding regimen was adopted consisting of once daily feeding with a mixture of *Chaetoceros* sp. and *Isochrysis* sp. (150,000 cells/mL for each microalgae species).

While *Parvocalanus* sp. nauplii could be intermittently harvested from these mixed copepod populations, optimal nauplius harvest frequencies and sustainable harvest quantities were not known using this culture method. A two-phase production cycle was therefore implemented in the 400-L culture tanks to enable complete harvesting of the required copepod size class as prey for fish larvae. Tanks were initially stocked with a single cohort of *Parvocalanus* sp. nauplii, then fed daily and harvested and restocked at 5-day intervals until the copepods reached maturity. Once the tanks began generating copepod nauplii, these were harvested at approximately 3-day intervals as prey for fish larvae. During harvests, the nauplii were separated from copepodids and adults by rinsing through a 70-µm screen onto a 28-µm nylon screen. Those copepods retained on the 70-µm screen were then restocked into a clean tank, and the sub-70 µm fraction offered to fish larvae.

Effects of Prey Type on Survival and Growth of Red Snapper Larvae

An experiment was carried out to compare the efficacy of *Parvocalanus* sp. nauplii versus ss-type rotifers, *Brachionus rotundiformis* (Hagiwara et al. 1995) as prey for first-feeding red snapper larvae. Larvae were reared in 1-L glass beakers located in a temperature-controlled water bath maintained at 26 ± 0.5°C. The glass beakers were acid washed, then filled with "conditioned" seawater from a 300-L reservoir tank containing polyethylene media ("Bio Ball"; Aquatic Ecosystems Inc., Apopka FL, USA), through which water was exchanged at a rate of approximately 100% of tank volume per day. The beakers were kept static, without aeration, for the 8-day duration of each experiment. A 12-hour light/12-hour dark photoperiod was provided by means of overhead fluorescent lighting, providing approximately 800-lux illumination at the water surface.

On day 0, 18 glass beakers were filled with 1-L conditioned seawater and inoculated with *Isochrysis* sp. at a concentration of 100,000 cells/mL. Each beaker was then stocked with 50 red snapper embryos (18 hours after fertilization) and assigned at random to the following three diet groups (six beakers per group):

Group 1: *Isochrysis* sp. (negative control)
Group 2: *Isochrysis* sp. and *Parvocalanus* sp. nauplii (5/mL)
Group 3: *Isochrysis* sp. and ss-type rotifers (5/mL)

Mean lorica dimensions of the ss rotifers used in the experiment were 143.4 μm × 96.4 μm (from A. Hagiwara, Nagasaki University). All rotifers were enriched for 12 hours using Algamac 2000 (Aquafauna Biomarine Inc., Hawthorne, CA, USA). This enrichment product was applied at a concentration of 300 mg per million rotifers, for a rotifer population density of 250/mL.

Two beakers from each group were classified as "sentinels" and were sampled daily to measure physical water parameters (temperature, pH, and dissolved oxygen concentration) and residual microalgae and zooplankton densities. Total ammonia nitrogen, nitrite, and nitrate concentrations were additionally measured at the end of the experiment on day 7 after hatching. Zooplankton rations for the remaining four beakers in groups 2 and 3 were calculated on the basis of mean residual values from their respective sentinels. This monitoring approach was adopted due to previous observations that physical disturbance to the 1-L rearing containers is detrimental to larvae.

Hatch rate was estimated for all beakers on day 1 after hatching by counting the number of hatched larvae and unhatched embryos *in situ*. *Parvocalanus* sp. nauplii and rotifers were introduced to larvae in groups 2 and 3 on day 2 after hatching, at a density of 5/mL. Further prey was subsequently added once per day when required to regain a copepod/rotifer density of 5/mL. One red snapper larva per beaker was sampled for size measurement on day 1 after hatching and 5 larvae per beaker were sampled on days 3 and 5 after hatching for size measurement and gut contents analysis (excluding sentinel beakers). Numbers of larvae were estimated daily by *in situ* observations, and total numbers of surviving larvae were counted at the end of the experiment on day 7 after hatching. Up to 15 larvae were sampled for size measurement per beaker, excluding sentinels. Larvae size (myotome height) was calculated from images recorded with an Olympus Camedia C-2000Z digital camera attached to an Olympus SZX-12 stereo-zoom microscope. Images were analyzed using Sigmascan software (SPSS Inc.).

Data Analysis

Repeated measures analysis of variance (ANOVA) was used to determine the effects of diet on copepod survival, growth, and fecundity with time and the effects of diet on red snapper larvae survival and growth with time. All percentage data were arc-sine–transformed before analysis. ANOVAs were carried out using the General Liner Model procedure on SPSS version 11.0 (SPSS Inc.). A significance level of $P < 0.05$ was applied throughout.

RESULTS

Stage-Specific Effects of Different Microalgae Combinations

In diet experiment I, percentage of survival declined significantly over the 4-day experimental period. Adult copepods from group 1 (fed *Chaetoceros* sp.) displayed the highest mean percent-

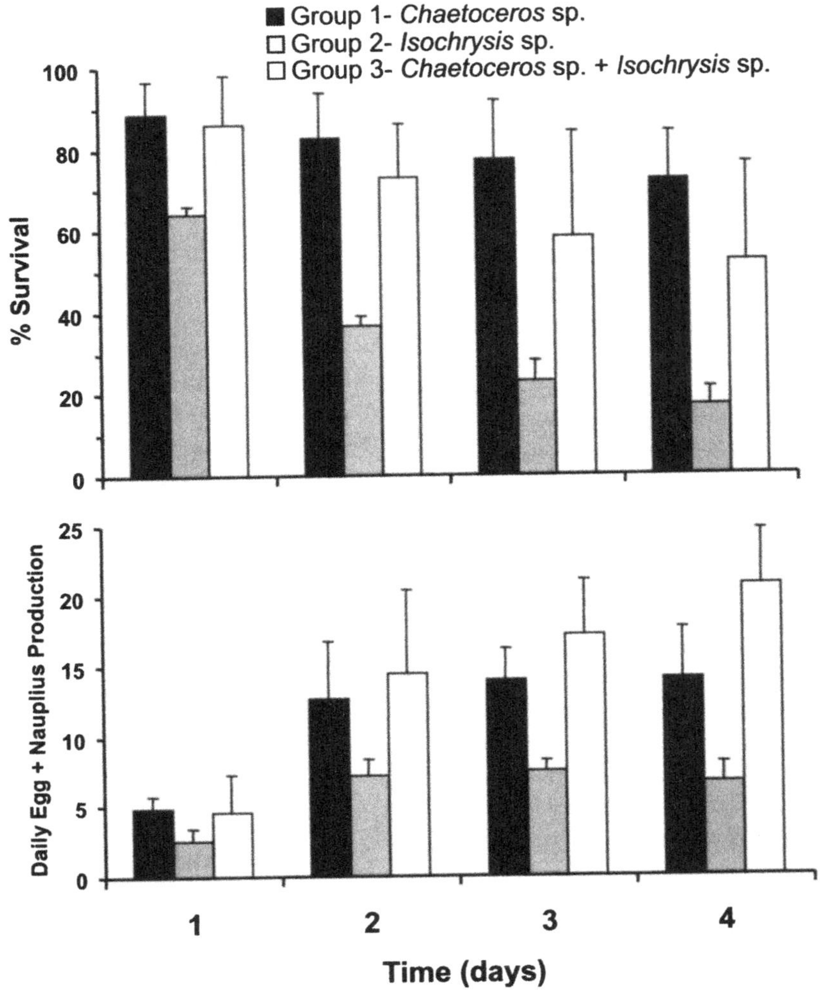

Figure 16.1. Mean percentage of survival (top) and fecundity (bottom) of adult *Parvocalanus* sp. fed *Chaetoceros* sp. or *Isochrysis* sp. singly or in combination (diet experiment I). Values are mean ± S.D. of three replicates.

age of survival to day 4 (Fig. 16.1A), while those receiving only *Isochrysis* sp. (group 2) exhibited significantly lower survival than either of the other two groups. Fecundity was also significantly lower among copepods from group 2 (Fig. 16.1B). Mean fecundity was highest among copepods fed a combination of *Isochrysis* sp. and *Chaetoceros* sp. from day 2 onward (group 3), and increased progressively throughout the experiment, to a maximum of 20 offspring per adult per day (Fig. 16.1B).

In diet experiment II, adult copepod diet and nauplius diet displayed significant independent and interactive effects on the survival and growth of nauplii. One day after stocking, nauplii from group 1 (*Chaetoceros* sp. parental diet) suffered approximately 40% greater mortality than those from group 2 (*Chaetoceros* sp. + *Isochrysis* sp. parental diet), irrespective of the diet provided to the nauplii (Fig. 16.2). A similar mortality differential was observed 2 days after stocking. Thereafter, survival within groups 1 and 2 were strongly affected by nauplius diet, with *Isochrysis* sp.–fed nauplii (groups 1b and 2b) exhibiting higher survival than those fed *Chaetoceros* sp. (Fig. 16.2). The interactions between parental and offspring nutrition were further illustrated in terms of copepod growth (Fig. 16.3A and Fig.

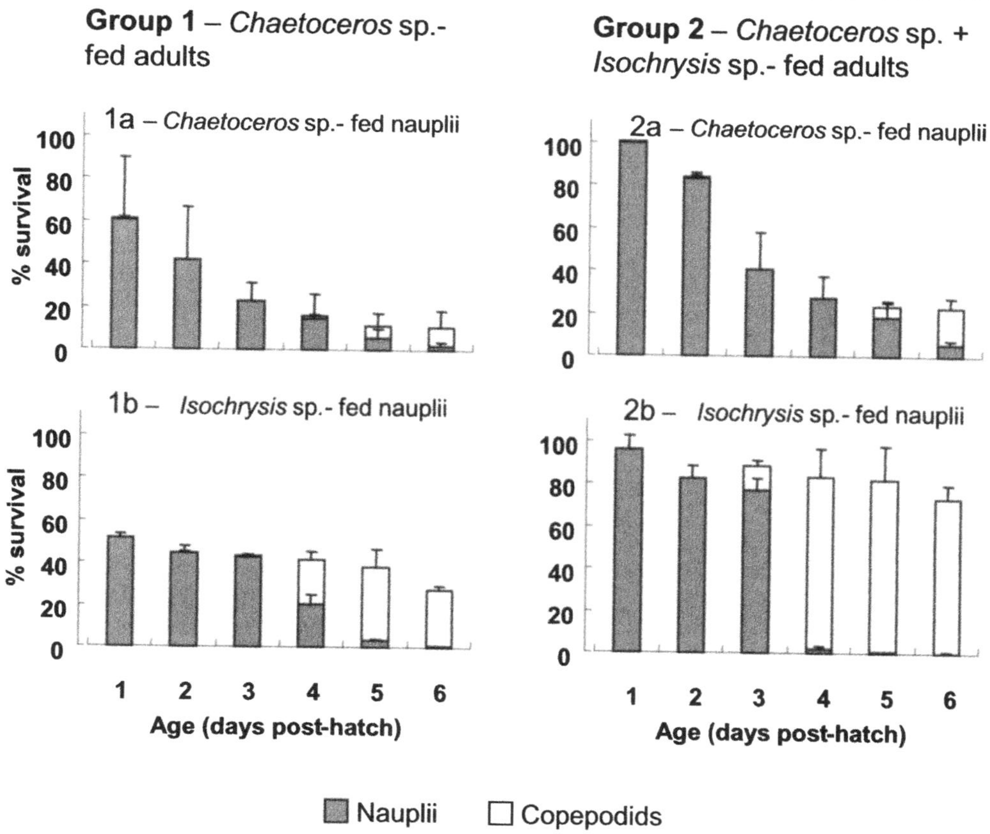

Figure 16.2. Combined effects of parental and offspring diet on mean percentage of survival and development from nauplius to copepodid of *Parvocalanus* sp. (diet experiment II). Values are mean ± S.D. of three replicates.

16.3B). *Chaetoceros* sp.–fed offspring from group 1 exhibited significantly slower growth than those fed *Isochrysis* sp., whereas microalgae species had comparatively little effect on growth rate of offspring from group 2.

In diet experiment III, copepodids fed only *Isochrysis* sp. (group 2) exhibited the highest mean percentage of survival throughout the 6-day experiment, while those fed *Chaetoceros* sp. in combination with *Isochrysis* sp. (group 3) experienced the lowest survival (Fig 16.4A). Survival differences among groups were not statistically significant, but *Chaetoceros* sp.–fed copepodids (group 1) grew significantly slower than those in the other two diet groups (Fig. 16.4B).

Parvocalanus sp. Population Dynamics in Pilot Scale Cultures

Parvocalanus sp. populations (mixed size classes) were successfully maintained in 400-L culture tanks when fed only *Chaetoceros* sp or a combination of *Chaetoceros* sp. and *Isochrysis* sp. (Table 16.1). As a mixed diet of *Chaetoceros* sp. and *Isochrysis* sp. produced the highest mean densities of *Parvocalanus* sp. nauplii, copepodids, and adults, this feeding regimen was adopted as standard.

Figure 16.5 summarizes *Parvocalanus* sp. population densities and quantities of nauplii harvested from two culture tanks operated using a two-phase production cycle. Both tanks were inoculated with copepod nauplii at a density of 5/mL. Adult copepods were first observed 2 days after stocking, and nauplius production was underway by 4 days after stocking. Nauplius population density had risen significantly by day 7, after which time harvesting started. Total copepod numbers, population size distribution, and numbers of nauplii harvested fluctuated cyclically during the nauplius production phase. Ten har-

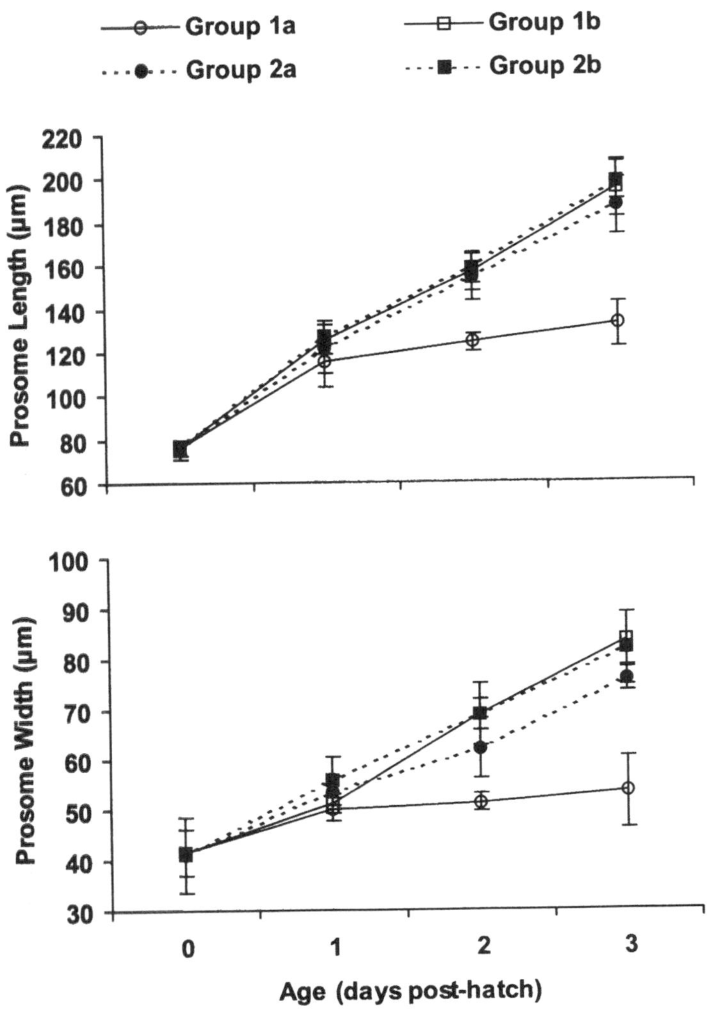

Figure 16.3. Combined effects of parental and offspring diet on growth of *Parvocalanus* sp. (diet experiment II). Prosome length and width values are mean ± S.D. of five copepods per tank.

vests were collected from tank 1 and 11 harvests from tank 2 over the 29–30 day nauplius production phase, producing a total of 40.2 million and 49.0 million nauplii, respectively.

Effects of Prey Type on Survival and Growth of Red Snapper Larvae

Mean hatch rate of red snapper larvae in 1-L glass beakers was 60.0% ± 2.1%. Cumulative survival after hatching declined significantly with time through the experiment (Fig. 16.6A). Larvae fed *Isochrysis* sp. and *Parvocalanus* sp. nauplii (group 2) exhibited significantly higher survival over the 7-day experimental period than those receiving either *Isochrysis* sp. alone (group 1) or *Isochrysis* sp. in combination with ss-type rotifers (group 3; Fig. 16.6A). For group 2, the mean survival to day 7 after hatching was 50.3% of larvae hatched, versus 2.6% for group 3, and 0% for group 1.

Larvae size exhibited a similar response to survival data with regard to diet type (Fig. 16.6B). Snapper larvae from group 2 (receiving *Isochrysis* sp. and *Parvocalanus* sp. nauplii) increased significantly in size during the experiment and were significantly larger than those in the other two diet groups from day 5 after hatching onwards. Larvae fed *Isochrysis* sp. and ss-type rotifers decreased in size during the experiment, but

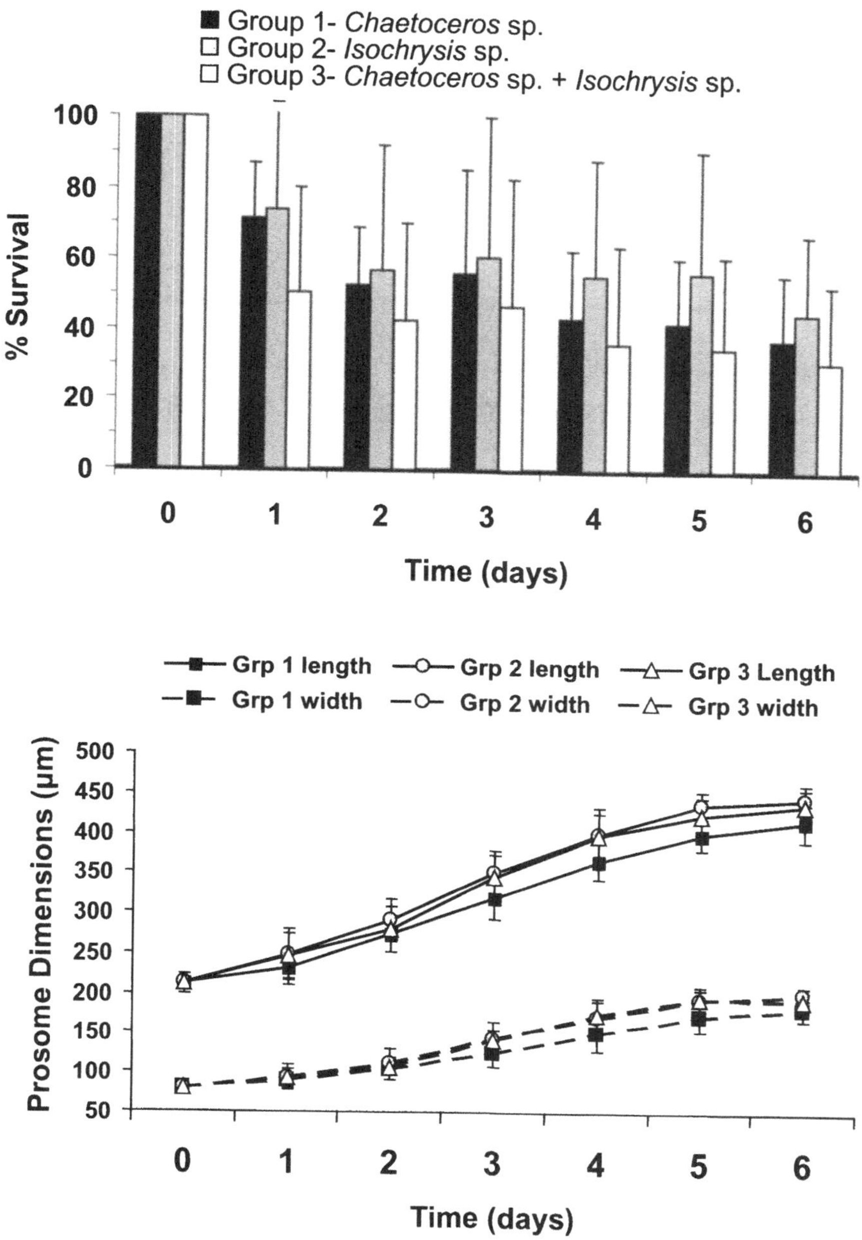

Figure 16.4. Mean percentage of survival (top) and growth (bottom) of *Parvocalanus* sp. copepodids fed *Chaetoceros* sp. or *Isochrysis* sp. singly or in combination (diet experiment III). Survival values are mean ± S.D. of three replicates; prosome length and width values are mean ± S.D. of five copepods per tank. Grp, group.

were significantly larger than larvae receiving only *Isochrysis* sp. on day 5 after hatching.

DISCUSSION

The findings from this study support the recent assertion by McKinnon et al. (2003) that paracalanid copepods are well suited for intensive culture as prey for small marine fish larvae. *Parvocalanus* sp. from Hawaiian coastal waters were maintained at high culture densities using the prymnesiophyte, *Isochrysis* sp., and the diatom, *Chaetoceros* sp., as feed. Nauplius densities attained in 400-L cultures were higher than previ-

Table 16.1. Mean population densities and minimum/maximum densities of *Parvocalanus* sp. adults, copepodids, and nauplii recorded in four 400-L culture tanks over a 50-day period

Diet group	Tank number	Adults + copepodids (individuals/mL)		Nauplii (individuals/mL)	
		Mean	Min.–max.	Mean	Min.–max.
Chaetoceros sp. (300,000/mL)	1	2.3 ± 1.4	0.4–6.0	6.0 ± 3.3	1.8–16.9
	2	2.1 ± 0.9	0.7–4.3	7.6 ± 4.4	1.1–23.1
Chaetoceros sp. + *Isochrysis* sp.	3	3.6 ± 1.3	1.1–9.0	9.4 ± 3.6	3.3–19.4
(200,000/mL + 100,000/mL)	4	3.1 ± 1.6	1.0–6.9	8.7 ± 5.7	1.2–29.9

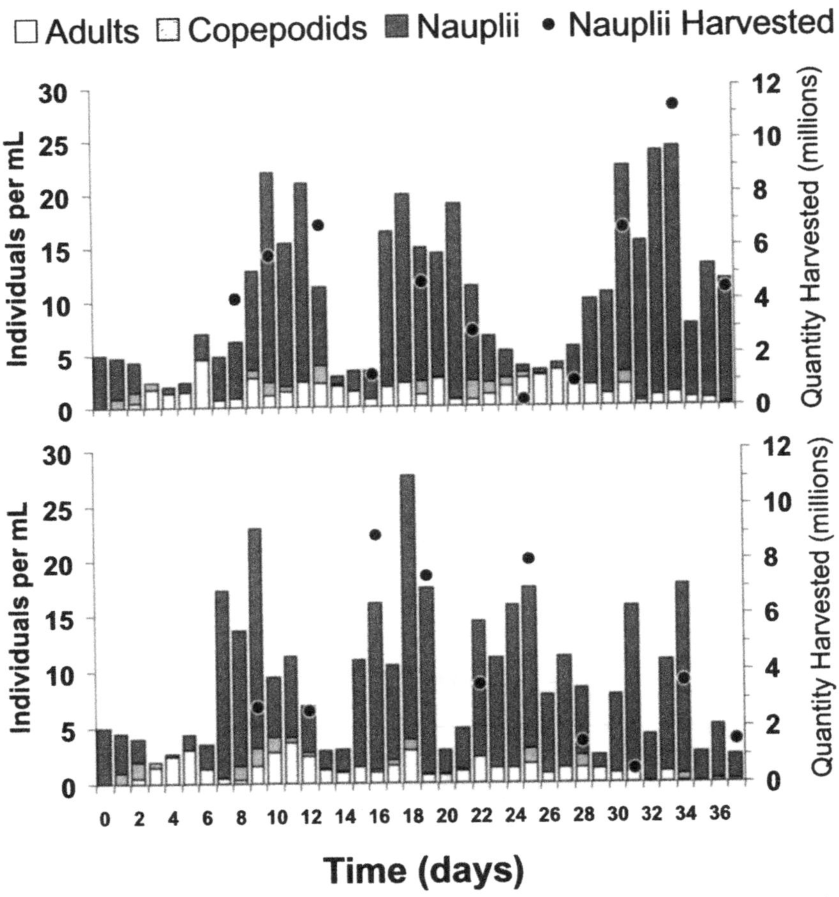

Figure 16.5. *Parvocalanus* sp. population densities (bars) and nauplius harvest quantities (circles) for two 400-L culture tanks operated using a two-phase production cycle. Values are mean of two counts from five pooled subsamples per tank per day.

ous reports for *Acartia* sp. (Schipp et al. 1999) and *G. imparipes* (Payne and Rippingale 2001) and approached those reported by Lipman et al. (2001) for *Oithona* sp. reared in 40-L bags. Although the mean dimensions of NI-stage *Parvocalanus* sp. nauplii were slightly larger than those reported by McKinnon et al. (2003) for *P. crassirostris* (75.6 × 41.3 μm versus 62.1 × 38.7 μm), they were successfully ingested by the small larvae of red snapper, *L. campechanus*.

Numerous earlier studies have examined the effects of microalgae species on survival, growth,

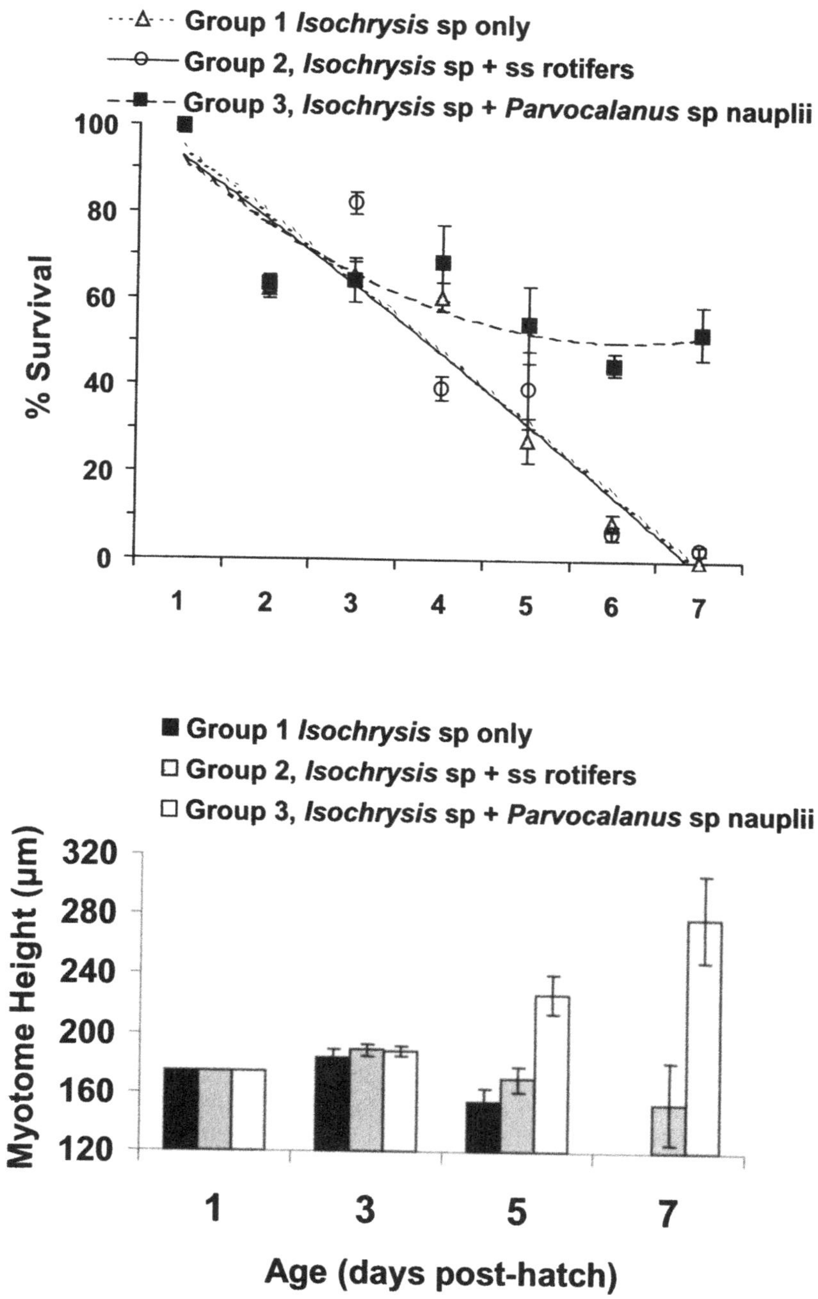

Figure 16.6. Survival (top) and growth (bottom) of first-feeding red snapper larvae, *Lutjanus campechanus*, in relation to diet. Values are mean ± S.D. of four replicates.

and fecundity of calanoid copepods, either from a plankton ecology perspective (for example, Berggreen et al. 1988; Kleppel et al. 1998; Rey et al. 2001) or for culture optimization (Støttrup and Jensen 1990; Semmens et al. 1999; Payne and Rippingale 2000; Murray and Marcus 2000; McKinnon et al. 2003). Microalgae cell size is known to be important in this respect, affecting the energy acquisition and feeding efficiency of different copepod developmental stages (Mullin 1963; Berggreen et al. 1988). Several applied culture studies have reported high survival rates and fecundity among adult copepods offered large-celled microalgae, in particular, dinoflagellates (Semmens et al. 1999; Murray and Marcus 2000; McKinnon et al. 2003). While the *Chaetoceros* sp. and *Isochrysis* sp. used in this study do not span such a broad range of size and energetic content as those tested elsewhere, their combined use enabled favorable production levels of *Parvocalanus* sp. nauplii with adequate nutritional value for rearing small marine fish larvae. This fits well with the goal of mass-producing copepods using microalgae species that are easy to culture. Schipp et al. (1999) similarly reported a benefit of using a mixed algal diet (*Rhodomonas* sp. plus *Tetraselmis* sp. plus *Isochrysis* sp.) for mass production of a tropical *Acartia* species.

The current findings illustrate a practical advantage to the aquaculturist of catering to stage-specific microalgae requirements of the copepod, with smaller developmental stages of *Parvocalanus* sp. benefiting from an *Isochrysis* sp.–rich diet (small Prymnesiophyte) and adults favoring *Chaetoceros* sp. (larger dinoflagellate). Although the two-microalgae diets were not balanced isoenergetically in the current feeding experiments, the rations applied should have provided sufficient carbon to meet the feeding requirements of the copepods (Kiørboe et al. 1985).

For adult *Parvocalanus* sp. fed a unialgal diet, survival was much higher when offered *Chaetoceros* sp., rather than *Isochrysis* sp. This may reflect better feeding efficiency by the adult copepods on diatoms, although several authors have documented high egg-production rates among *Isochrysis*-fed adults of other copepod species, including *A. tonsa* (Støttrup and Jensen, 1990) and *G. imparipes* (Payne and Rippingale 2000).

While *Chaetoceros* sp.–fed adult *Parvocalanus* sp. exhibited higher mean fecundity than those receiving *Isochrysis* sp., fecundity was increased significantly by providing a mixture of these two algal diets. Based on a sex ratio of 1:1 within populations of cultured adult *Parvocalanus* sp. (T. Kotani, unpublished observations), this equates to more than 40 offspring per female per day. In contrast, McKinnon et al. (2003) did not discern any significant difference in fecundity between female *P. crassirostris* fed *Chaetoceros muelleri* versus those fed *Isochrysis aff galbana*. Both of these diets yielded approximately 10 eggs per female per day, while the highest egg-production rates from *P. crassirostris* females (31 per female per day) were obtained using the dinoflagellate, *Heterocapsa niei* (McKinnon et al. 2003).

Although *Chaetoceros* sp.–fed adult *Parvocalanus* sp. exhibited high survival rates and high fecundity, their offspring displayed very low viability compared with those from adults fed a mixture of *Chaetoceros* sp. and *Isochrysis* sp. It is possible that nauplii produced by *Chaetoceros* sp.–fed adult copepods were nutritionally deficient. Alternatively, the strain of *Chaetoceros* sp. offered may have contained antimitotic compounds of the type isolated from other diatom species and known to arrest copepod embryonic development (Miralto et al. 1999; Ianora 2001). This topic deserves further attention, with a view to identifying an alternative diatom species that retains high fecundity of adults without impairing egg or nauplius viability.

Chaetoceros sp. was also an inferior diet for nauplii of *Parvocalanus* sp., based on low survival, growth, and development rates compared with *Isochrysis* sp.–fed nauplii. This was the case irrespective of parental nutritional history and may reflect a mismatch between the dimensions of the algae offered and nauplius prey-size requirements. Payne and Rippingale (2000) similarly reported more rapid maturation among *Isochrysis*-fed versus *Chaetoceros*-fed *G. imparipes*. Copepodids of *Parvocalanus* sp. exhibited an intermediate dietary response compared with adults and nauplii, surviving at similar rates on either of the two microalgae diets, although growing significantly faster on *Isochrysis* sp.

The benefits of feeding a combined *Chaetoceros* sp./*Isochrysis* sp. diet were also evident in 400-L pilot scale *Parvocalanus* sp. cultures, in which mixed diets yielded higher mean population densities than *Chaetoceros* sp. alone. Data from the pilot-scale culture tanks also confirmed recent findings (Schipp et al. 1999; Payne and Rippingale 2001) that warm-water calanoid copepods can attain much higher population densities under intensive culture than temperate species such as *A. tonsa* (Støttrup 2000). The maximum *Parvocalanus* sp. nauplius densities recorded during this study were, furthermore, higher than those reported by Schipp et al. (1999) for a tropical *Acartia* sp. (circa 30/mL versus 5/mL), supporting the observation by McKinnon et al. (2003) that the predatory feeding habit of adult

Acartia may prevent high culture densities being achieved.

System productivity in 400-L cultures, expressed as mean nauplii harvested per liter per day, was also higher in this study for *Parvocalanus* sp. than for *Acartia* sp. (Schipp et al. 1999) or *G. imparipes* (Payne and Rippingale 2001). For *Parvocalanus* sp. cultures harvested at 5-day intervals, nauplius production was equivalent to 3,750/L/day, compared with approximately 440/L/day for *Acartia* sp. (1,000-L batch cultures) and 878/L/day for *G. imparipes* (500-L batch cultures). Follow-up trials have confirmed that these *Parvocalanus* sp. nauplius production levels can be achieved routinely and that the method is scalable to at least 800-L without any significant loss in productivity (data not shown). Further research is required on *Parvocalanus* sp. harvesting strategies to stabilize nauplius output per harvest and minimize the interval between harvests.

Intensively cultured *Parvocalanus* sp. nauplii proved to be a suitable food source for first-feeding larvae of red snapper, *L. campechanus*, based on higher growth and survival rates among nauplius-fed larvae compared with those receiving enriched ss-type rotifers. In addition to the laboratory scale findings presented here, the authors have successfully reared larvae of red snapper, flame angelfish, *Centropyge loriculus*, and bluefin trevally, *Caranx melampygus*, to metamorphosis using cultured *Parvocalanus* sp., either as the sole prey or in combination with rotifers and *Artemia* (data not shown). Prey dimensions are clearly not the only variable determining diet efficacy in this case, since larvae of each of these species can ingest ss-type rotifers, yet exhibit very high mortality and slow growth following yolk exhaustion (R. Shields, unpublished observations). McKinnon et al. (2003) recently reported that intensively cultured *P. crassirostris* contain a high proportion of polar lipid classes and favorable *n*-3 highly unsaturated fatty acid (HUFA) ratios. It is the expectation of the authors that the cultured *Parvocalanus* sp. nauplii used in these studies also contain essential *n*-3 fatty acids in a readily digestible form. Further work should be carried out to confirm this, and to ascertain the effects of diet manipulation on *Parvocalanus* sp. biochemical composition.

There is considerable scope for future refinement of *Parvocalanus* sp. mass-culture techniques, especially in relation to feeding regimen and copepod-harvesting strategy (frequency and method of harvest, maximum sustainable yield). The current 1:1 *Chaetoceros* sp./*Isochrysis* sp. diet combination provided satisfactory preliminary results at pilot scale, but provision of a higher proportion of *Isochrysis* sp. during early developmental stages should be tested in the future, with the aim of accelerating copepod growth and maturation. Similarly, the fixed ration level chosen for this study was intended to avoid underfeeding of the copepod populations, but it would be more efficient in the future to adjust feed quantity in response to copepod population density. Alternative microalgae species, such as large dinoflagellates, should also be considered for future use, subject to meeting appropriate cost-benefit criteria. As a longer-term goal, the successful development of live algae replacement diets would greatly facilitate industry uptake of intensively cultured calanoid copepods.

Although the productivity levels currently achievable for *Parvocalanus* sp. and other tropical calanoids are much lower than for rotifers, it is important to distinguish the methods of administering these prey organisms. Under intensive larviculture conditions, copepod nauplii provide a crucial dietary input only during early developmental stages when water exchange rates are low, which translates to limited quantities administered daily over a short time period (approximately 3–7 days). Furthermore, larvae-rearing tanks may be inoculated with adult copepods in advance, providing an endogenous source of nauplii during first feeding.

CONCLUSIONS

The authors support the view of McKinnon et al. (2003) that tropical paracalanids possess advantageous life-cycle traits for intensive culture relative to copepods of the genus *Acartia*, which have received greater attention as candidate aquaculture feeds. Within the family Paracalanidae, however, the authors do not consider small nauplius size to be the sole or primary determinant of diet quality. The species of *Parvocalanus* isolated for the current work is both well suited to intensive culture conditions and is likely to provide adequate nutrition for a broad range of warm-water marine fish larvae with small mouth sizes. Mass culture of this and similar copepod species offers an important technique for the controlled, biosecure production of fish species that are not amenable to conventional rotifer/*Artemia*-based feeding methods. The currently developed copepod culture methods are sufficient for pilot scale juvenile production of "difficult-to-rear" marine fish species, although increased system productivity and scale of operation will certainly be required in the future.

ACKNOWLEDGMENTS

The current research on copepod culture techniques was funded by an award from the U.S. National Oceanic and Atmospheric Administration, NOAA, grant number NA16FV1491 (Hawaiian Sustainable Fisheries Development project). The red snapper feeding experiment was funded within the U.S. Gulf of Mexico Marine Stock Enhancement Programme, NOAA grant number NA96FL0358. The authors gratefully acknowledge the skilled husbandry assistance provided by Noland Wong and Andrea Suzuki.

REFERENCES

Berggreen, U., Hansen, B., and Kiørboe, T. 1988. Food size spectra, ingestion and growth of the copepod *Acartia tonsa* during development: implications for determination of copepod production. *Marine Biology* 99:341–352.

Dhert, P., Divanach, P., Kentouri, M., and Sorgeloos, P. 1997. Development of rearing techniques using large enclosed ecosystems in the mass production of marine fish fry. *Reviews in Fisheries Science* 5:367–390.

Doi, M., Ohno, A., Taki, Y., Singhagraiwan, T., and Kohno, H. 1997a. Nauplii of the calanoid copepod, *Acartia sinjiensis* as an initial food organism for larval red snapper *Lutjanus argentimaculatus*. *Suisan Zoshoku* 45:31–40.

Doi, M., Toledo, J.D., Golez, M.S.N., and Ohno, A. 1997b. Preliminary investigation of feeding performance of larvae of early red-spotted grouper, *Epinephelus coiodes*, reared with mixed zooplankton. In *Live Food in Aquaculture, Proceedings of the Live Food and Marine Larviculture Symposium*, edited by A. Hagiwara, T.W. Snell, E. Lubzens, and C.S. Tamaru. Japan, 1–4 September 1996. *Developments in Hydrobiology* 124:259–263.

Evjemo, J.O. and Olsen, Y. 1997. Lipid and fatty acid content in cultivated live feed organisms compared to marine copepods. *Hydrobiologia* 358:159–162.

Hagiwara, A., Kotani, T., Snell, W., Assava-Aree, M., and Hirayama, K. 1995. Morphology, reproduction, genetics and mating behavior of small, tropical marine rotifer *Brachionus* strains(Rotifera). *Journal of Experimental Marine Biology and Ecology* 194: 25–37.

Heath, P.L. and Moore, C.G. 1997. Rearing Dover sole larvae on *Tisbe* and *Artemia* diets. *Aquaculture International* 5:29–39.

Ianora, A. 2001. Do diatoms act as birth control pills for marine copepods? *Journal of Phycology* 37:26.

Kiørboe, T., Møhlenberg, F., and Hamburger, K. 1985. Bioenergetics of the planktonic copepod *Acartia tonsa*: relation between feeding, egg production and respiration, and composition of specific dynamic action. *Marine Ecology Progress Series* 26:85–97.

Kleppel, G.S., Burkart, C.A., and Houchin, L. 1998. Nutrition and regulation of egg production in the calanoid copepod *Acartia tonsa*. *Limnology and Oceanography* 43:1000–1007.

Liao, I.C., Su, H.M., and Chang, E.Y. 2001. Techniques in finfish larviculture in Taiwan. *Aquaculture* 200:1–31.

Lipman, E.E., Kao, K.R., and Phelps, R.P. 2001. Production of the copepod *Oithona* sp. under hatchery conditions. In *Book of Abstracts. Aquaculture 2001*, p. 379. Lake Buena Vista, Florida, USA, 21–25 January 2001. Baton Rouge, Louisiana, USA: World Aquaculture Society.

Marshall, A.J. and Purser, G.J. 1999. The use of temperate harpacticoid copepod nauplii as a first feed for flounder larvae (*Rhombosolea tapirina*). In *Book of Abstracts. World Aquaculture '99*, p. 493. Sydney, Australia, 26 April–2 May 1999. Baton Rouge, Louisiana, USA: World Aquaculture Society.

McKinnon, A.D., Duggan, S., Nichols, P.D., Rimmer, M.A., Semmens, G., and Robino, B. 2003. The potential of tropical paracalanid copepods as live feeds in aquaculture. *Aquaculture* 223:89–106.

Miralto, A., Barone, G., Romano, G., Poulet, S.A., Ianora, A., Russo, G.L., Buttino, I., Mazzarella, G., Laabir, M., Cabrini, M., and Giacobbe, M.G. 1999. The insidious effect of diatoms on copepod reproduction. *Nature* 402:173–176.

Mullin, M.M. 1963. Some factors affecting the feeding of marine copepods of the genus *Calanus*. *Limnology and Oceanography* 8:239–250.

Murray, M. and Marcus, N.H. 2000. Maximising copepod egg production: the effect of different algal diets on survival and fecundity of the copepod *Centropages hamatus*. In *Book of Abstracts*.

Aquaculture America 2000, p. 235. New Orleans, Louisiana, USA, 2–5 February 2000. Baton Rouge, Louisiana, USA: World Aquaculture Society.

Ogle, J., Lotz, J., Nicholson, C., and Barnes, D. 2000. Larval culture of red snapper *Lutjanus campechanus* using copepod nauplii for first feeding. In *Book of Abstracts. Aquaculture America 2000*, p. 249. New Orleans, Louisiana, USA, 2–5 February 2000. Baton Rouge, Louisiana, USA: World Aquaculture Society.

Payne, M.F. and Rippingale, R.J. 2000. Evaluation of diets for culture of the calanoid copepod *Gladioferens imparipes*. *Aquaculture* 187:85–96.

Payne, M.F. and Rippingale, R.J. 2001. Intensive cultivation of the calanoid copepod *Gladioferens imparipes*. *Aquaculture* 201:329–342.

Rey, C., Harris, R., Irigoien, X., Head, R., and Carlotti, F., 2001. Influence of algal diet on growth and ingestion of *Calanus helgolandicus* nauplii. *Marine Ecology Progress Series* 216:151–165.

Sato, V. 1991. Development of a phytoplankton production system as a support base for finfish larval rearing research. In *Rotifer and Microalgae Production Systems. Proceedings of a U.S.-Asia Workshop*, edited by W. Fulks and K.L. Main, pp. 257–274. Honolulu, Hawaii, USA: The Oceanic Institute.

Schipp, G.R., Bosmans, J.M.P., and Gore, D.J. 2001. A semi-intensive larval rearing system for tropical marine fish. In *Larvi 2001, 3rd Fish and Shellfish Larviculture Symposium, Ghent, Belgium*, edited by C.I. Hendry, G. van Stappen, M. Wille, and P. Sorgeloos. *European Aquaculture Society Special Publication* 30:536–539.

Schipp, G.R., Bosmans, J.M.P., and Marshall, A.J. 1999. A method for hatchery culture of tropical calanoid copepods, *Acartia* spp. *Aquaculture* 174:81–88.

Semmens, G.L., Stewart, M.D., Fry, P.A., Rimmer, M.A., and McKinnon, D. 1999. Comparison of yield produced by the copepod *Acartia* fed on different microalgae. In *Book of Abstracts. World Aquaculture '99*, p. 685. Sydney, Australia, 26 April–2 May 1999. Baton Rouge, Louisiana, USA: World Aquaculture Society.

Støttrup, J. 2000. The elusive copepods: their production and suitability in marine aquaculture. *Aquaculture Research* 31:703–711.

Støttrup, J.G. and Jensen, J., 1990. Influence of algal diet on feeding and egg-production of the calanoid copepod *Acartia tonsa* Dana. *Journal of the Marine Biological Association of the United Kingdom* 141:87–105.

Støttrup, J.G. and Norsker, N.H. 1997. Production and use of copepods in marine fish larviculture. *Aquaculture* 155:231–247.

Støttrup, J.G., Richardson, K., Kirkegaard, E., and Pihl, N.J. 1986. The cultivation of *Acartia tonsa* Dana for use as a live food source for marine fish larvae. *Aquaculture* 52:87–96.

Toledo, J.D., Golez, M., Doi, M., and Ohno, A. 1999. Use of copepod nauplii during early feeding stage of grouper *Epinephelus coioides*. *Fisheries Science* 65:390–397.

Urup, B., 1994. Methods for the production of turbot fry using copepods as food. In *Turbot Culture: Problems and Prospects. Proceedings of the Satellite Workshop of World Aquaculture '93*, edited by P. Lavens and R.A.M. Remmerswaal. Torremolinos, Spain, 25–27 May 1993. *European Aquaculture Society Special Publication* 22:47–53.

17
Characterization of an Extensive Zooplankton Culture System Coupled with Intensive Larval Rearing of Red Snapper *Lutjanus campechanus*

John T. Ogle, Jason T. Lemus, L. Casey Nicholson, Donald N. Barnes, and Jeffrey M. Lotz

ABSTRACT

As part of a program to produce red snapper (*Lutjanus campechanus*), we have developed a low-cost extensive zooplankton-production system. The system uses natural estuarine water with no supplemental feed, phytoplankton, or fertilizer. This method, which was originally used to rear oyster larvae and later adapted to produce the copepod *Acartia tonsa,* has been successfully scaled up to produce zooplankton in 70-m^3 tanks. The estuarine water used to fill the tanks provides the initial inoculum of zooplankton but is coarsely filtered to remove large predators. The tanks are left undisturbed for 48 hours, at which time 40% of the water is removed to harvest zooplankton. The zooplankton is sieved out, and the zooplankton tanks are refilled with new coarsely filtered estuarine water. Naupliar harvests during 2001 and 2002 exceeded 4×10^6 nauplii per day 40% and 71% of the time, respectively. Mean daily production during the past 2 years has been 2.4×10^6 copepods per tank per day and 6.6×10^6 copepods per tank per day. Copepod dynamics of two tanks was monitored for a 30-day period, and the work indicates a resident density of copepods that is 30 times greater than the inoculum. An average of 5.2×10^6 copepods were harvested daily, while only 411,000 copepods were added with the replacement water from those two tanks.

Copepods harvested from the system have been successfully used to rear red snapper during the past five years. Production has increased from 3,574 fish in 1998 to 28,545 fish in 2002. Larvae of the red snapper were initially reared in a variety of containers and locations during 1998. Cumulative advances during the following 3 years consisted of concentrating initial prey density at first feeding, eliminating *Artemia* as a feed, and moving the culture indoors under artificial light. During 2002, recirculation and biofiltration of the culture water was instituted, which, in consort with the previous changes, has resulted in a standard protocol.

INTRODUCTION

The most satisfactory food for larval marine fish is wild zooplankton consisting mostly of copepods (May 1970; Turner 1984; Tucker 1998).

Corresponding author: John T. Ogle, University of Southern Mississippi–Gulf Coast Research Laboratory, P.O. Box 7000, Ocean Springs, MS 39566-7000, USA; Phone: (228) 872-4675; Fax: (228) 872-4204; E-mail: john.ogle@usm.edu.

Although modern filtration technology has made it possible to harvest sufficient quantities of wild zooplankton for production of marine fish larvae (Barnabe 1990), the cost is prohibitive, and the spatial and temporal fluctuations in their abundance preclude routine use.

Traditionally, larval marine fish culture has entailed stocking larvae into ponds that have been fertilized to facilitate natural mixed-species zooplankton blooms (Horvath et al. 1984; Lannan et al. 1986; Harrell et al. 1990; Marcel 1990; Anderson and Tave 1993). Although larval marine fish are still cultured in ponds of mixed-species zooplankton blooms, the modern era of controlled marine fish culture began after the discovery by Rollefsen (1939) that plaice (*Pleuronectes platessa*) could be reared on brine shrimp nauplii (*Artemia* spp.). The next advance in feed for the culture of marine fish larvae occurred when rotifers (*Brachionus* spp.) were used to successfully culture species with larvae too small to consume brine shrimp nauplii. Most likely the work of Ito (1955) on mortalities in eel ponds led to the development of his system for rotifer culture (Ito 1960). Although Ito (1960) wrote that rotifers at the time were "demanded and expected much as food for the larval cultures of fish, lobsters, and crabs," he offered no details, and the first published account of rotifers as feed for larval fish was that of Vasil'eva and Okuneva (1961).

The use of brine shrimp and rotifers increased the number of marine species that could be cultured. There is a group of marine fish, however, whose larvae cannot be reared on either brine shrimp or rotifers, but can be reared only on marine copepod nauplii (Doi et al. 1994). In fact, it has been suggested that even those fish species that can eat brine shrimp or rotifers at first feeding grow and survive better on a diet of copepods (Pedersen 1984; Whitt et al. 1984; Nellen 1985; Shields et al. 1999; Ogle and Lotz 2000). Presently, copepods are not used routinely in marine aquaculture because they are difficult to culture compared with brine shrimp and rotifers, and ensuring sufficient quantities in a timely manner is problematic. Nonetheless rearing these copepod-dependent marine fish species commercially will occur only after the development of routine culture methods for copepods.

REVIEW OF COPEPOD PRODUCTION SYSTEMS

The culture of copepods is developing along several lines from extensive culture to intensive culture. On the extensive end lie production methods that produce low densities of mixed species, typically in ponds that are fertilized to provide blooms of food for the copepods. Typically, fish larvae are cocultured in the same pond as the bloomed copepods; a combination referred to as "endogenous" production (Shields 2001). On the intensive end lie methods that produce high densities of a single species in a controlled environment using separately cultured sources of nutrition that are supplied to the copepods in a prescribed feeding regimen. Typically, fish larvae are cultured separately from the copepods, and the partitioning of copepod production from the rearing of larvae is referred to as "exogenous" production (Shields 2001). For the most part, as culture intensifies, both the density of copepods and the effort required to operate the system increase. The appropriate choice is dependent on the resources and goals of a facility and should optimize the yield per unit of effort.

Copepod Density and Yield

The number of copepods that a system can produce is dependent on the potential copepod density that can be achieved, the volume of water used, and the duration and nature of the production cycle (Table 17.1). Extensive methods generally produce large numbers of copepods, because although the densities may be low, the volumes are large. Ohno and Okamura (1988) using organic fertilizer, and Raymont and Miller (1962) using inorganic fertilizer, successfully cultured calanoid copepods in large outdoor tanks (20 m^3 and 24 m^3). The large tanks and ponds were necessary to produce large numbers of nauplii because typical copepod densities were usually fewer than 300 copepods per liter (Raymont and Miller 1962; Ohno and Okamura 1988; Doi et al. 1994; Singhagraiwan et al. 1994), with one or two peaks as high as 1,150 copepods per liter (Doi et al. 1994). The need for large volumes may exclude extensive systems from areas where coastal lands are at a premium and the more compact footprint of intensive systems is desirable.

Table 17.1. Characteristics and yields of some calanoid copepod production systems

Type of system	General components (complexities)	Production cycle	Culture vessel volume	Density (/L)	Mean copepod yield per culture vessel
Intensive, Static, *Gladioferens imparipes*[a]	Algae production (1 or 2 spp.), Broodstock production, Naupliar production	42-day batch, Daily harvest for 21 day beginning on day 21	2 × 60-L maintained concurrently on asynchronous cycles	—	Nauplii 6.7×10^4/day
Intensive, Intermittent flow-through, *G. imparipes*[a]	Algae production (2 spp.) Broodstock production, Naupliar production	Semicontinuous, 21-day copepod rinsing and tank cleaning intervals; Harvest every 2nd day for 121 day	2 × 1000-L maintained concurrently on asynchronous cycles	—	Nauplii 5.2×10^5/day
Intensive, Automated, Intermittent recirculation, *G. imparipes*[a]	Algae production (1 sp.), Broodstock production Naupliar production	42-day batch; Daily harvest for 21 day beginning on day 21	2 × 500-L maintained concurrently on asynchronous cycles	—	Nauplii 4.4×10^5/day
Intensive, Static, *Acartia tonsa*[b]	Algae production (2 spp.), Copepod maintenance ("basis") culture Copepod production ("harvest") culture Egg hatching and growth tank	Continuous daily harvest	200-L basis tank, 450-L harvest tank, Nursery/grow-out tank	~50–100 adults	Nauplii 8.3×10^4/day
Intensive, Static, *Acartia* spp.[c]	Algae production (3 spp.), Copepod stock culture maintenance, Broodstock production	8-day batch production with single harvest on day 8	60-L stock culture, 1,000-L production	Mean: Adults 319 Copepodids 752 Nauplii 2,000	Adults 3.19×10^5 Copepodids 7.52×10^5 Nauplii 2.0×10^6 every 8 days
Intensive, static, *Acartia* sp.[d]	Algae production, copepod working culture	7- to 10-day batch production with daily harvest	400 L	—	Nauplii 5.0×10^5/day

(continued)

Table 17.1. Characteristics and yields of some calanoid copepod production systems (*continued*)

Type of system	General components (complexities)	Production cycle	Culture vessel volume	Density (/L)	Mean copepod yield per culture vessel
Intensive, static, *A. tonsa*[e]	Naupliar production coupled with broodstock	Semicontinuous	170–1,500 L	Mean: Adults 679 Nauplii 1,225	No harvest: production potential studied
Semi-intensive, intermittent flow-through, *A. tonsa*[f]	Production, mass water exchanges, Broodstock production, Naupliar production	Batch	1,900 L	Mean: Adults 232 Nauplii 820	No harvest: production potential studied
Static, batch (continuous cyclic production), *Acartia tsuensis*[g]	Broodstock collection, multiple fertilizations	Continuous production with production peaks every 12–15 days	24,000 L	Mean: Adults 94 Copepodids 101 Nauplii 374 Peak: Adults 280 Copepodids 588 Nauplii 1,136	No harvest: production potential studied
Extensive, Intermittent flow-through, multiple species[h]	Mass water exchanges	Semicontinuous with daily harvest; production peaks every 10–15 day	8 × 70,000-L maintained concurrently	Mean: Copepodids + adults 66 Nauplii 120	Copepodids + adults 1.9×10^6/day Nauplii 3.4×10^6/day

[a]Payne and Rippingale 2001.
[b]Støttrup et al. 1986.
[c]Schipp et al. 1999.
[d]Knuckey et al. 2001.
[e]Turk et al. 1981.
[f]Ogle 1979.
[g]Ohno and Okamura 1988.
[h]Current study.

Intensive systems, on the other hand, can produce high densities of copepods (Schipp et al. 1999; Lippman et al. 2001; Payne and Rippingale 2001). These systems, however, are constrained by the need for intense management, restricting the scale that can be achieved to obtain the required number of copepods even at high copepod densities.

Production cycle

Extensive and intensive zooplankton production systems differ in production consistency primarily as a result of their management principles. Extensive systems have little control over the copepod population dynamics while intensive systems focus on maintaining a consistent population. Pure extensive endogenous systems are successional and thus may not produce copepods for extended periods (Lannan et al. 1986; Harrell et al. 1990; Anderson and Tave 1993). The prospects for longer production cycles are greater for intensive systems and some more extensive systems that do not coproduce fish larvae (Ohno and Okamura 1988; Doi et al. 1994; Singhagraiwan et al. 1994). The extensive systems that focus on continuous copepod production are cyclic in nature, producing peaks lasting for periods of only a few to several days (Ohno and Okamura 1988; Doi et al. 1994; Singhagraiwan et al. 1994). The blooms that occur in extensive systems are only loosely predictable and vary in magnitude. This can cause problems with scheduling larval stocking, which should coincide with peak bloom, particularly if fish larvae are coproduced with zooplankton. The semicontinuous intensive system of Payne and Rippingale (2001) can produce large numbers of copepod nauplii for 121 days, thus reducing the risk of mistiming fish larvae production with naupliar production. The intensive system of Schipp et al. (1999), however, relies on coproduction and thus must be managed more strictly in terms of timing the production with fish larvae production.

Culture Effort

As copepod density intensifies, the culture system increases in complexity due to increased reliance on specialized equipment and expertise. This results in a greater number of critical points for failure, and therefore increased effort is required to operate the system effectively. The critical points and concomitant increased effort are associated with water quality, phytoplankton production, broodstock maintenance, and the need for monitoring feed during larval rearing.

Water Quality

The high copepod density in an intensive system often causes water quality to deteriorate, unlike extensive systems, and may require treatment. Remediation has been implemented through intermittent or regular water exchange or recirculation with biofiltration and foam fractionation (Schipp et al. 1999; Payne and Rippingale 2001). Støttrup et al. (1986) maintained water quality in 450-L semicontinuous cultures of *Acartia tonsa* at low adult densities (50–100/L) by siphoning the tank bottom each day to remove detritus and eggs and replacing the removed water with clean water.

Phytoplankton

Perhaps the most important contributor to complexity is the addition of laboratory-cultured phytoplankton for feeding to copepods (Støttrup et al. 1986; Schipp et al. 1999; Payne and Rippingale 2001). Extensive copepod production systems for endogenous production systems are driven by fertilization that promotes phytoplankton blooms (Raymont and Miller 1962; Ohno and Okamura 1988; Doi et al. 1994; Singhagraiwan et al. 1994). They require much less labor and equipment than intensive systems; the reduced control, however, may compromise the consistency of the copepod quality and species composition. This concern can be somewhat ameliorated by fertilization schemes that promote growth of desired zooplankton species (Geiger 1983).

Prey Size

An additional concern of intensive copepod culture systems is that they focus production efforts on only one size class of prey. In the case of Lutjanid culture, that prey is the early naupliar stages (Schipp et al. 1999). Because fish larvae outgrow a prey size (copepod nauplii in particular) within several days, additional effort must be directed toward production of larger prey such as rotifers during the early days of larviculture (Schipp et al. 1999) and brine shrimp during the later days of larviculture or by maintaining copepod nauplii in grow-out vessels until they reach a larger size (Støttrup et al. 1986). Extensive systems can be harvested for a variety of prey sizes simultaneously; however, the quantity of each prey size may not be predicted easily due to their cyclic nature.

Broodstock

Maintenance of copepod broodstock is necessary for intensive cultures because those systems are typically batch rather than continuous (Schipp et al. 1999; Payne and Rippingale 2001). Operation of batch intensive cultures requires regular inoculation with early life-cycle stages such as eggs (Støttrup et al. 1986), or the first naupliar instar (Payne and Rippingale 2001), or with adults (Schipp et al. 1999).

Extensive copepod culture systems, on the other hand, do not require laboratory-maintained copepod broodstock because copepods are present in the raw fill water. Supplementation, however, with additional copepods in extensive culture may enhance copepod production (Ohno and Okamura 1988; Doi et al. 1994; Singhagraiwan et al. 1994).

DESCRIPTION, OPERATION, AND PRODUCTION: EXTENSIVE ZOOPLANKTON PRODUCTION SYSTEM

Over the past several years the authors have developed an extensive zooplankton production system to generate live feed for larval red snapper (*Lutjanus campechanus*). Here the authors describe the system, its management, and its productivity. In addition they present results of rearing larval red snapper using the zooplankton produced in this system.

In contrast to most other extensive systems, the system developed by the authors requires no fertilization and is capable of production over several copepod generations. The lengthy production period allows copepods to be available over a longer time and provides greater leeway in timing of spawning broodstock fish. In addition, the overlap of several generations of copepods allows a variety of copepod stages (sizes) to be available simultaneously, and a single culture can be used to feed larval fish throughout their development. This system is a scaled-up version of the "brown-water technique" used by Ogle (1979, 1982) to culture *A. tonsa* and oyster larvae.

Description

Zooplankton Production Tanks

The walls of the zooplankton production tanks are constructed of metal wire panels and support a vinyl liner as described in Ogle (1991). The tanks are 11.5 m in diameter and have a surface area of just over 100 m^2.

Water

The source of water is Davis Bayou, an inlet off Mississippi Sound and an arm of the Gulf of Mexico. The water has a brown color, which is presumed to be due to suspended detritus and diatoms. Water from Davis Bayou is pumped to a tank by a dedicated 0.75-horsepower (hp) Jacuzzi pump (Jacuzzi USA; Little Rock, AK, USA). Water is removed from a tank for harvest of copepods by a dedicated 0.5-hp Sta-Rite submersible harvest pump (Sta-Rite Industries, Madison, WI, USA) positioned approximately 4 cm from the tank bottom. Harvest water is pumped through a 5-cm polyvinyl chloride harvest pipe from the tank to a sieve station in the sieve house.

Sieve House

The sieve house is a small, screened enclosure that serves as a central place for controlling copepod harvesting, sorting the harvest into size classes, and providing shelter from the elements for individuals performing the work. The house contains eight sieve stations, each located at the outflow of a harvest pipe. Electrical switches in the sieve house allow centralized control of the pumps for the various production tanks.

Sieve Stations

A sieve station consists of a fiberglass sieve tank (3.9 m × 0.6 m × 0.025 m) to hold water that is pumped from a production tank for harvest and in which sieving and size sorting occurs. Sieves are made of a piece of 20-μm mesh Nitex that is sewn into a rectangular-shaped box (90 cm × 53 cm × 9 cm) and supported by a stainless steel frame. A sieve is positioned in the sieve tank so that the sieve is partially submerged, and the harvest pipe opening is beneath the surface of the water to minimize damage to copepods during harvest.

Operation

Copepod production tanks are filled to a depth of 0.7 m with estuarine water from Davis Bayou of

Table 17.2. Relative abundance of copepod species determined from surveys in zooplankton tanks during three production seasons (1999, 2000, 2002)

	Acartia tonsa	*Oithonia* sp.	*Pseudodiaptomus pelagicus*	Other Calanoid spp.
n	11	11	11	11
Mean proportion (SD)	0.52 (0.22)	0.23 (0.18)	0.19 (0.22)	0.03 (0.44)
Proportion range	0.12–0.77	0.00–0.56	0.10–0.74	0.00–0.14

Mississippi Sound. Fill water is pumped through a 300-μm mesh bag to remove fish larvae, crab larvae, and other potential predators of copepods prior to flowing into a production tank. Wild zooplankton that passes through the mesh bag (primarily copepods) serves to inoculate the tank and initiate a production cycle. Harvest is achieved by removing 40% of the tank volume, passing it through a 500-μm mesh bag to remove debris and aquatic insects, and then passing it through a 20-μm sieve to remove the desired zooplankton. Harvested material retained on the 20-μm sieve is rinsed into a 70-L aerated container. Rinse water is pumped from the sieve tank by a small pump (Little Giant, 1EUAA-MD; Little Giant Pump Co., Oklahama City, OK, USA). Sieves are rinsed every 10–15 minutes. Harvested zooplankton is passed through a variety of sieve sizes to collect the size fraction of zooplankton required to feed fish larvae. After harvest, water is returned to the bayou. Nothing is added to the water during zooplankton production, therefore the quality of the discharge water is not different from that of the natural water. As many as four production tanks can be harvested simultaneously. Approximately 2 hours are required to harvest 40% of the volume of all the tanks each day. Tanks are recharged with new brown water from the estuary after harvest.

Over time and many recharges the tanks may become fouled with bivalves, barnacles, jellyfish, ctenophores, and sediment, all of which cause the productivity to decline. Tanks can be totally harvested, cleaned, and restarted in less than 4 hours. The zooplankton harvested during cleaning can be returned to the fresh tank, thus maintaining the production schedule. The authors routinely clean a tank every 4 weeks to prevent tanks from becoming fouled.

Production

Species Composition

The zooplankton harvested from the system is a mixed-species assemblage and includes representatives of the Phyla Nematoda, Rotifera, Annelida, Mollusca, and Arthropoda. Nematodes, polychaetes, and bivalve larvae are usually present in low numbers, with an occasional large bloom. Incidental arthropods include small numbers of barnacle, crab, and palaemonid shrimp larvae. Rotifers are rarely seen at salinities above 5 ppt.

The dominant component of the harvest is the Copepoda, and of that, the calanoid species are most abundant. The relative proportions of the copepods estimated from surveys taken during several years are presented in Table 17.2. *A. tonsa* is the most common calanoid copepod, but *Pseudodiaptomus pelagicus* is also abundant at times. Other calanoid copepods observed in low numbers are *Tortanus* sp. and *Eurytemora affinis*. The most common of the cyclopoid copepods, *Oithona* spp., are sometimes present in high numbers. Several unidentified cyclopoid and harpacticoid copepods are present in lower numbers.

Copepod Dynamics

During August 2001, the naupliar populations and copepodid (including adult stage) populations of two zooplankton tanks (tanks A and B) were monitored for 30 days during operation. The two tanks were harvested and recharged on alternating days. The number of copepods harvested from the tanks and the number of copepods added to the tanks during recharge were recorded (Table 17.3). Temperature was recorded using a minimum-maximum recording thermometer. Dissolved oxygen was recorded twice daily using a meter (YSI Model 95 oxygen meter). The salinity was determined for the harvest water and the recharge water using a refractometer.

The mean morning low temperature was

Table 17.3. Copepod population abundance, copepod harvest, and number of copepods introduced with recharge water of tanks A and B (2001 tank dynamics study)

		Nauplii			Copepodids + Adults		
	Tank	Mean (SD)	/L	Population (%)	Mean (SD)	/L	Population (%)
Tank population	A	6,175,342 (6,412,015)	88	—	4,860,304 (2,564,806)	69	—
	B	10,511,400 (6,805,311)	150	—	4,389,554 (2,458,890)	63	—
	Combined	8,343,395 (6,928,412)	120	—	4,624,929 (3,504,517)	66	—
Copepods harvested	A	2,470,137 (2,564,806)	88	40	1,944,122 (1,728,134)	69	40
	B	4,204,579 (2,722,324)	150	40	1,755,822 (983,556)	63	40
	Combined	3,337,358 (2,771,364)	120	40	1,849,972 (1,401,807)	66	40
Copepods introduced with recharge	A	232,878 (240,352)	3.3	3.7	173,010 (222,642)	2.5	3.6
	B	220,015 (217,323)	3.1	2.1	196,204 (314,745)	2.8	4.5
	Combined	226,427 (228,010)	3.2	2.7	184,608 (271,423)	2.6	3.9

27.8°C for both tanks, the mean afternoon high temperatures were similar (33.3°C in tank A and 32.2°C in tank B). The mean salinity for both tanks was 12 ppt. The mean morning and afternoon oxygen levels were 6.4 and 7.5 ppm for tank A and 6.3 and 7.4 for tank B.

The mean number of copepods added during recharge was 405,888 (232,878 nauplii and 173,010 copepodids) to tank A and 416,219 (220,015 nauplii and 196,204 copepodids) to tank B (Table 17.3). The mean harvest was 4,414,259 (2,470,137 nauplii and 1,944,122 copepodids) for tank A and 5,960,401 (4,204,579 nauplii and 1,755,822 copepodids) for tank B (Table 17.3). The mean harvest for both tanks was 5,187,330 (3,337,358 nauplii and 1,849,972 copepodids). The mean harvest in tank A represented a 987% net copepod yield (961% in nauplii and 1,024% in copepodids), and the mean harvest in tank B was a 1,350% net copepod yield (1,810% nauplii and 795% copepodids).

The total population of copepods in the tanks A and B combined (12,968,000) was over 30 times larger than the number of copepods introduced with each recharge of brown water (411,000) (Table 17.3). In fact, the mean number of copepods introduced into the tanks with each recharge represented less than 5% of the total population of copepods in the tanks (Table 17.3). The number of females introduced into the tanks during recharge was estimated, because instar stages were not distinguished in this study. We estimated that 1/6 (17%) of the copepodids introduced into the tanks were adults because there are six copepodid instars, including the adult instar. If 50% of the adults are assumed to be female, then approximately 15,700 mature females were introduced with each tank recharge. If each introduced female produced 25 eggs (Parrish and Wilson 1978), it could be expected that 392,500 nauplii would be available the next day, of which 157,000 (40%) would be harvested. In fact the mean harvest is 3,337,358 nauplii per day, indicating an actively reproducing and growing population of copepods within the tanks. This supports the conclusion of previous reports that brown water is an effective nutrition source for the growth of copepod populations (Ogle 1979; Lemus 2001).

Similar copepod dynamics over time were found in both the tanks, where two distinct peaks in naupliar density were each followed by a peak in copepodid density (Fig. 17.1). This seems to

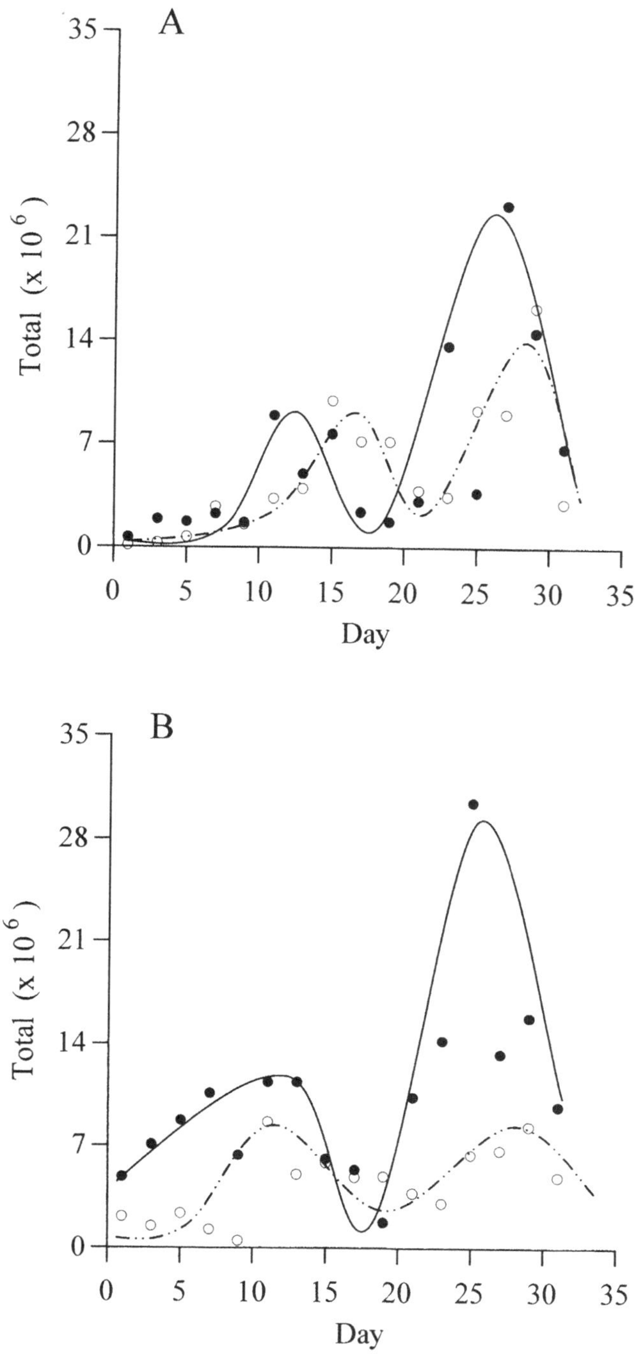

Figure 17.1. Copepod populations in tank A (A) and tank B (B) during the copepod dynamics study of 2001. Naupliar populations are designated by filled circles and solid lines. Copepodid populations are designated by open circles and dashed lines. Lines are fit to the data points by eye.

Table 17.4. Daily copepod harvests by size category during 2001 and 2002 production seasons

Year		Size category		
		Small	Medium	Large
2001[a]	Mean harvest	4,517,129 (4,102,971)	4,108,114 (4,256,682)	2,634,553 (2,566,295)
	(SD) Range	600,000–20,980,000	140,000–15,620,000	140,000–9,640,000
2002[b]	Mean harvest	8,313,830 (7,242,084)	14,260,200 (5,379,294)	5,874,812 (5,375,683)
	(SD) Range	1,016,666–40,130,000	7,860,000–23,140,00	506,000–34,080,000

[a]2001 size categories: small = 20–100 μm; medium = 100–200 μm; large = 200–300 μm.
[b]2002 size categories: small = 20–100 μm; medium = 68–200 μm; large = > 200 μm.

represent three overlapping generations of adult copepods, wherein the first generation of adult copepods was introduced with the fill water and the second and third generations resulted from reproduction of an established population inside the tanks. Very few nauplii and copepodids were introduced with the recharge water relative to the total abundance of nauplii during the entire study (Table 17.3). Therefore, the majority of the production of the tanks can be attributed to the growth of the population in the tanks. The major naupliar productions were separated by approximately 12 days, which is in agreement with an estimation of the generation time of *A. tonsa* by Ogle (1979). The continuous presence of copepodids and nauplii indicated that although there were periods of reproduction, the generations did overlap, and therefore the systems provided continuous production of nauplii despite a 40% harvest of the entire population every other day.

Harvests

During the production season of 2001, six tanks were used for zooplankton production (three harvested each day), and in 2002 eight tanks were used (four harvested each day). On average, 7×10^6 copepods were harvested each day in 2001 (2.4×10^6/tank/day) and 2.6×10^7 million copepods were harvested each day in 2002 (6.5×10^6/tank/day). After harvest, zooplankton was separated into three size categories, and each category was concentrated into 10 L of water for feeding to red snapper larvae. In 2001 the size categories were 20–100 μm (small), 100–200 μm (medium), and 200–300 μm (large). In 2002 the size categories were 20–100 μm (small), 68–200 μm (medium), and more than 200 μm (large).

The mean and range of copepods harvested each day from each size class for 2001 (three tanks per day) and 2002 (four tanks per day) are presented in Table 17.4. The mean density of each size category per tank was estimated using the harvest data and the volume of water from which the harvest was obtained (Table 17.4). In 2001 87%, 86%, and 68% of the mean daily copepod harvests were greater than or equal to 1.0×10^6 in the small (n = 62), medium (n = 35), and large (n = 47) size categories, respectively. In 2002, 100%, 100%, and 96% of the mean daily copepod harvests were $\geq 1.0 \times 10^6$ in the small (n = 42), medium (n = 10), and large (n = 48) size categories, respectively. Greater than 4×10^6 small-sized copepods (nauplii) were harvested 40% and 71% of the time in 2001 and 2002, respectively.

LARVAL FISH PRODUCTION

Rearing red snapper *Lutjanus campechanus* through the larval stages has only recently been accomplished. Work on its culture, however, was begun in the 1970s. In 1978, red snapper were shown to ovulate in response to hormone injection (Trimble and Tatum 1978; Minton et al. 1983), and to spawn spontaneously in captivity (Arnold et al. 1978; Chesney and Filippo 1994). No larvae, however, survived for longer than 10 days, even though they were offered rotifers as prey (Rabalais et al. 1980). Minton, who was working on larval rearing in the 1980s, improved the quantity and quality of rotifers offered, but again, the larvae survived for only 5–7 day (V. Minton, Claude Peteet Mariculture Facility, personal communication).

It was not until 1997 that Bootes (1998), working at the Claude Peteet Mariculture Center (Gulf Shores, AL, USA), reported the successful rearing of 283 red snapper larvae in a tank of bloomed zooplankton. Recent attempts to rear red snapper using the harpactoid copepod *Euterpina* sp. have not been encouraging (Shields et al. 2001). In the same study by Shields et al. (2001), it was demonstrated that although snapper larvae consumed a small strain of rotifer (ss-rotifer), growth and survival were no better than unfed controls. By contrast, Lemus et al. (2002) found that survival of red snapper to 7 days after hatching was 15% in a single tank when fed laboratory-cultured *A. tonsa*. Another study (performed at the Gulf Coast Research Laboratory) with laboratory-cultured *A. tonsa* as food for red snapper larvae was performed with replication. Seven hundred day 0 red snapper larvae were stocked into each of four 25-L containers with clean 5-μm filtered 30 ppt seawater inoculated with 10,000 adult *A. tonsa* and the flagellate *Isochrysis* sp. at a cell density of 2.0×10^5 cells/mL. Seven days after hatching, mean larval survival was 16.5% (errors, 10.82%). Others have also demonstrated the value of the laboratory-cultured calanoid copepods, that is, *Parvocalanus* sp. in small containers and in a 4-ton tank (R. Shields, The Oceanic Institute, personal communication) and *Centropages hamatus* in small containers (Lemus et al. 2002) as food for red snapper larvae. Later work by Shields (2005) demonstrated the production of large numbers of nauplii of the calanoid copepod *Parvocalanus* sp. under intensive conditions that were used to culture red snapper larvae.

Red snapper have also been reared successfully through the larval stages over the past 5 years using the mixed-species zooplankton produced from the system described above (Ogle and Lotz 2000, 2002). Production of postlarval juvenile fish has increased yearly from 3,574 fish in 1998 to 28,545 fish in 2002 (Table 17.5).

In 1998 and 1999, most rearing trials were in tanks protected from rain and direct sun by shade cloth–covered clear plastic sheets supported by a wooden pole frame. Preliminary trials to culture larvae indoors in 1998 were unsuccessful (Ogle et al. 2000). In 2000, indoor culture was revisited, and a single tank of larvae was cultured indoors using a metal halide lamp for illumination (Ogle et al. 2001). This success led to greater effort on indoor culture in 2001, when over half of the tanks were indoors (Ogle and Lotz 2002). In 2002, one system design and one management protocol was adopted and used to produce 28,545 postlarval red snapper indoors under artificial light with recirculation and biofiltration.

Table 17.5. Red snapper larvae production by year

Year	Number of spawn batches	Number of larvae stocked	Fish produced	Mean larval survival
1998	3	31,500	3,574	11.3%
1999	4	190,000	4,008	2.1%
2000	5	152,000	5,504	3.6%
2001	7	198,000	5,862	3.0%
2002	6	200,000	28,545	14.3%

Observations

1998

In 1998, it became apparent that larvae were very sensitive to disturbance. For example, of the tanks that produced fish, the lowest survival (2.2%) was in a tank (tank 9, Table 17.6) with water recirculation (water was pumped out of the tank, passed through a biofilter, and returned to the tank). Tanks using internal filters and being manually cleaned by daily siphoning had 7.5% and 7.6% survival (tanks 10 and 11, Table 17.6). The tank receiving no disturbance other than feeding had 12.5% survival (tank 12, Table 17.6).

There was evidence that high feed rates might be important to survival. One 200-L tank (tank 13, Table 17.6) was fed 20 copepods/mL (compared with 5 copepods/mL for the other tanks) and had the highest survival (30.4%). In recent years, the authors have realized that transfer of copepods from low-salinity production tanks to high-salinity larval-rearing tanks may result in copepod mortalities. Therefore, it is possible that the increased larval survival in the tank fed 20 copepods/mL was achieved because the high number compensated for deaths of copepods after their transfer to high salinity from the lower-salinity zooplankton production system.

There was also a hint that red snapper larvae might do better at lowered salinities. Two 200-L tanks with reduced salinities had higher survival (tanks 20 and 21; Table 17.6) than two matched tanks at higher salinities (tanks 18 and 19; Table 17.6).

Table 17.6. Larval rearing of red snapper in 1998

Spawn batch	Tank #	Location	Comment	Culture volume (L)	Larvae stocked per L[a]	Harvest age (day)	% Survival
1	1	Shade[b]	—	1,000	2	26	5.1
1	2	Porch[c]	—	1,000	1	26	3.3
1	3	Shade	—	200	5	26	2.3
1	4	Porch	—	200	5	26	3.7
1	5	Shade	—	200	5	26	1.3
1	6	Indoors[d]	Dead day 7	200	5	—	0
1	7	Indoors	Dead day 7	100	10	—	0
1	8	Porch	Dead day 7	100	10	—	0
2	9	Shade	Recirculation	200	5	26	2.2
2	10	Shade	Filtered[e]	200	5	26	7.5
2	11	Shade	Siphoned	200	5	26	7.6
2	12	Shade	Undisturbed	200	5	26	12.5
2	13	Shade	20 copepods/mL	200	5	26	30.4
3	14	Shade	—	200	2.5	—	13.2
3	15	Shade	—	200	5	—	7.9
3	16	Shade	—	200	10	—	8.7
3	17	Shade	—	1,000	10	—	20.9
3	18	Shade	35 ppt	200	5	30	7.7
3	19	Shade	30 ppt	200	5	30	4.4
3	20	Shade	25 ppt	200	5	30	9.7
3	21	Shade	20 ppt	200	5	30	12.9

[a]All larvae were fed copepods from day 2–12 followed by *Artemia*.
[b]Shade was 80% shade cloth cover over clear plastic roof.
[c]Porch was a solid cover open to the south.
[d]Indoors was a room with eight 40-W fluorescent lamps and no windows.
[e]Filtered was an in-tank sponge filter.

1999

Total mortality occurred in all six tanks located outdoors under the shade at day 20 after hatching (Table 17.7). Total mortality occurred at day 9 for larvae in six additional tanks located outdoors under the same shade (Table 17.7). The mass mortalities of both groups occurred on the same calendar day, suggesting an external event that was unrelated to conditions in the tanks.

2000

Mean survival for the tanks fed exclusively on copepods was 5.7%, while survival was 1.8% for the eight tanks that were transitioned to brine shrimp. A tank placed indoors under an artificial light had 7.7% survival (tank 16; Table 17.8).

2001

Survival for the 15 tanks located outdoors in the shade (tanks 7, 8, 9, 17, 18, and 19; Table 17.9) or in the greenhouse (tanks 21–27; Table 17.9) averaged 0.9%. Survival for the 24 tanks located indoors under artificial light averaged 4.9% (Table 17.9). Average survival of fish in three tanks held under constant light (tanks 31–33; Table 17.9) was 1.4%, which was not different from a mean of 1.1% for three additional tanks held under the standard photoperiod of 14 hours light/10 hours dark (tanks 28–30; Table 17.9). Larvae of spawn batch six were used to compare the effect of aeration on survival. The mean survival in three tanks with air flow of 50 mL/min (tanks 37–39; Table 17.9) and three tanks with air flow of 250 mL/min (tanks 34–36; Table 17.9) were both 3.8%.

2002

The cumulative successes over the past years led to the development of one system design and one management protocol in 2002. It was used to produce 28,545 postlarval red snapper indoors under artificial light with recirculation and biofiltration.

Table 17.7. Larval rearing of red snapper in 1999

Spawn batch	Tank number	Location[a]	Comment[b]	Culture volume (L)[c]	Larvae stocked per L	Harvest age	% Survival
1	1	Shade	Died day 20	1,000	10	20	0
1	2	Shade	Died day 20	1,000	10	20	0
1	3	Shade	Died day 20	1,000	10	20	0
1	4	Shade	Died day 20	1,000	10	20	0
1	5	Shade	Died day 20	1,000	10	20	0
1	6	Shade	Died day 20	1,000	10	20	0
1	7	Greenhouse	—	1,000	10	20	8.3
2	8	Shade	Died day 9	1,000	10	—	0
2	9	Shade	Died day 9	1,000	10	—	0
2	10	Shade	Died day 9	1,000	10	—	0
2	11	Shade	Died day 9	1,000	10	—	0
2	12	Shade	Died day 9	1,000	10	—	0
2	13	Shade	Died day 9	1,000	10	—	0
3	14	Shade	Density, Selco	500–1,000	20	23	17.6
3	15	Shade	Density	500–1,000	20	23	12.9
3	16	Shade	Density	500–1,000	20	23	6.8
3	17	Shade	Density	500–1,000	20	23	0.3
3	18	Shade	Density	500–1,000	20	23	1.3
3	19	Shade	Density	250–1,000	40	23	6.2
3	20	Greenhouse	Density	250–1,000	40	23	11
3	21	Greenhouse	Density, Selco	250–1,000	40	23	2.9
4	22	Greenhouse	Density	250–1,000	40	23	1.4

[a]Shade was 80% shade cloth cover over clear plastic roof; Greenhouse was a plastic Quonset style greenhouse with a 40% shade cloth.

[b]Density refers to larval density at stocking in the initial culture volume and then the larval density following an increase in the water volume; Selco indicates Selco enrichment of brine shrimp.

[c]Culture volume ranges indicate a tank's initial and final volumes.

One 4,000-L tank was reared under the standard protocol to evaluate the potential for using larger tanks (tank 17; Table 17.10).

DISCUSSION

The dominant zooplankter in the above-described extensive production system as well as the dominant species in the source water is *A. tonsa*, a small neritic copepod (Stein 1981). Ikeda (1973) concluded that small neritic brackish-water copepods would be ideal for mass culture. In fact *Acartia* spp. are the dominant zooplankters in most neritic water of the United States (Roman 1977) and usually come to dominate large fertilized tanks as well (Raymont and Miller 1962). *Acartia* spp. may be able to dominate these estuarine areas due to their wide salinity tolerance (Raymont and Miller 1962; Stein 1981; Cervetto et al. 1999). Furthermore, *A. tonsa* is an opportunistic, omnivorous, particle grazer and suspension feeder (Poulet 1978; Lonsdale et al. 1979). As such it may be able to use the detrital component of the brown water as well as the algal component.

Studies of natural populations of a number of copepod species have suggested that copepods cannot meet their energy demands by grazing exclusively on phytoplankton (Heinle and Flemer 1975), and other sources of energy are required. Pafenhoffer and Strickland (1970) and Poulet (1983) have demonstrated that copepods use detritus, and indeed detritus may provide the additional energy. In fact, there is five times more detritus than phytoplankton biomass in the ocean (Poulet 1983). Much of the brown color of "brown water" appears to be from detritus and may be an important source of energy for copepods in the above-described system.

The extensive zooplankton production system

Table 17.8. Larval rearing of red snapper in 2000

Spawn batch	Tank number	Location[a]	Comment[b]	Culture volume (L)[c]	Larvae stocked per L	Harvest age (d)	% Survival
1	1	West shade	Copepods and *Artemia*	350–1,000	28	24	0.3
1	2	West shade	Copepods and *Artemia*	350–1,000	28	24	0
1	3	West shade	Copepods and *Artemia*	350–1,000	28	24	0
1	4	West shade	Copepods and *Artemia*	350–1,000	28	24	0
2	5	East shade	Copepods	350–1000	28	29	11.7
2	6	East shade	Copepods	350–1,000	28	29	3.5
2	7	East shade	Copepods	350–1,000	28	29	2.8
3	8	East shade	Copepods	350–1,000	28	24	0.9
3	9	East shade	Copepods	350–1,000	28	24	0.7
3	10	East shade	Copepods	350–1,000	28	24	5.2
3	11	West Shade	Copepods	350–1,000	28	24	1.6
4	12	Greenhouse	Copepods	350–1,000	28	22	15.4
4	13	Greenhouse	Copepods	350–1,000	28	22	11.5
4	14	Greenhouse	Copepods	350–1000	28	22	0
4	15	Greenhouse	Copepods	350–1,000	28	22	0
5	16	Indoors	Copepods	200	10	24	7.7

[a]West and east shade were west and east ends of the shaded structure—80% shade cloth cover over clear plastic roof; greenhouse was a plastic Quonset-style greenhouse with a 40% shade cloth; indoors was a room with eight 4-W fluorescent lamps and no windows.

[b]Copepods and *Artemia*: Copepods fed from day 2–12 followed by brine shrimp from day 12–24; Copepods: Copepods fed day 2–24.

[c]Culture volume ranges indicate a tank's initial and final volumes.

described here is simple, reliable, inexpensive to build and operate, and can easily be scaled-up. On the other hand, use of the system may be restricted to localities with productive estuarine water and where land use is not a limitation.

Comparison with Extensive Endogenous Production Systems

By coupling an estuarine-based zooplankton system with an intensive larval fish-rearing system, the authors have produced marine fish in an estuarine setting. Coproduction of zooplankton with marine fish in ponds (coproduction or endogenous zooplankton production) requires large volumes of high-salinity water, which limits those production systems to sites with ready access to that water.

In addition, because outdoor ponds or tanks are subject to dilution by rain and concentration by evaporation, endogenous production systems may not be as suitable for the larval fish. The zooplankton production system developed by the authors has not been adversely affected by the fluctuating salinities and high temperatures characteristic of estuarine areas. In fact, the system has been successfully operated during several tropical storms when estuarine salinity was reduced to nearly 0 ppt. Salinity changes in the estuary are not transmitted directly to the tanks because water exchange occurs only every other day, and then only at 40% of the tank volume.

During periods of extremely low salinity in the copepod production tanks, however, copepods are placed into high salinity to feed the fish larvae. Such a salinity change can cause copepod mortality. Although copepods were acclimated to the higher salinity of the fish larvae tanks in 1998–2001, laboratory studies performed here have indicated that survival of copepods transferred directly to larval production tanks was not appreciably different from that of copepods acclimated over several hours. The threshold of salinity shock of *A. tonsa* collected from Davis Bayou was between 15 ppt and 20 ppt 1 hour after shock (Zimmerman et al. 2004), which is in good agreement with the salinity shock tolerance of *A. tonsa* determined by Cervetto et al. (1999). This information has lead to the practice of directly transferring copepods to larval fish tanks to avoid depletion of their nutritional quality. During the

Table 17.9. Larval rearing of red snapper in 2001[a]

Spawn batch	Tank number	Location[b]	Comment[c] Treatment, study, or other comment	Culture volume (L) day	Larvae stocked per L	Harvest age	% Survival
1	1	Indoors	—	200	10	18	14.2
1	2	Indoors	—	200	10	18	8.3
1	3	Indoors	—	200	10	18	10.8
1	4	Indoors	—	500–1,000	20	18	7.4
1	5	Indoors	—	500–1,000	20	18	8.3
1	6	Indoors	—	500–1,000	20	18	8.3
2	7	Shade	—	200	10	18	0.2
2	8	Shade	—	200	10	18	0
2	9	Shade	—	200	10	18	0
2	10	Indoors	—	200	10	18	2.7
2	11	Indoors	—	200	10	18	6.7
2	12	Indoors	—	200	10	18	4.2
2	13	Indoors	—	500–1,000	20	21	2.7
2	14	Indoors	—	500–1,000	20	21	2.5
2	15	Indoors	—	500–1,000	20	21	1.5
3	16	Shade	—	500–1,000	20	20	0.6
3	17	Shade	—	500–1,000	20	20	0.8
3	18	Shade	—	500–1,000	20	20	0.4
3	19	Shade	—	500–1,000	20	20	0.5
4	20	Greenhouse	Dead day 3	500	20	—	0
4	21	Greenhouse	Dead day 3	500	20	—	0
4	22	Greenhouse	Dead day 3	500	20	—	0
4	23	Greenhouse	Dead day 3	500	20	—	0
5	24	Greenhouse	—	500–1,000	20	27	4.3
5	25	Greenhouse	—	500–1,000	20	27	0.4
5	26	Greenhouse	—	500–1,000	20	27	1.1
5	27	Greenhouse	—	500–1,000	20	27	3.1
6	28	Indoors	14L:10D	200	10	20	1.6
6	29	Indoors	14L:10D	200	10	20	0
6	30	Indoors	14L:10D	200	10	20	2.7
6	31	Indoors	24L:0D	200	10	20	0.8
6	32	Indoors	24L:0D	200	10	20	0.8
6	33	Indoors	24L:0D	200	10	20	1.8
7	34	Indoors	250 mL/min	200	10	22	4.8
7	35	Indoors	250 mL/min	200	10	22	3
7	36	Indoors	250 mL/min	200	10	22	3.6
7	37	Indoors	50 mL/min	200	10	22	3.2
7	38	Indoors	50 mL/min	200	10	22	7.8
7	39	Indoors	50 mL/min	200	10	22	0.4

[a]All tanks were fed exclusively copepods.

[b]Indoors was a room with no windows. Each tank had a 100-W metal halide lamp; greenhouse was a plastic Quonset-style greenhouse with a 40% shade cloth; shade was 80% shade cloth cover over clear plastic roof.

[c]14L:10D is 14-h light followed by 10-h dark photoperiod; 24L:0D is continuous light photoperiod; 250 mL/min and 50 mL/min were the aeration rates.

[d]All larvae died at day 3 after stocking.

[e]Culture volume ranges indicate a tank's initial and final volumes.

Table 17.10. Larval rearing of red snapper in 2002

Spawn batch	Tank number	Comment[a]	Culture volume (L)[b]	Larvae stocked per L	Harvest age (day)	Survival (%)
1	1	Standard	500–1,000	20	24	1.0
1	2	Standard	500–1,000	20	24	6.1
1	3	Standard	500–1,000	20	24	5.5
1	4	Standard	500–1,000	20	24	23.6
2	5	Standard	500–1,000	20	24	7.7
2	6	Standard	500–1,000	20	24	27.4
3	7	Standard	500–1,000	20	24	33.7
3	8	Standard	500–1,000	20	24	18.6
4	9	Standard	500–1,000	20	24	27.1
4	10	Standard	500–1,000	20	24	18.5
4	11	Standard	500–1,000	20	24	28.1
4	12	Standard	500–1,000	20	24	18.7
5	13	Standard	500–1,000	20	24	6.2
5	14	Standard	500–1,000	20	24	31.5
5	15	Standard	500–1,000	20	24	25
5	16	Standard	500–1,000	20	24	6.8
6	17	Tank size	2,000–4,000	20	24	1.2

[a]Standard indicates that larvae were reared at 25 ppt salinity in a room with no windows. Each tank had a 100-W metal halide lamp; Spawn tank: 4,000-L tank under standard conditions.
[b]Culture volume ranges indicate a tank's initial and final volumes.

2000 production year, the salinity being used to rear the fish larvae was lowered from 30 ppt to 25 ppt to reduce the mortality of copepods associated with salinity shock during the transfer from zooplankton tanks to larval fish tanks. In addition to increasing prey numbers, decreasing the salinity from 30 ppt to 25 ppt provided a benefit to red snapper larvae (Zimmerman et al. 2003) The zooplankton production system operates well throughout the spring and summer, when temperatures may reach 40°C in the tanks. Although the authors have not operated the system continuously outside of spring and summer, they have occasionally checked operation at other times of the year, and production seems unaffected.

The above-described system is managed for semicontinuous production of zooplankton, and therefore it is not as affected by the bloom and crash cycles typical of fertilized coproduction ponds. In coproduction ponds, it is necessary to closely match larval stocking with peak zooplankton abundance in the ponds. In the system described here, zooplankton peaks in individual tanks are not as closely tied to larval fish availability because larvae are not reared in the zooplankton production tanks and harvests from several tanks helps alleviate within-tank variability in copepod density.

Comparison with Intensive Exogenous Systems

The zooplankton production system described herein requires only an inexpensive tank, a pump, and a sieve. It is simple and therefore more reliable per unit of effort than most intensive systems. In particular, it is not subject to collapse after a power failure, and the only labor required is for harvesting and cleaning of tanks. Although copepod densities are lower than those in intensive systems, it may be more efficient because of reduced cannibalism.

Intensive production of marine fish involves the production of several live feeds (e.g., algae, rotifers, brine shrimp) and therefore requires several production units, while the one described above requires only two, that is, one for fish and one for copepods. This system does not use supplemental phytoplankton cultures, while intensive copepod production systems described to date require one or two species of phytoplankton. Additional requirements for space, labor, and cost are incurred with phytoplankton culture. Although generally reliable, mass phytoplankton cultures are subject to crashes.

The system described here is designed to provide semicontinuous production of a variety of copepod sizes (nauplii, copepodids, and adult) simultaneously. Differential harvesting by size allows the producer to meet the changing prey-size requirements of fish larvae as they grow. On the other hand, intensive copepod systems target prey of a single size (usually the nauplius), and the systems must be supplemented or restarted regularly with copepods. The animals for supplementation come from a broodstock culture that is maintained separately. Because a narrow range of sizes is targeted, additional live feeds must be available to meet the requirements for larger prey as the fish larvae grow. Rotifers and brine shrimp are most commonly used for feeding larger fish larvae.

Any need for rotifers and brine shrimp brings additional requirements. Rotifer production requires phytoplankton as food, and although a common practice, rotifer culture is subject to failure (Dhert et al. 2002; Hagiwara et al. 2002). Brine shrimp can be readily hatched from stored cysts. The commercial supply of those cysts, however, is variable, and the price fluctuates and at times is excessive. Perhaps even more importantly, both rotifers and brine shrimp are known to be nutritionally deficient and require enrichment with essential fatty acids before feeding (Sorgeloos et al. 1993; McEvoy and Sargent 1998).

Expansion of Production

Production in the system developed by the authors can be augmented to meet commercial requirements by optimizing (a) the harvest rate, (b) the labor, (c) copepod use, or (d) production by fertilizing the system.

Harvest Rate

At present, the authors harvest the system at 40% every other day. Laboratory experiments, however, using 1,000-L tanks (Lemus 2001) suggest that the naupliar harvest rate can be as high as 75% every other day without affecting the cultures. Harvesting at 75% would increase production with no additional facilities and virtually no increase in labor.

Labor

Zooplankton is harvested twice daily from two tanks during each harvest period. Harvesting zooplankton currently takes 4 hours per day, but only a few minutes of that time are required for personnel involvement to concentrate the zooplankton from two tanks and wash sieves for each collection period. Despite the low level of physical labor required, a person must be present for the 4 hours to monitor the sieves to ensure that they do not clog. Little extra effort would be required for one person to operate as many as six or possibly eight sieves during the 4-hour harvest time. Therefore, expansion to 24 (12 tanks harvested daily) or 32 (16 tanks harvested daily) zooplankton tanks would increase production by three- or fourfold with no additional labor.

Copepod Use

The optimal feed rate has yet to be determined for red snapper. At present, the authors feed the entire daily harvest to four rearing tanks and have fed as many as 20 per milliliter. When optimal feed rates are determined, copepods can be used more efficiently and perhaps more fish can be fed.

Production

Ponds are not fertilized because it results in blooms of rotifers that interfere with copepod production. Lemus (2001) and Ogle et al. (2002), however, found that supplementing with rice bran increased copepod productivity. Rice bran has also been used by Turk et al. (1982) as a sole food source to produce *A. tonsa* and has been used to enhance zooplankton in ponds (Bishara 1979). Perhaps fertilizing with rice bran would increase copepod yields without the occurrence of rotifer blooms.

SUMMARY

Work over the past 5 years has resulted in a standardized system and management protocol for rearing marine fish larvae. That system has the following salient features: an extensive zooplankton outdoor production system coupled with an intensive larval-rearing system, exclusive use of size-graded copepods as food, use of clear rather than green water, a three-phase water management scheme, consisting of (a) static water, followed by (b) doubling the water volume, and subsequently (c) water recirculation with biofiltration, 25 ppt. water, and indoor operation at constant temperature (26–28°C) under artificial light.

ACKNOWLEDGMENTS

The authors acknowledge the Alabama Department of Natural Resources' Claude Peteet Mariculture Center for supplying larvae during 1998 and 1999, The Oceanic Institute finfish program for larvae during 2000 and 2001, Dr. Paul Chigbu of Jackson State University for help with production, and the cadre of University of Southern Mississippi graduate students and staff for their assistance. This work was funded in part through grant numbers NA86FL0476, NA96FL358, and NA06FL0501 from the National Oceanic and Atmospheric Administration, NOAA Fisheries.

REFERENCES

Anderson, R.O. and Tave, D. 1993. *Strategies and Tactics for Management of Fertilized Hatchery Ponds*. Binghamton, New York: The Haworth Press. 262 pp.

Arnold, C.R., Wakeman, J.M., Williams, T.D., and Treece, G.D. 1978. Spawning of red snapper (*Lutjanus campechanus*) in captivity. *Aquaculture* 15:301–302.

Barnabe, G. 1990. Harvesting zooplankton. In *Aquaculture,* edited by G. Barnabe, Vol. 1, pp. 264–272. New York, New York: Ellis Horwood.

Bishara, N.F. 1979. Fertilizing fish ponds III. Growth of *Mugil capito* in Egypt by pond fertilization and feeding. *Aquaculture* 16:47–55.

Bootes, K.L. 1998. Culture and description of larval red snapper, *Lutjanus campechanus*. Master's Thesis. Auburn University, Auburn, Alabama, USA. 142 pp.

Cervetto, G., Gaudy, R., Pagano, M. 1999. Influence of salinity on the distribution of Acartia tonsa (Copepoda, Calanoida). *Journal of Experimental Marine Biology and Ecology* 239:33–45.

Chesney, E.J. and Filippo, R.A.S. 1994. Size dependent spawning and egg quality of red snapper. MARFIN final report February 28, 1994, Project No. NA37FF0048-01. National Marine Fisheries Service, St. Petersburg, Florida, USA.

Dhert, P., Rombaut, G., Suantika, G., and Sorgeloos, P. 2002. Advancement of rotifer culture and manipulation techniques in Europe. *Aquaculture* 200:129–146.

Doi, M., Singhagraiwan, T., Singhagraiwan, S., and Ohno, A. 1994. An investigation of copepods being applied as initial food organisms for red snapper larvae. *Thai Marine Fisheries Research Bulletin* 5:21–26.

Geiger, J.G. 1983. A review of pond zooplankton production and fertilization for the culture of larval and fingerling striped bass. *Aquaculture* 35:353–369.

Hagiwara, A., Gallardo, W.G., Assavaaree, M., Kotani, T., and de Araujo, A.B. 2002. Live food production in Japan: recent progress and future aspects. *Aquaculture* 200:111–127.

Harrell, R.M., Kerby, J.H., and Minton, R.V. 1990. Culture and propagation of striped bass and its hybrids. Striped Bass Committee, Southern Division, American Fisheries Society, Bethesda, Maryland, USA. 323 pp.

Heinle, D.R. and Flemmer, D.A. 1975. Carbon requirements of a population of the estuarine copepod *Eurytemora affinis*. *Marine Biology* 31:235–247.

Horvath, L., Tamas, G., and Tolg, I. 1984. *Special Methods in Pond Fish Husbandry*. Seattle, Washington: Halver Corporation. 148 pp.

Ikeda, T. 1973. On the criteria to select copepod species from mass culture. *Bulletin of the Plankton Society of Japan* 20:41–48.

Ito, T. 1955. Studies on the "mizukawari" in eel-culture ponds I. The feeding activity of *Brachionus plicatilis* on phytoplankton (as a cause of "mizukawari"). *Report of the Faculty of Fisheries Prefecture University of Mie* 2:162–167.

Ito, T. 1960. On the culture of the mixohaline rotifer *Brachionus plicatilis* O.F. Muller in sea water. *Report of the Faculty of Fisheries Prefecture University of Mie* 3:708–709.

Knuckey, R., Semmens, G., and Della-Rodolfa, B. 2001. Research in progress at the Live Prey Research Unit, QDPI Northern Fisheries Centre, Cairns. In *Hatchery Feeds: Proceedings of a Workshop Held in Cairns*, pp. 34–41. Cairns, Australia, 9–10 March 2000. Australian Institute of Marine Science.

Lannan, J.E., Smitherman, R., and Tchobanoglous, G. 1986. *Principles and Practice of Pond Aquaculture*. Corvallis, Oregon: Oregon State Press. 252 pp.

Lemus, J.T. 2001. Development of a protocol for the mass production of copepods. Master's Thesis. University of Southern Mississippi, Hattiesburg, Mississippi, USA. 62 pp.

Lemus, J.T., Ogle, J.T., Marcus, N.H., and Lotz, J.M. 2002. Potential calanoid copepods for the larviculture of the red snapper *Lutjanus campechanus*. In *Book of Abstracts, Aquaculture America 2002*, p. 68. San Diego, California, USA, 27–30 January 2002. Baton Rouge, Louisiana: World Aquaculture Society.

Lippman, E.E., Kao, K.R., and Phelps, R.P. 2001. Production of the copepod *Oithona* sp. under hatchery conditions. In *Book of Abstracts. Aquaculture 2001*, p. 379. Lake Buena Vista, Florida, 21–25

January 2001. Baton Rouge, Louisiana: World Aquaculture Society.

Lonsdale, D., Heinle, D., and Siegfried, C. 1979. Carnivorous feeding behaviour of the adult calanoid copepod *Acartia tonsa*, Dana. *Experimental Marine Biology and Ecology* 36:235–248.

Marcel, J. 1990. Fish culture in ponds. In *Aquaculture*, edited by G. Barnabe, Vol. 2, pp. 593–627. New York, New York: Ellis Horwood.

May, R.C. 1970. Feeding larval marine fishes in the laboratory: a review. California Marine Resource Commission. *California Cooperative Oceanic Fisheries Investigations Report* 14:76–83.

McEvoy, L.A. and Sargent, J.R. 1998. Problems and techniques of live prey enrichment. *Bulletin of the Aquaculture. Association of Canada* 98:12–16.

Minton, R.V., Hawke, J.P., and Tatum, W.M. 1983. Hormone induced spawning of red snapper, *Lutjanus campechanus*. *Aquaculture* 30:363–368.

Nellen, W. 1985. Live animal food for larval rearing in aquaculture. Non-*Artemia* organisms. In *Realism in Aquaculture. Proceedings of the World Conference on Aquaculture,* edited by M. Bilio, H. Rosenthal, and C.J. Sinderman, pp. 215–248. Venice, Italy, 21–25 September 1981. Bredene, Belgium: European Aquaculture Society.

Ogle, J.T. 1979. Adaptation of a brown water culture technique to the mass culture of the copepod *Acartia tonsa*. *Gulf Research Reports* 6:291–292.

Ogle, J.T. 1982. Operation of an oyster hatchery utilizing a brown water culture technique. *Journal of Shellfish Research* 2:153–156.

Ogle, J.T. 1991. A method to construct low-cost ponds for aquaculture. *Gulf Research Reports* 8:299–300.

Ogle, J.T. and Lotz, J.M. 2000. Culture of red snapper. *Global Aquaculture Advocate* 3(5):23–26.

Ogle, J.T. and J.M. Lotz. 2002. Larval culture of the red snapper *Lutjanus campechanus*. In *Program and Abstracts, Joint Annual Meeting Mississippi and Louisiana Chapters of the American Fisheries Society*, p. 47. Biloxi, Mississippi, USA, 6–8 February 2002. Ocean Springs, Mississippi: Gulf Coast Research Laboratory.

Ogle, J.T., Lotz, J.M., Nicholson, L.C., and Barnes, D.N. 2000. Larval culture of red snapper *Lutjanus campechanus* using copepod nauplii for first feeding. In *Book of Abstracts. Aquaculture America 2000*, p. 249. New Orleans, Louisiana, USA, 2–5 February 2000. Baton Rouge, Louisiana: World Aquaculture Society.

Ogle, J.T., Nicholson, L.C., Barnes, D.N., Shields, R.J., and Lotz, J.M. 2001. Recent developments in the culture of larval red snapper *Lutjanus campechanus*. In *Book of Abstracts Aquaculture 2001*, p. 490. Lake Buena Vista, Florida, USA, 21–25 January 2001. Baton Rouge, Louisiana: World Aquaculture Society.

Ogle, J.T., Nicholson, L.C., and Lotz, J.M. 2002. Culture of the Copepod *Acartia tonsa* utilizing various artificial feeds. *Proceedings of the Fifty-Third Annual Gulf and Caribbean Fisheries Institute*, pp. 234–240. Biloxi, Mississippi, USA, 6–10 November 2000. Fort Pierce, Florida: Gulf and Caribbean Fisheries Institute.

Ohno, A. and Okamura, Y. 1988. Propagation of the calanoid copepod, *Acartia tsuensis*, in outdoor tanks. *Aquaculture* 79:39–51.

Pafenhoffer, G.A. and Strickland, J.D.H. 1970. A note on the feeding of *Calanus helgolandicus* on detritus. *Marine Biology* 5:97–99.

Parrish, K. and Wilson, D. 1978. Fecundity studies on *Acartia tonsa* (Copepoda: Calanoida) in standardized culture. *Marine Biology* 46:65–81.

Payne, M.F. and Rippingale, R.J. 2001. Intensive cultivation of the calanoid copepod *Gladioferens imparipes*. *Aquaculture* 201:329–342.

Pedersen, B.H. 1984. The intestinal evacuation rates of larval herring (*Clupea harengus* L.) predating on wild plankton. *Dana* 3:21–240.

Poulet, S.A. 1978. Comparison between five coexisting species of marine copepods feeding on naturally occurring particulate matter. *Limnology and Oceanography* 23:1126–1143.

Poulet, S.A. 1983. Factors controlling utilization on non-algal diets by particle grazing copepods: a review. *Oceanologica Acta* 69:221–234

Rabalais, N.N., Rabalais S.C., and Arnold, C.R. 1980. Description of eggs and larvae of laboratory reared red snapper *Lutjanus campechanus*. *Copeia* 4:704–708.

Raymont, J.E.G. and Miller, R.S. 1962. Production of marine zooplankton with fertilization in an enclosed body of sea water. *Hydrobiologia* 47:169–209.

Rollefsen, G. 1939. Artificial rearing of the fry of sea water fish. Preliminary communication. *Rapports el Proces Verbaux des Conseil International Pour l'Exploration de la Mer* 30:204–221.

Roman, M.R. 1977. Feeding of the copepod *Acartia tonsa* on the diatom *Nitzschia closterium* and brown algae (*Fucus vesiculosus*) detritus. *Marine Biology* 42:149–155.

Schipp, G.R., Bosmans, J.M.P., and Marshall, A.J. 1999. A method for hatchery culture of tropical calanoid copepods, *Acartia* spp. *Aquaculture* 174:81–88.

Shields, R.J. 2001. Larviculture of marine finfish in Europe. *Aquaculture* 200:55–88.

Shields, R.J., Bell, J.G., Luizi, F.S., Gara, B., Bromage, N.R., and Sargent, J.R. 1999. Natural copepods are superior to enriched *Artemia* nauplii as feed for halibut larvae (*Hippoglossus hippoglossus*) in terms of survival pigmentation, and retinal morphology: relation to dietary essential fatty acids. *Journal of Nutrition* 12:1186–1194.

Shields, R.J., Molnar, A., Ostrowski, A.C., and Ogle, J. 2001. Development of intensive larviculture techniques for red snapper *Lutjanus campechanus*. In *Book of Abstracts. Aquaculture 2001*, p. 587. Lake Buena Vista, Florida, USA, 21–25 January 2001. Baton Rouge, Louisiana: World Aquaculture Society.

Shields, R.J., Kotani, T., Molnar, A., Marion, K., Kobashigawa, J., and Tang, L. 2005. Intensive culture of a calanoid copepod, *Parvocalanus* sp., as prey for small sub-tropical marine fish larvae. In *Culture of Copepods and Applications to Marine Finfish Larval Rearing*, edited by C.-S. Lee, P.J. O'Bryen, and N.H. Marcus. Ames, Iowa: Blackwell Publishing Professional.

Singhagraiwan, S., Charaoenwat, J., Suwannachote, P., Singhagraiwan, T., Doi, M., and Ohno, A. 1994. Effect of fertilization on the propagation and mortality patterns of the calanoid copepod, *Acartia sinjiensis*. *Thai Marine Fisheries Research Bulletin* 5:37–45.

Sorgeloos, P., Dehasque, M., Dhert, P., and Lavens, P. 1993. Review of present status of marine fish larviculture. In *Fish Farming Technology. Proceedings of the First International Conference on Fish Farming Technology*, edited by H. Reinertsen, L.A. Dahle, L. Jorgensen, and K. Tvinnereim, p. 3. Trondheim, Norway, 9–12 August 1993. Rotterdam, Netherlands: AA Balkema.

Stein, J.P. 1981. Spatial and temporal distribution of zooplankton in a low-salinity Mississippi bayou system. Ph.D. dissertation, University of Mississippi, University, Mississippi, USA. 183 pp.

Støttrup, J.G., Richardson, K., Kirkegaard, E., and Pihl, N.J. 1986. The cultivation of *Acartia tonsa* for use as a live food source for marine fish larvae. *Aquaculture* 52:87–96.

Trimble, W.C. and Tatum, W.M. 1978. Experimental pond production of selected marine species, completion report for NOAA (National Oceanic and Atmospheric Administration), NFS, PL88-309 Project 2-263-R, U.S. Dept. of Commerce, Washington, D.C., USA. 127 pp.

Tucker, J.W. 1998. *Marine Fish Culture*. Boston, Massachusetts: Kluwer Academic Publishers. 750 pp.

Turk, P.E., Krejci, M.E., and Yang, W.T. 1982. A laboratory method for the culture of *Acartia tonsa* using rice bran. *Journal of Aquariculture and Aquatic Sciences* 3(2):25–27.

Turner, J.T. 1984. The feeding ecology of some zooplanktors that are important prey items of larval fish. NOAA (National Oceanic and Atmospheric Administration) Technical Report NMFS-7, U.S. Dept. of Commerce, Washington, D.C., USA. 28 pp.

Whitt, U., Quantz, G., Kuhlmann, O., and Kattner, G. 1984. Survival and growth of turbot larvae, *Scophthalmus maximus* L. reared on different food organisms with special regard to long chain polyunsaturated fatty acids. *Aquacultural Engineering* 3:177–190.

Vasil'eva, G.L. and Okuneva, G.L. 1961. Experiments on rearing the rotifer *Brachionus rubens* Ehrbg as food for young fish. *Voprosy Ikhtiologii* 4:752–761.

Zimmerman, J.K., Lotz, J. M., and Ogle, J.T. 2003. Salinity tolerance of red snapper *Lutjanus campechanus* eggs and larvae. In *Book of Abstracts, Aquaculture America 2003*, p. 345. Louisville, Kentucky, USA, 18–21 February 2003. Baton Rouge, Louisiana: World Aquaculture Society.

Zimmerman, J.K., Lotz, J.M., and Ogle, J.T. 2004. Salinity tolerance of *Acartia tonsa*. In *Abstracts Book, Aquaculture 2004. Aquaculture — An Ecologically Sustainable and Profitable Venture*, p. 673. Honolulu, Hawaii, USA, 1–5 March 2004. Baton Rouge, Louisiana: World Aquaculture Society.

18
Culture of Copepods and Applications to Marine Finfish Larval Rearing Workshop Discussion Summary

Patricia J. O'Bryen and Cheng-Sheng Lee

INTRODUCTION

A workshop hosted by The Oceanic Institute was held in Honolulu, Hawaii, USA, on May 5–8, 2003, under a grant from the Oceanic and Atmospheric Research Program of the National Oceanic and Atmospheric Administration (NOAA). Academic and industry representatives consisting of copepod biologists and fish culturists met to present papers reviewing the current status of research in their respective fields and to discuss the challenge of first feeding in various marine finfish species. Before the workshop, participants sent their suggestions for topics to be discussed, which were grouped into nutritional aspects and utilization of copepods, disease issues, and issues of production and transport. A draft discussion summary was sent to all of the invited speakers for comment, and it was clear that, had there been more time for discussion at the workshop, topics would have been covered in more depth and more opinions would have been expressed on the various issues. To avoid duplication, this chapter therefore presents a sample of the comments on the major topics discussed at the workshop, but is not intended to represent the views of NOAA or of all of the participants on all of the topics discussed.

SOURCES OF COPEPODS

Collecting wild plankton is partly an engineering problem and partly an issue of economics. A plankton net would be needed for collecting them, but the efficiency of collection would vary widely according to location. In estuaries it would be difficult to collect using plankton nets because of the detrital mass. The unique plankton wheel culture system has been developed for large-scale harvesting of copepods in natural, enriched environments within a relatively enclosed body of water. A vast pumping and screening system for zooplankton was designed and commercialized in Norway for use in enclosed fertilized fjords systems.

A pathogen entry route certainly exists when using wild-collected organisms. Cultured cod (*Gadus morhua*) larvae that were fed untreated wild-collected copepods became infected with parasitic copepods. *Epistylus* sp., a stalked ciliate that negatively impacts productivity, has been observed on both adults and nauplii of *Acartia* sp. reared as feed for larval red snapper (*Lutjanus campechanus*). Low-dose copper sulfate might be used to remove some of the contaminants that are introduced with the copepods. Changing the salinity was another possible way to control cer-

Corresponding author: Patricia O'Bryen, The Oceanic Institute, 41-202 Kalanianaole Hwy., Waimanalo, HI 96795, USA; Phone: (808) 259-3176; Fax: (808) 259-8395; E-mail: pobryen@oceanicinstitute.org

tain pathogens on wild-caught copepods. One participant reported that treating harpacticoid copepods with a small amount of chlorine bleach or formalin right before they molted successfully removed ciliates and bacteria that had attached to the copepods.

It has been shown that nodaviruses were introduced into pond culture systems with wild copepods. Introduction of viral pathogens is especially problematic in a hatchery environment. Once a virus is introduced and it begins replicating, a mass mortality event develops quickly. Treatment with ultraviolet light is a relatively easy method of reducing the number of pathogens introduced with the incoming water.

Commercial harvesting of wild copepods is not expected to meet the needs of aquaculture in terms of quality, quantity, and constant supply. For these reasons, culture methods to produce large numbers of copepod nauplii need to be developed further.

CULTURE METHODS FOR COPEPODS

Copepods are raised as live feed by a variety of methods and under different culture conditions. Part of the strategy for culturing copepods is growing the appropriate phytoplankton. Like rotifer production, continuous or batch culture methods can be used to produce copepods.

In the 1990s, copepod production farms began to appear in Taiwan. Typically, copepods are grown at low salinity in ponds with no water exchange. Some producers prefer ponds with clay bottoms because nutrients that are beneficial for plankton blooms can be concentrated in these types of ponds. Other producers prefer ponds with a sandy bottom for a better environment for the copepods. Initially, copepods were grown in the ponds for 10–20 days, after which the entire population was harvested at one time. Later, farmers found it was better to harvest about 25% of the total every day for about 1 month. By doing interval harvests, farmers increased the total yield by up to 30% from the same volume of water.

Initial studies indicate that harpacticoids are comparatively easy to culture. Among the calanoid copepods, *Acartia* spp. are much easier to cultivate than *Calanus* spp. or *Temora* spp. Culture techniques may be considerably different between even closely related species. For example, techniques for raising *Calanus helgolandicus* were not transferable to *C. glacialis*.

FEED FOR COPEPODS

It is important to provide an optimal diet for the copepod. Useful information about what to feed copepods may be gained by identifying what wild copepods eat and the types of plankton existing in their surroundings. Gut analysis or examination of the fecal pellets can show if the copepods are feeding on diatoms, which have silicious frustules. Microflagellates leave no obvious evidence, but identifying the plankton in the natural surroundings may give an indication of the types of microflagellates being eaten. This approach, however, may not always result in an accurate representation of this component of the diet.

Food particle size is an important factor to consider in selecting the food for copepods. For the copepods species discussed at the workshop, for instance, it was estimated that the food particle should be smaller than 5 μm. It was pointed out that *Isochrysis* sp. and other very small algae may be suitable for smaller copepods such as some *Acartia* spp., but may be too small for copepods such as the calanoid copepods *Centropages hamatus* or *Labidocera aestiva*. Good results were obtained by feeding copepods with *Isochrysis* sp., *Chaetoceros* sp., and *Tetraselmis* sp. *Nannochloropsis oculata* (~2–5 μm) was difficult for the copepods to digest. Digestibility of the algae, therefore, is equally as important as size.

Variations in the chemical composition of an algal species may produce variable effects. Comparisons of the effects of different microalgae on the calanoid copepod *Gladioferens imparipes* showed that *N. oculata* was an undesirable feed. At higher temperatures, *G. imparipes* grew to maturity with *Nannochloropsis* sp., but had no survival at lower temperatures. Among the algal species tested, *N. oculata* was the worst diet for *G. imparipes* in terms of nauplius and egg production. For each of the parameters studied, *Isochrysis* sp. was a better food for *G. imparipes*. Similarly, *Isochrysis* sp. was found to be the best of three species of microalgae used in large-scale culture of the cyclopoid copepod, *Apocyclops royi*. No significant differences in nauplius stage development were observed in laboratory cultures of *A. royi* fed *Isochrysis* sp., *Chaetoceros* sp. (~7 μm), or *Tetraselmis* sp. (~10 μm), while *Nannochloropsis* sp. completely inhibited development of the copepod.

Participants referred to studies conducted by Australian researchers on the effect of algal cell size on female fecundity in Paracalanid copepods. For example, a larger alga, *Heterocapsa* sp. (20 μm), resulted in higher copepod fecundity than the smaller *Chaetoceros* sp. Participants pointed out that some *Chaetoceros* spp. produce chemicals that reduce the viability of copepod eggs. Some *Isochrysis* spp. or larger-size algae do not support high fecundity.

Practical guidance on which groups of phytoplankton to investigate further and which to avoid for different life stages of the copepods has yet to be developed. Dinoflagellates and diatoms are both excellent diets for copepods, but dinoflagellates are more difficult to culture, very delicate, and seem to be negatively affected by aeration, making them a greater challenge for mass culture. Some diatom species are rich in fatty acids that are considered excellent for copepod growth and produce very high egg production, yet hatching success is zero, because these diatoms produce unwanted unsaturated aldehydes. Reports in the literature indicate that freshwater Chrysophyte algae produce these same molecules during the course of blooms. Several species of *Alexandrium*, a genus of dinoflagellates known to produce nontoxic strains of saxitoxins, seem to produce antiproliferative compounds, but the nature of these compounds is not yet known. While algal pastes were unsuccessful with a calanoid copepod species and the cyclopoid copepod *Apocyclops royi*, harpacticoid copepods, such as *Nitokra lacustris* and *Robertsonia* sp., have been successfully raised on a locally supplied algal paste. *Chaetoceros* sp. and *Skeletonema* sp. are relatively easy phytoplankton to grow, but *Skeletonema* is not a good food for copepods. Participants were generally in favor of using *Isochrysis* sp. as a feed for copepods unless another suitable algal species is found.

It is possible that more than one algal species is required to culture a copepod species, but whether it has to be provided throughout the life span is not clear. For example, the calanoid copepod *Centropages hamatus* can be raised to adulthood on the dinoflagellate *Prorocentrum* sp., but it does not produce eggs on this species alone. It is necessary to switch to another dinoflagellate, *Gymnodinium* sp. Egg production of the calanoid *Acartia tonsa* being fed *Rhodomonas lens* and *R. salina* was poor, relative to reports in published studies. Switching to *Gymnodinium* sp. resulted in four to five times more eggs per female.

Hatchery operators are well aware of the advantages of using multiple algal species. If the diet for the copepod contains a mixed culture of different species, there is a greater chance of getting the desired nutrients. *Oxyrrhis marina*, a heterotrophic dinoflagellate, was used successfully with the microalgae *Isochrysis* sp. to culture *Acartia* sp. The microalgae is good for the nauplii, and the larger, heterotrophic dinoflagellate is good for the adults. Thus far, this approach has been used only to grow on an experimental scale, but it seems promising for larger-scale cultures. Another participant mentioned that a mix of four dinoflagellates, *Gymnodinium nelsoni*, *Gonyaulax pallida*, *Prorocentrum micans*, and *Scrippsiella trochoidia*, was found to be effective for feeding copepods that produce diapause eggs. A mixed diet of algae *Tetraselmis* sp. and *Nannochloropsis* sp. plus *Chaetoceros* sp. or *Skeletonema* sp. was used in a 200-L recirculating system for culturing copepods in the Philippines. Production of eggs reached about 1 million per day, but decreased after about 6–7 days, and viability decreased from 80–90% to about 35%, possibly due to unsaturated aldehydes present in the *Skeletonema* sp. and *Chaetoceros* sp.

The microbial food web is thought to play a major role in the production of copepods in ponds, where the application of organic fertilizer not only stimulates the production of phytoplankton but also that of heterotrophic bacteria. Some species of calanoid copepods (e.g., *Acartia* spp.) and harpacticoid copepods have been cultured with bacteria. Harpacticoid copepods are able to scavenge bacteria off surfaces, and the nauplii have been found to develop well on bacteria. Copepods, however, may lack the necessary enzymes to detoxify a variety of noxious compounds contained in the bacteria, thus bacteria may take over the culture and cause a number of water quality problems.

Harpacticoid copepods are nonselective feeders, ingesting both phytoplankton and detritus. A formulated feed consisting of yeast, flax seed oil, and a vegetable juice produced good culture results with the harpacticoid copepods, *Nitokra lacustris*, *Tisbe holothuriae*, and *Tigriopus californicus*. The generation time of *T. californicus* was shorter when it was fed the formulated feed than with live feeds. It was proposed that the formulated feed stimulated bacterial growth and that the copepods fed on bacteria. There is some concern about the bacterial buildup in the culture system when using artificial feeds. It was recommended that the amount of formulated feed be adjusted to result in less debris so that the bacterial buildup in the culture system could be controlled.

An important key to selecting an artificial food for copepods is finding one with the appropriate chemosensory signal, which may be species-specific. The suggested approach to identifying the appropriate feed was to put different artificial feeds into natural seawater and see if any of the candidate copepod species proliferates. These species should be harvested and fed the artificial feed to see if they can reproduce solely on the feed.

Besides the right attractants and nutrients, the artificial feed must possess physical characteristics that are detectible by copepods. Making artificial feed leaf-shaped to suspend it in the water column may make it more attractive to copepods. Slight aeration in the rearing tanks causes vibrations of leaf-shaped particles that may attract copepods.

GROWTH AND DENSITY OF COPEPODS

Development in copepods varies across genera. The nonfeeding developmental stages of certain symbiotic harpacticoid copepods develop very quickly; that is, they complete a development stage almost every 6 hours and complete their development up to the CI stage in 48 hours. The calanoid copepod *Gladioferens imparipes* takes about 11–12 days at 25°C from hatching to reproductive maturity and is capable of reproducing for at least 4 weeks.

There is some published information available about the maximum longevity of copepods. The life span of adult female calanoid copepods ranges from 8 to 80 days, depending on living conditions, during which time they continue to produce eggs. Survivorship of subtropical copepods being grown in the laboratory for use in aquaculture has been good. Percentage of survivorship is likely to be a combination of husbandry practices and senescence within the population. In a system in which the nauplii were cropped out at regular intervals without recruitment within the system, the life span of the adults was estimated at about 2 weeks.

To meet the needs of aquaculture, a system designed for copepod production must be expandable, with an ability to produce multiple millions of the appropriate-sized stage of copepod per day on demand. Participants expressed some doubt if such an economically viable system would ever exist. Scaling up to commercial production also depends on the type of larval fish being produced. Commercial scale production can be achieved in the types of systems already in use for ornamental fish, but would severely limit production of food fish.

For some copepod species, there seems to be a certain obligate domestication period, i.e., when given a hatchery diet of two or three algal species, wild-caught copepods do not perform well in culture for several generations. After a 1- or 2-month period, their production slowly increases and then takes off. In some harpacticoid copepod species, separating the broodstock from the production has been shown to have a positive effect on production.

There seems to be some sort of self-regulation occurring in copepod populations. Hatchery operators have not been able to increase the density of their copepod cultures beyond a certain level, even though the water conditions were favorable. Selecting a suitable species of copepod for high-density culture is essential, since some copepod species are very sensitive to crowding. In at least one species of calanoid copepod, *Centropages typicus*, egg-production rates are highly density dependent. If certain species of copepods are in close proximity to each other, it interferes with their feeding behavior and decreases their potential productivity. Swarming copepods, which include some *Acartia* spp. in tropical areas, and the cylclopoid copepod, *Dioithona* sp., adapt well to being cultured at high densities. Some harpacticoid copepods, such as *Nitokra lacustris*, thrive at densities of > 150/mL, even in large-scale cultures.

If the objective is to have more nauplii, it is better to lower the stocking density of females and to remove the females after spawning, since this would decrease cannibalistic behavior and minimize the effect of crowding. Nevertheless, with current aquaculture systems, there is a limitation on biomass for the production of nauplii. For example, a 500-L tank stocked at a fairly large biomass of adults can produce a million copepods, but the number of 500-L tanks needed to produce copepods to meet the needs at a commercial hatchery is astounding. The amount of scrubbing and cleaning that would be necessary because of the increased number of culture vessels was generally regarded as too labor-intensive to be cost-effective. These problems involve both engineering and biological issues.

BEHAVIOR OF COPEPODS

The nauplii of many, but not all, copepod species swim in the water column. While most harpacticoid naupliar stages are nonswimming, nauplii of some symbiotic poecilostomatoid copepods are swimming. Some copepod species swim at naupliar stages, but are nonswimming at later life stages. For example, during its copepodid stages, the calanoid copepod *Gladioferens imparipes* develops structures with which it attaches itself to surfaces and may then spend extended periods of time attached rather than swimming. Some copepod species produce extremely small copepodids that might be of a favorable size and have appropriate behavioral characteristics for culture. Copepodids, however, are generally more sensitive to disturbances and harder for larval fish to catch.

Under culture conditions, some copepod species will prey on their own eggs or nauplii. Cannibalism may occur very severely in some calanoid copepod species when the population density is high. When the calanoid copepod *Acartia* sp. is grown in culture vessels, regardless of being in a food-rich environment or a tank of the right design, adult copepods eat the eggs that accumulate on the bottom. Calculations of the clearance rates of the total population, however, have shown that adults consume a certain percentage of the eggs even if the eggs are evenly distributed in the water. Cannibalism of copepods on eggs or nauplii may be a behavioral response to captivity in the laboratory, as there is little evidence of this behavior occurring in nature.

The calanoid copepod *Sulcanus* sp. is omnivorous, and at high densities the populations are uniformly adult. It is not known whether they no longer reproduce after they have overexploited their food resources or if they have consumed all their nauplii. After the generation is entirely adult, it crashes. The same population structure of more adults and fewer nauplii was reported in *Parvocalanus* sp. On the other hand, no evidence of adults preying upon nauplii in even very small containers has been found in the calanoid copepods, *Eurytemora affinis*, *Gladioferens imparipes,* and *Pseudodiaptomus elongatus*.

One strategy for minimizing cannibalism in cultures is to separate the adult copepods from the nauplii using light. In one laboratory study, *G. imparipes* nauplii were collected by attracting them to light through a mesh, since adults made no response to light. Even if all of the life stages of copepods in a culture system were phototactic, they could be separated out through different sizes of mesh into different size categories.

REPRODUCTION OF COPEPODS

There is evidence in the literature that certain environmental or dietary variables skew the sex ratio in favor of females, but the optimal sex ratio for good egg production is not yet known. Some genera of copepods (e.g., *Paracalanus*, *Oithona,* and *Dioithona*) have a high female-to-male ratio, while others, for example, *Gladioferens*, produce an exact 50:50 sex ratio. Manipulating the sex ratio of culture populations would be one way of increasing productivity. In a laboratory culture of symbiotic poecilostomatoid copepods, the same females continued producing eggs for 800 days from one mating. In these species of copepods, fecundity is related to the age of the adult. It is likely that among most species of copepods, there is an optimal age among females for maximum reproduction. Removing part of the population and renewing younger females would be one strategy for sustaining fecundity of the population.

Reports of the effects of temperature on egg production are inconsistent. Some researchers have found that fewer eggs are produced per female per day at higher temperatures and suggested that a slightly lower culture temperature would produce a slightly larger copepod, which would produce more eggs. In contrast, although raising *G. imparipes* at higher temperatures produced a smaller female with a slightly smaller number of embryos per clutch, eggs were produced more frequently. Others report that egg production is greater at higher temperatures, but survival is lower. In general, however, there was not enough information for aquaculturists to be able to maximize the reproductive capacity of a particular copepod species by precisely manipulating the culture temperature.

Diapause eggs produced by copepods were discussed at the workshop and are presented in Chapter 1. An inability to provide a predictable number of nauplii when they are needed would be a serious problem for hatchery operators. Through a manipulation of environmental conditions, it may be possible to change the hatching time, but further studies are needed to establish the conditions. Using diapause eggs would allow hatchery operators to synchronize their culture and avoid the risk of having a naupliar culture suddenly crash just when it was needed for the fish larvae. A lot of basic research still needs to be done, because it is not yet known whether the biochemistry of diapause eggs changes nutritionally over that period of time. Ultimately, however, whatever is done with diapause eggs needs to be done on an economically feasible basis.

TRANSPORT OF COPEPOD PRODUCTS

A suitable density for transporting copepods is likely to be species-specific. When being transported, harvested copepods should be kept cool, with the water temperature at about 18–20°C. For transferring copepods over short distances, less water, with highly dissolved oxygen, results in better survival. Anesthetizing copepods to reduce swimming activity is not recommended.

In Taiwan, 10 kg of copepods can be transported to a nearby hatchery (reached within 2 hours) at 18°C, in 100 L of oxygenated water. Temperature seems to be the most critical factor for survival. The method and timing of the harvest are important for controlling the density of the copepods and, in turn, reducing the amount of injury to them during harvest and transport.

One participant reported keeping *Gladioferens imparipes* at 1,000 animals per liter in an unopened container for 3 weeks at low temperatures (6–8°C) and at 20°C for 3 or 4 days. Various species of harpacticoid copepods have been sent by overnight express at densities of 20–30 per milliliter without any problems. The swarming cylclopoid copepod, *Dioithona oculata*, has been transported from the Caribbean to Texas at similar high densities (20–30 per milliliter) in a sealed container for 24 hours without additional oxygen.

The importance of ensuring that exporting certain copepod taxa all around the world would not be contributing to the introduction of exotic species was briefly discussed. One suggested strategy was to choose species from among genera that were found locally, for example selecting an *Apocyclops* sp. for culturing in an area where another *Apocyclops* sp. was already found.

Participants also discussed the potential for introducing pathogens with transported copepods. Nodaviruses seem to be transferred on the outside of the eggs, rather than within the egg itself. In Taiwan, vertical transmission of a fish pathogen, viral nervous necrosis (VNN), was confirmed in grouper from broodstock, to the eggs, and to the larvae. It was found that VNN came from copepods harvested from fishponds, but not from the copepods that were raised in shrimp ponds. There may also be a potential for introducing pathogens that can be transferred up the food chain, although there does not seem to be any conclusive evidence that the virus is active within the copepod itself. The copepod may be ingesting water, for example, in which the virus is contained, and acting as a passive vector.

Ultraviolet light (UV) has been used successfully (100% reduction in *Vibrio* sp. contamination) for disinfecting copepod nauplii. With UV treatment, however, only the outside of the nauplius is being treated, not the internal surfaces. Other experiments have investigated the effects of UV treatment on copepod eggs and nauplii. Nauplius stage I and stage II are the most susceptible to UV irradiation. UV-C, which is often used in aquaculture to kill viruses, is very harmful, and flow rates through the UV light need to be carefully adjusted. UV-B has also been shown to have some harmful effects on copepod eggs and nauplii. Hatchery operators in Taiwan routinely treat copepods they buy with formalin. Ozone, which has been used to disinfect rotifers and *Artemia*, may be effective for disinfecting copepods. Although rotifers are quite susceptible to damage at higher concentrations and exposure periods, *Artemia* are very robust. Copepods are expected to have an intermediate response. Ozone treatment has perhaps greater potential as a way of obtaining clean stocks of diapause eggs.

NUTRITIONAL VALUE OF COPEPODS

More information is needed about the nutritional value of individual nauplii to fish larvae, since it relates to how many nauplii will be needed for large-scale fish production. Larvae of red snapper, a subtropical fish, have been successfully reared in a controlled environment using copepod nauplii as feed, but with considerable variability in survival, even within the same batch of larvae. Four- to six-day-old mahi mahi (*Coryphaena hippurus*) larvae, a tropical fish species, were reared successfully on nauplii of the harpacticoid copepod *Euterprina acutifrons*, but when *E. acutifrons* nauplii were tested on earlier stages of four species of smaller tropical finfish, all of the larvae died very quickly. Participants also pointed out that larval fish may ingest copepod nauplii, but the nauplii must be nutritionally rewarding. In the calanoid copepod *Gladioferens imparipes*, the lipid reserves of a freshly hatched nauplius are in the form of large droplets, which are very high in food value. In contrast, another calanoid copepod, *Acartia* sp., has prodigious daily egg production, but the lipid reserves of an individual nauplius when it first hatches are quite low.

The biochemical composition of copepods varies within and between copepod species, in terms of what they are fed and the biochemical profile of the nauplius produced, but there is little data available on the specific nutritional composition of the nauplius at first hatching. In general, nauplii incorporate proteins most rapidly for their somatic growth. Information on the lipids, fatty acids, and other composition in response to life-cycle stage and starvation, etc., is available for some Northern temperate marine copepods, but little is available for the tropical and subtropical copepod species. There is a gap between the needs of the aquaculture industry and the research that has been done on basic copepod biology. Increased knowledge of nauplius nutritional value is essential.

The long-chain omega-3 highly unsaturated fatty acids (sometimes referred to as *n*-3 HUFAs) play a major role as sources of energy in fish nutrition. Very little information, however, is available on lipids in copepod nauplii; the vast majority of research is on adult stages. Generally, copepods with large lipid reserves are found in higher latitudes or in deep sea or estuarine habitats where food may be unavailable for certain periods of time. Copepod species that undergo diapause also tend to have large lipid reserves to carry them through a season with very low productivity. In some species of calanoid copepods, up to 30% of the body weight is lipids. In contrast, most coastal water copepods, such as the calanoid copepods *Acartia* spp. and *Centropages* spp., tend not to have many lipid reserves because they convert food very quickly into eggs, but they have a rapid growth rate and a high biomass-production rate. Researchers have found a greater total lipid concentration in harpacticoid copepods at lower temperatures. If a copepod species from a high-latitude area (with large lipid reserves) could also tolerate higher temperatures, it would be ideal for rearing tropical and subtropical fish larvae. Further research is needed to discover whether such a species exists.

Retention of the nutritive values of copepods during storage varies widely among groups and species. Some starved harpacticoid copepod adults have been shown to retain most of their lipids for up to 5 days. In contrast, the lipid content of the calanoid copepods *Acartia* spp. decreases rapidly when they are starved. The form in which lipids are stored may also vary greatly by genera and species of copepod. Generally, lipids seem to be stored in copepods mainly as polar lipids, in contrast to enriched rotifers, which store lipids mainly as triglycerides ("storage lipids"). Polar lipids may be important for larval fish development. A study on the effects of algae with different chemical compositions on copepods showed that the omega-3 fatty acid, eicosapentaenoic acid (EPA), might not be important for copepod growth. It may be more important for some fish species that the omega-3 fatty acids are available as phospholipids. Many fish species seem to have phospholipids at first hatching, but may not have the requisite lipases to break them down into fatty acids.

It is likely that copepods provide digestive enzymes that may serve as cues for the secretion of digestive enzymes in larval fish. Preliminary results from studies on the ontogeny of digestive enzymes in larval grouper showed evidence of a higher level of enzyme activity in larvae fed copepod nauplii only, compared with those fed with rotifers. It was mentioned that Norwegian researchers found better digestive enzyme production by halibut (*Hippoglossus hippoglossus*) larvae fed copepods than those fed *Artemia* and that the fatty acid profiles that occurred in the copepod-fed halibut larvae during development were better. Further research is needed to study the enzymes that fish larvae are producing and to determine whether those enzymes are capable of digesting particular types of copepods.

The diet can be used to fine-tune the nutritional value of copepods for a particular developmental stage of fish species. A 24-hour response to food changes has been observed in the Calanoid copepod *Acartia tonsa*. The composition of the egg of the target finfish species may indicate the need for a particular copepod species. If a nutrient were not in the endogenous reserve of the egg, it would probably not be needed in the exogenous feed that is supplied. Some harpacticoid copepods have enzymes that convert the *n*-3 fatty acids to longer-chain fatty acids. Thus, these types of copepods may be able to create an optimal fatty acid profile on a less than an optimal diet.

APPLICATION OF COPEPODS IN FISH LARVAL REARING

The prospects of fish larvae encountering and capturing prey through the critical first-feeding period can be increased through prey density and a light regime. Increasing the density to about 2–5 nauplii per milliliter increases their encounter rate with larval fish. There is some evidence that swarming behavior makes it more difficult for fish to feed on individual nauplii due to a confusion effect. On the other hand, it may not be necessary to nurture fish larvae on semisessile or standard swimming organisms such as the rotifer. The marine fish larvae of many of the fish of interest for aquaculture are evolutionarily adapted to feed on copepod nauplii, among other organisms. A nauplius with a quick escape response may result in healthier fish, especially if the nutrition from the nauplius that is eventually consumed is highly beneficial to the fish. Light can be used to increase the encounter rates of larval fish with nauplii that are photopositive. For example, swarms of *Acartia* sp. nauplii and larval red snapper were both attracted to hot spots created under the high-intensity lights.

A small amount of turbulence may also help to reduce the evasive ability of copepods and increase encounter rates with larval fish. One of the biggest problems in aquaculture is having some larvae growing much faster than others and having size gradients develop among the fish. By dispersing the food and the larvae with turbulence, it may be possible to solve this problem. For example, the growth of cultured striped bass (*Morone saxatilis*) larvae that were in a very highly turbulent environment was more uniform. Red snapper larvae in a 1-m^3 tank, with one airstone and one slight current, have been observed to cluster and orient themselves into that current, then strike at nauplii as they come by. These reactions, however, are species-specific, and it is difficult to generalize about the effect of turbulence.

CANDIDATE COPEPOD SPECIES

At the end of the workshop, participants identified a number of promising candidate genera with which to proceed. There are two major requirements for a candidate copepod species for first feeding: It must be small, and it must be moving.

Among the calanoid copepods, the genera *Acartia*, *Pseudodiaptomus*, *Sinocalanus*, *Eurytemora*, *Gladioferens*, *Parvocalanus*, *Bestiolina*, *Temora*, *Centropages*, and *Labidocera* were proposed for first feeding. Nauplii of some *Acartia* spp. are as small as 100 μm in length and 50–60 μm in width, making them suitable for first feeding as well. *Labidocera* sp. was found to grow successively larger and produced more eggs in an established laboratory culture. Several Paracalanid copepod species are larger, and thus are more suitable for feeding to larger fish larvae. Paracalanid copepods are very common in coastal areas, while *Acartia* spp. are found almost everywhere.

Harpacticoid copepod species tend to have high production rates, are generally not cannibalistic, and are found all over the world. In addition, it is likely that many genera among this group can be raised on formulated feeds. Two species that can be mass cultured at high densities are *Tisbe holothuriae* and *Nitokra lacustris*. Although *T. holothuriae* may sometimes be cannibalistic, it is somewhat unusual among harpacticoid copepods in having pelagic nauplii. The nauplii of *N. lacustris* are benthic, but the copepodid stages swim and are of a suitable size (~90 μm length, 30–40 μm width) for first-feeding fish larvae. Results of short-term feeding of *N. lacustris* to red drum (*Sciaenops ocellatus*) have been positive. Adult *N. lacustris* is about the size of a large rotifer, with a slower escape response than many other species of copepods. Two exceptions were the harpacticoid genera, *Tigriopus* and *Euterpina*, since the nauplii of *Tigriopus* are not good swimmers, and *Euterpina* nauplii could not be digested by early larval stages of several tropical fish species. Nauplii of harpacticoid copepod species, however, may be somewhat difficult to separate from the culture, and harvesting adults from sediments may be quite difficult. Symbiotic poecilostomatoid copepods, which have swimming but nonfeeding nauplii, may be considered if the target finfish species can be co-cultured with mussels, the obligate host of the adult stages of the copepod.

Among the cyclopoid copepods, the genera *Oithona* and *Dioithona* were listed as candidates for first-feeding larvae. Some *Dioithona* species have been shown to produce robust nauplii in laboratory cultures and some brood eggs. Many of the swarming copepods are found in this genus. Adults of swarming copepods grow to about 800 μm and have shown very little escape behavior. Nauplii are less than 100 μm in length and are negatively phototactic, but they can be collected in plankton nets at night and will enter the water column with slight turbulence. Nauplii of the cyclopoid copepod *Apocyclops royi* develop from eggs in 4–5 days and are used for first-feeding practices in Taiwan.

EPILOGUE

Culturists and copepod biologists expressed their appreciation for the opportunity to learn from each other during the 3-day information exchange and discussion. Dialogue between experts from diverse disciplines is sure to accelerate the improvement of copepod culture technology. An ideal copepod species for aquaculture was not identified at the workshop, but several potential species were recommended for further study. More research is needed to explore their potential. As one participant pointed out, an ideal copepod species would be one that produced 100–200 nauplii per milliliter. Although such a species of copepod may never be found, and production levels are unlikely to match that of rotifers, the potential contribution of copepods to hatchery production can be enhanced through the combined efforts of biologists, nutritionists, and engineers. Copepods can play an increasingly important role in the production of new aquaculture species in the years to come.

Index

Printed and bound by CPI Group (UK) Ltd, Croydon, CR0 4YY
16/09/2026
14963448-0006